D1749209

Nucleic Acids from A to Z

Edited by
Sabine Müller

Related Titles

N. Sewald, H.-D. Jakubke

Peptides from A to Z

A Concise Encyclopedia

2008
ISBN 978-3-527-31722-6

P. Herdewijn (Ed.)

Modified Nucleosides

in Biochemistry, Biotechnology and Medicine

2008
ISBN 978-3-527-31820-9

C. S. Tsai

Biomacromolecules – Introduction to Structure, Function and Informatics

2006
ISBN 978-0-470-08012-2

R. K. Hartmann, A. Bindereif, A. Schön, E. Westhof (Eds.)

Handbook of RNA Biochemistry

2005
ISBN 978-3-527-30826-2

G. Kahl

The Dictionary of Gene Technology

Genomics, Transcriptomics, Proteomics
2 Volumes

2004
ISBN 978-3-527-30765-4

Nucleic Acids from A to Z

A Concise Encyclopedia

Edited by
Sabine Müller

WILEY-VCH Verlag GmbH & Co. KGaA

The Editor

Prof. Dr. Sabine Müller
Ernst-Moritz-Arndt-Universität
Greifswald
Institut für Biochemie
Felix-Hausdorff-Strasse 4
17487 Greifswald
Germany

All books published by Wiley-VCH are carefully produced. Nevertheless, authors, editors, and publisher do not warrant the information contained in these books, including this book, to be free of errors. Readers are advised to keep in mind that statements, data, illustrations, procedural details or other items may inadvertently be inaccurate.

Library of Congress Card No.: applied for

British Library Cataloguing-in-Publication Data
A catalogue record for this book is available from the British Library.

Bibliographic information published by the Deutsche Nationalbibliothek
Die Deutsche Nationalbibliothek lists this publication in the Deutsche Nationalbibliografie; detailed bibliographic data are available in the Internet at http://dnb.d-nb.de

© 2008 WILEY-VCH Verlag GmbH & Co. KGaA, Weinheim

All rights reserved (including those of translation into other languages). No part of this book may be reproduced in any form – by photoprinting, microfilm, or any other means – nor transmitted or translated into a machine language without written permission from the publishers. Registered names, trademarks, etc. used in this book, even when not specifically marked as such, are not to be considered unprotected by law.

Printed in the Federal Republic of Germany
Printed on acid-free paper

Composition   Aptara, New Delhi, India
Printing   Strauss GmbH, Mörlenbach
Bookbinding   Litges & Dopf GmbH, Heppenheim

ISBN 978-3-527-31211-5

Preface

With the discovery of the DNA double helix in 1953, the structure and function of nucleic acids became one of the central subjects of research in molecular biology. As a result, we have today a more exact knowledge of the structure–function relation of nucleic acids and of the many parameters that are involved in their regulation. Since then, the field of nucleic acids research has exploded, particularly over the past two decades. Major discoveries include that of ribozymes, the phenomenon of RNA interference as well as the many different functions of small noncoding RNAs in the living cell. Oligonucleotides show promise as therapeutic agents and oligonucleotide conjugates play an important role as nonradioactive probes in diagnostic medicine. At the same time, nucleic acid chemistry has developed to a degree that allows the synthesis of long DNA and RNA oligonucleotides of any desired sequence from the microgram to multi-gram scale. This provides scientists with important tools for studies in molecular biology, biochemistry and medicine. Moreover, a number of other disciplines make use of synthetic oligonucleotides, e.g. as tethers in positioning functional groups at specific locations in molecular assemblies. In the field of material science, research is directed toward creating novel materials from nanoparticle oligonucleotide conjugates. Obviously, nucleic acids nowadays have entered into many research areas. The present book is an attempt to consolidate the knowledge that has been collected, together with definitions and explanations of nucleic acid relevant terms. Thus, *Nucleic Acids from A to Z* is envisioned to serve as a concise encyclopedia covering all aspects of nucleic acids research in the different areas of science.

When Frank Weinreich from Wiley-VCH first suggested that I work on this project, I was excited and thought that it would be a good idea to put all the information on nucleic acids together into one book. Later, when I had started developing a concept and collecting keywords, I understood what a tremendous plan that was. Looking at terms covering the field of RNA alone would deliver sufficient material to fill a book. The same holds for DNA, as genetic material, as drug, as building block for molecular devices, nanostructures or functional units. Further on, the synthesis of nucleic acids, enzymes working on and molecules interacting with them, application of nucleic acids as well as methods for structural and functional analysis should be considered in a nucleic acid encyclopedia. And not to forget, structures derived from and related to nucleic acids also belong to it.

I am far from claiming that the book is complete. Nevertheless, I hope that I have succeeded in collecting a large variety of nucleic acid-related terms coming from different research fields. It was my aim to accumulate expertise from nucleic acid analytics, biochemistry, genetics, molecular medicine, nanotechnology, supramolecular chemistry, synthetic chemistry, theoretical and structural biology, all combined into one book. I may have overlooked important keywords and likely the one or the other term is missing. Please, keep me informed about such cases.

Nucleic Acids from A to Z: A Concise Encyclopedia. Edited by Sabine Müller
Copyright © 2008 WILEY-VCH Verlag GmbH & Co. KGaA, Weinheim
ISBN: 978-3-527-31211-5

Of course, such a book is not written by one person. I wish to thank all participating authors and sub authors, whose efforts and high-quality contributions made this book possible. I am particularly grateful to Mauro Santos, Milan Stojanovic and Stefan Vörtler, who all have done a lot more then I had required. The deepest thanks, however, is to the people of my group, Bettina Appel, Valeska Dombos, Irene Drude, Slawomir Gwiazda, Matthäus Janczyk, Denise Strohbach, Jörn Wolf and Stéphanie Vauléon, who supported me throughout the project and helped a lot to bring the manuscript into it's final shape. I thank Fritz Eckstein for helpful advice in the initial stage. Last, but not least, I wish to thank the team from Wiley-VCH, Frank Weinreich for encouraging me to take the challenge of editing this book, Bettina Bems, Tim Kersebohm and Stefanie Volk for their assistance during manuscript preparation, Ray Loughlin for copyediting, Claudia Grössl for the good collaboration during completion of the book, and all of them for their endless patience and help.

I hope many scientists, university teachers and students, and all other people who are interested in this fascinating field will find this book a valuable source of information showing the many faces of nucleic acids in more than one area of science.

Greifswald, January 2008
Sabine Müller

How to Use this Encyclopedia

There are two kinds of cross-references in this mini-encyclopedia:

- An arrow (→) denotes synonymous or closely related entries.

- An asterisk (*) indicates that a separate entry is available for further reading.

List of Contributors

Dr. Bettina Appel
Ernst-Moritz-Arndt-Universität
Greifswald
17487 Greifswald
Germany

Prof. Dr. Bernd-Joachim Benecke
Ruhr-Universität Bochum
44780 Bochum
Germany

Dr. Yaakov Benenson
Harvard University
Cambridge, MA 02138
USA

Dipl.-Chem. Lucas Bethge
Humboldt-Universität zu Berlin
12489 Berlin
Germany

Dr. Susanne Brakmann
Universität Dortmund
44227 Dortmund
Germany

Dr. Nina G. Dolinnaya
Lomonossov Moscow State University
119991 Moscow
Russia

Dipl.-Chem. Valeska Dombos
Ernst-Moritz-Arndt-Universität
Greifswald
17487 Greifswald
Germany

Dipl.-Biochem. Irene Drude
Max-Planck-Institut für
Molekulare Physiologie
44227 Dortmund
Germany

Dr. Pierre Fechter
UPR 9002-ARN, ULP, CNRS
Institute de Biologie Moléculaire
et Cellulaire du CNRS
67084 Strasbourg
France

Prof. Dr. Hans-Joachim Fritz
Georg-August-Universität
Göttingen
37077 Göttingen
Germany

Dr. Boris Fürtig
Johann-Wolfgang-Goethe-Universität
Frankfurt
60438 Frankfurt
Germany

Prof. Dr. Roser González-Duarte
Universitat de Barcelona
08028 Barcelona
Spain

Dipl.-Chem. Tom Grossmann
Humboldt-Universität zu Berlin
12489 Berlin
Germany

List of Contributors

Dipl.-Biochem. Slawomir Gwiazda
Ernst-Moritz-Arndt-Universität
Greifswald
17487 Greifswald
Germany

Prof. Dr. Uli Hahn
Universität Hamburg
20146 Hamburg
Germany

Prof. Dr. Roland Hartmann
Philipps-Universität Marburg
35032 Marburg
Germany

Prof. Dr. Christian Herrmann
Ruhr-Universität Bochum
44780 Bochum
Germany

Dr. Ivo Hofacker
Universität Wien
1090 Wien
Austria

Dipl.-Biochem. Matthäus Janczyk
Ernst-Moritz-Arndt-Universität
Greifswald
17487 Greifswald
Germany

Prof. Dr. Andres Jäschke
Universität Heidelberg
69120 Heidelberg
Germany

Dr. Hans-Werner Junghans
Universität Konstanz
78457 Konstanz
Germany

Prof. Dr. Evgeny Katz
Clarkson University
Potsdam, NY 13699-5810
USA

Dr. Sven Klussmann
NOXXON Pharma AG
10589 Berlin
Germany

Prof. Dr. Eric T. Kool
Stanford University
Stanford, CA 94305
USA

Dr. Andrew T. Krueger
Stanford University
Stanford, CA 94305
USA

Dr. Elena A. Kubareva
Lomonossov Moscow State University
119991 Moscow
Russia

Dr. Jens Kurreck
Universität Stuttgart
70569 Stuttgart
Germany

Dr. David Loakes
MRC Laboratory of Molecular Biology
Cambridge CB2 2QH
United Kingdom

Dr. Gemma Marfany
Universitat de Barcelona
08028 Barcelona
Spain

Prof. Dr. Andreas Marx
Universität Konstanz
78457 Konstanz
Germany

Dr. Valeriy G. Metelev
Lomonossov Moscow State University
119991 Moscow
Russia

List of Contributors

Prof. Dr. Ronald Micura
Leopold-Franzens-Universität
6020 Innsbruck
Austria

Prof. Dr. Mario Mörl
Universität Leipzig
04103 Leipzig
Germany

Prof. Dr. Sabine Müller
Ernst-Moritz-Arndt-Universität
Greifswald
17487 Greifswald
Germany

Prof. Dr. Franz Narberhaus
Ruhr-Universität Bochum
44780 Bochum
Germany

Prof. Dr. Wolfgang Nellen
Universität Kassel
34109 Kassel
Germany

Mark Olah
University of New Mexico
Albuquerque, NM 87131
USA

Prof. Dr. Tatiana S. Oretskaya
Lomonossov Moscow State University
119991 Moscow
Russia

Dr. Tina Persson
Lund University
22100 Lund
Sweden

Dr. Nicolas Piganeau
Universität Hamburg
20146 Hamburg
Germany

Prof. Dr. Tobias Restle
Universität zu Lübeck
23538 Lübeck
Germany

Dipl.-Chem. Lars Röglin
Humboldt-Universität zu Berlin
12489 Berlin
Germany

Dr. Pascale Romby
UPR 9002-ARN, ULP, CNRS
Institute de Biologie Moléculaire
et Cellulaire du CNRS
67084 Strasbourg
France

Dr. Paul W. K. Rothemund
California Institute of Technology
Pasadena, CA 91125
USA

Prof. Dr. Mauro Santos
Universitat Autónoma de Barcelona
08193 Bellaterra (Barcelona)
Spain

Dr. Olav Schiemann
University of St. Andrews
St. Andrews, KY 169ST
United Kingdom

Prof. Dr. Harald Schwalbe
Johann-Wolfgang-Goethe-Universität
Frankfurt
60438 Frankfurt
Germany

Prof. Dr. Georg Sczakiel
Universität zu Lübeck
23538 Lübeck
Germany

List of Contributors

Prof. Dr. Nadrian C. Seeman
New York University
New York, NY 10003
USA

Prof. Dr. Claus Seidel
Heinrich-Heine-Universität
Düsseldorf
40225 Düsseldorf
Germany

Prof. Dr. Oliver Seitz
Humboldt-Universität zu Berlin
12489 Berlin
Germany

Dr. Petr Sergiev
Lomonossov Moscow State University
119991 Moscow
Russia

Prof. Dr. Roland K. O. Sigel
Universität Zürich
8057 Zürich
Switzerland

Prof. Dr. Snorri Thor Sigurdsson
University of Iceland
101 Reykjavik
Iceland

Dr. Adam Silvermann
Stanford University
Stanford, CA 94305
USA

Dr. Rajesh Kumar Singh
Ernst-Moritz-Arndt-Universität
Greifswald
17487 Greifswald
Germany

Dr. Andreas Springer
Freie Universität Berlin
14195 Berlin
Germany

Prof. Dr. Peter Stadler
Universität Leipzig
04107 Leipzig
Germany

Prof. Dr. Darko Stefanovic
University of New Mexico
Albuquerque, NM 87131
USA

Prof. Dr. Milan Stojanovic
Columbia University
New York, NY 10032
USA

Dr. Denise Strohbach
Ruhr-Universität Bochum
44780 Bochum
Germany

Dr. Beatrix Süß
Johann-Wolfgang-Goethe-Universität
Frankfurt
60438 Frankfurt
Germany

Prof. Dr. Eörs Szathmáry
Collegium Budapest
1014 Budapest
Hungary

Dr. Stéphanie Vauléon
Ernst-Moritz-Arndt-Universität
Greifswald
17487 Greifswald
Germany

Dr. Stefan Vörtler
Universität Bayreuth
95447 Bayreuth
Germany

Prof. Dr. Hans-Achim Wagenknecht
Universität Regensburg
93053 Regensburg
Germany

Dr. Markus Wahl
Max-Planck-Institut für
Biophysikalische Chemie
37077 Göttingen
Germany

Prof. Dr. Elmar Weinhold
Rheinisch-Westfälische Technische
Hochschule Aachen (RWTH)
52056 Aachen
Germany

Prof. Dr. Klaus Weisz
Ernst-Moritz-Arndt-Universität
Greifswald
17487 Greifswald
Germany

Dr. Rüdiger Welz
Yale University
New Haven, CT 06520-8120
USA

Prof. Dr. Jesper Wengel
University of Southern Denmark
5230 Odense
Denmark

Dr. Annegret Wilde
Humboldt-Universität zu Berlin
10115 Berlin
Germany

Dr. Dagmar Willkomm
Philipps-Universität Marburg
35032 Marburg
Germany

Prof. Dr. Erik Winfree
California Institute of Technology
Pasadena, CA 91125
USA

Dipl.-Chem. Jörn Wolf
Ernst-Moritz-Arndt-Universität
Greifswald
17487 Greifswald
Germany

Prof. Dr. Ada Yonath
Weizmann Institute of Science
76100 Rehovot
Israel

General References

R. Besch, C. Giovannangeli, K. Degitz, Triplex-forming oligonucleotides – sequence specific DNA ligands as tools for gene inhibition and for modulation of DNA-associated functions. *Curr Drug Targets* **2004**, *5*, 691–703.

G. M. Blackburn, M. J. Gait, D. Loakes, D. M. Williams (Eds.), *Nucleic Acids in Chemistry and Biology*. RSC Publishing, Cambridge, **2006**.

X. Cheng, R.M. Blumenthal (Eds.), *S-Adenosyl-methionine-dependent Methyltransferases: Structures and Functions*. World Scientific Publishing, Singapore, **1999**.

D. R. Corey, RNA learns from antisense. *Nat Chem Biol* **2007**, *3*, 8–11.

S. T. Crooke, Progress in antisense technology. *Annu Rev Med* **2004**, *55*, 61–95.

V. V. Didenko (Ed.), *Fluorescent Energy Transfer Nucleic Acid Probes, Design and Protocols*. Humana Press, Totowa, NJ, **2006**.

A. Ducruix, R. Giege (Eds.), *Crystallization of Nucleic Acids and Proteins: A Practical Approach*, 2nd edn. Oxford University Press, Oxford, **1999**.

A. Fallert-Müller (Ed.), *Lexikon der Biochemie*. Spektrum Akademischer Verlag, München, **1999/2000**.

A. R. Ferre-D'Amare, J. A. Doudna, Crystallization and structure determination of a hepatitis delta virus ribozyme: use of the RNA-binding protein U1A as a crystallization module. *J Mol Biol* **2000**, *295*, 541–556.

E. Freisinger, R. K. O. Sigel, From nucleotides to ribozymes – a comparison of their metal ion-binding properties. *Coord Chem Rev* **2007**, *251*, 1834–1851.

R. Garret, S. R. Douthwaite, A. Liljas, A. T. Matheson, P. B. Moore, H. F. Noller (Eds.), *The Ribosome: Structure, Function, Antibiotics and Cellular Interaction*. ASM Press, Washington, DC, **2000**.

S. F. Gilbert, *Developmental Biology*, 8th edn. Sinauer Associates, Sunderland, MA, **2006**.

H. Grosjean, R. Benne (Eds.), *Modification and Editing of RNA*. ASM Press, Washington, DC, **1998**.

R. K. Hartmann, A. Schön, E. Westhof (Eds.), *Handbook of RNA Biochemistry*. Wiley-VCH, Weinheim, **2005**.

A. Herbert, A. Rich, The biology of left-handed Z-DNA. *J Biol Chem* **1996**, *271*, 11595–11598.

I. Hofacker, P. F. Stadler, RNA secondary structures, in: *Bioinformatics – From Genomes to Therapies*, Vol. 1, T. Lengauer (Ed.): Wiley-VCH, Weinheim, **2007**.

G. Kahl, *The Dictionary of Gene Technology*, 2nd edn. Wiley-VCH, Weinheim, **2001**.

J. M. Kalish, P. M. Glazer, Targeted genome modification via triple helix formation. *Ann NY Acad Sci* **2005**, *1058*, 151–161.

E. Katz, I. Willner, Integrated nanoparticle–biomolecule hybrid systems: synthesis, properties and applications. *Angew Chem Int Ed* **2004**, *43*, 6042–6108.

E. Katz, I. Willner, Biomolecule-functionalized carbon nanotubes: applications in nano-bioelectronics. *Chem Phys Chem* **2004**, *5*, 1194–1104.

E. Katz, B. Willner, I. Willner, Amplified electrochemical and photoelectrochemical analysis of DNA, in: *Perspectives in Bioanalysis*, Vol. 1, *Electrochemistry of Nucleic Acids and Proteins – Towards Electrochemical Sensors for Genomics and Proteomics*, E. Palecek, F. Scheller, J. Wang (Eds.). Elsevier, Amsterdam, **2005**.

J. S. Kieft, R. T. Batey, A general method for rapid and nondenaturing purification of RNAs. *RNA* **2004**, *10*, 988–995.

R. Knippers (Ed.), *Molekulare Genetik*. Thieme, Stuttgart, **2001**.

E. T. Kool, Modified DNA bases: probing base pair recognition by polymerases, in: *Modified Nucleosides in Biochemistry, Biotechnology, and Medicine*, P. Herdewijn (Ed.): Wiley-VCH, Weinheim, **2007**.

A. Krueger, H. Lu, A. H.-F. Lee, E. T. Kool, Synthesis and properties of size-expanded DNAs: toward designed, functional genetic systems. *Accts Chem Res* **2007**, *40*, 141–150.

A. L. Lehninger, D. L. Nelson, M. M. Cox, *Principles of Biochemistry*. Worth, New York, **1993**.

B. Lewin, *Genes VII*. Oxford University Press, New York, **2000**.

B. Lewin, *Genes IX*. Jones & Bartlett, Boston, MA, **2007**.

X. Li, D. R. Liu, DNA-templated organic synthesis: nature's strategy for controlling chemical reactivity applied to synthetic molecules. *Angew Chem Int Ed* **2004**, *43*, 4848–4870.

D. Loakes, The applications of universal DNA base analogues. *Nucleic Acids Res* **2001**, *29*, 2437–2447.

K. H. Nierhaus, D. E. Wilson (Eds.), *Protein Synthesis and Ribosome Structure: Translating the Genome*. Wiley-VCH, Weinheim, **2004**.

W. H. Pan, G. A. Clawson, Antisense applications for biological control. *J Cell Biochem* **2006**, *98*, 14–35.

A. M. Pingoud (Ed.), *Restriction Endonucleases*. Springer, Berlin, **2004**.

A. Pingoud, M. Fuxreiter, V. Pingoud, W. Wende, Type II restriction endonucleases: structure and mechanism. *Cell Mol Life Sci* **2005**, *62*, 685–707.

W. Saenger, *Principles of Nucleic Acid Structure*. Springer, New York, **1984**.

U. Schepers, *RNA Interference in Practice: Principles, Basics, and Methods for Gene Silencing in C. elegans, Drosophila, and Mammals*. Wiley-VCH, Weinheim, **2004**.

L. J. Scherer, J. J. Rossi, Approaches for the sequence-specific knockdown of mRNA. *Nat Biotechnol* **2003**, *21*, 1457–1465.

C. Schmuck, H. Wennemers (Eds.), *Highlights in Bioorganic Chemistry*. Wiley-VCH, Weinheim, **2004**.

N. C. Seeman, DNA in a material world. *Nature* **2003**, *421*, 427–431.

N. C. Seeman, P. S. Lukeman, Nucleic acid nanostructures. *Rep Prog Phys* **2005**, *68*, 237–270.

N. C. Seeman, From genes to machines: DNA nanomechanical devices. *Trends Biochem Sci* **2005**, *30*, 119–125.

N. C. Seeman, Nanotechnology and the double helix. *Sci Am* **2006**, *290*, 64–75.

R. K. O. Sigel, A. M. Pyle, Alternative roles for metal ions in enzyme catalysis and the implications for ribozyme chemistry. *Chem Rev* **2007**, *107*, 97–113.

L. Snyder, W. Champness (Ed.), *Molecular Genetics of Bacteria*. ASM Press, Washington, DC, **2003**.

D. Söll, U. L. RajBhandary, *tRNA: Structureaa, Biosynthesis and Function*. ASM Press, Washington, DC, **1994**.

C. E. Thomas, A. Ehrhardt, M. A. Kay, Progress and problems with the use of viral vectors for gene therapy. *Nat Rev Genet* **2003**, *4*, 346–358.

H. A. Wagenknecht (Ed.) *Charge Transfer in DNA: From Mechanism to Application*. Wiley-VCH, Weinheim, **2005**.

B. Vester, J. Wengel, LNA (locked nucleic acid): high affinity targeting of complementary RNA and DNA. *Biochemistry* **2004**, *43*, 13233–13241.

A

A → adenine

AA-platform Structural motif in RNA consisting of two adjacent *adenosine nucleobases in a coplanar arrangement.

aaRS → aminoacyl-tRNA synthetase

aa-tRNA → aminoacyl-tRNA

abasic site *Monomer residue within a contiguous nucleic acid strand (normally DNA) that does not carry a *nucleobase. Depending on the nucleobase lost (discernible in the case of *double-stranded DNA) an abasic site is either an apurinic or an apyrimidinic site. An abasic site arises from hydrolysis of a *glycosidic bond, which can either occur spontaneously or as the result of enzymatic catalysis. The latter reaction is carried out on a number of chemically damaged DNA residues and constitutes the first step of *base excision repair. With respect to the pentose moiety, abasic sites exist in equilibrium between an open-chain and two furanose forms (having β- or α-configuration). In DNA, the free aldehyde function of the open-chain form lends C/H acidity to carbon center C2' that, in turn, facilitates base-induced strand cleavage on the 3'-side of the abasic site by β-elimination of a *phosphomonoester with the formation of a 2',3' double bond and vinylogous shift of C/H acidity to center C4'. In DNA, spontaneous base loss is promoted by low pH and single-stranded structure; it is faster for *purines than for *pyrimidines. Enzymes catalyzing hydrolysis of glycosidic bonds in DNA are called *DNA glycosylases (in connection with their substrate base, e.g. uracil DNA glycosylase). For historical reasons, abasic sites are also called AP-sites (for apurinic or apyrimidinic site). HANS-JOACHIM FRITZ

absorbance Characterized in spectroscopy as:

$$A = \log \frac{I_0}{I}$$

where A is the measured absorbance and equals the $\log_{10}$ of the intensity I_0 of the incident light at a given wavelength divided by the transmitted intensity I. Absorbance is often also referred to as *extinction or *optical density (→ Beer–Lambert law).
VALESKA DOMBOS

absorption Physical process of absorbing light. Nucleic acids show an absorption maximum near 260 nm. → absorbance, → absorption spectra of nucleobases.
VALESKA DOMBOS

absorption coefficient (Synonym: extinction coefficient) → absorptivity

absorption spectra of nucleobases Used to determine the concentration of DNA and RNA in solution. The *nucleobases of DNA and RNA strands absorb ultraviolet light of 250–270 nm wavelength with the absorption maximum at 260 nm. According to the *Beer–Lambert law there is a linear relationship between the concentration and the absorption of light. The higher the *optical density, the higher the nucleic acid concentration of the sample. To determine the concentration of nucleic acid solutions, the sample is exposed to ultraviolet light at 260 nm and a photodetector measures the *absorbance. An example spectrum of a 41mer RNA is given below.

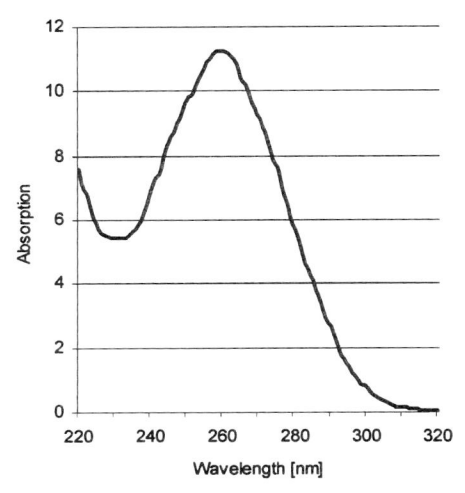

Absorption spectrum of a 41mer RNA.

The absorptivity is wavelength dependent:

$$a(\lambda) = \frac{A}{cL}$$

where A is the *absorbance, L is the length of the light path and c is the concentration of the analyte. If the concentration is expressed on a molar basis, the absorptivity $a(\lambda)$ becomes *molar absorptivity ε, which is a fundamental molecular property in a given solvent at a particular temperature and pressure. In the literature, the *molar absorptivity is also referred to as the *molar absorption coefficient or *molar extinction coefficient. The unit given is usually $L\,mol^{-1}\,cm^{-1}$, but the SI unit is $m^2\,mol^{-1}$. VALESKA DOMBOS

Knowing the *molar absorptivity of the sample, its concentration can be calculated. For an *optical density value of 1, there are some benchmarks corresponding to nucleic acids: (a) for *double-stranded DNA it is equivalent to a concentration of 50 $\mu g\,mL^{-1}$, (b) for single-stranded DNA to 37 $\mu g\,mL^{-1}$ and (c) for RNA to 40 $\mu g\,mL^{-1}$. Furthermore, the spectrophotometric measurement allows the purity of nucleic acids to be determined. The absorption value at 230 nm reflects impurities such as urea, aromatic compounds or peptides, although nucleic acid itself shows little absorption at 230 nm due to the aromatic bases. Proteins have an absorption maximum at 280 nm. Therefore, the absorbance ratio 260/280 can be used to estimate the contamination of the nucleic acid sample with proteins. A ratio value greater than 1.9 for DNA or greater than 1.8 for RNA is considered an indicator of protein-free nucleic acid samples. BETTINA APPEL

absorptivity (Synonym: absorption coefficient, extinction coefficient) Proportionality constant $a(\lambda)$ of the *Beer–Lambert law.

AC arm → anticodon arm, → transfer RNA

acceptor arm Amino acid accepting branch of the L-shaped three-dimensional structure of a *transfer RNA (tRNA) containing the conserved *CCA-tail at its 3'-end. It is formed by *coaxial stacking of the dihydrouridine (→ DHU loop) and acceptor RNA helices of a tRNA cloverleaf. The *anticodon arm forms the second branch. STEFAN VÖRTLER

acceptor end Conserved single-stranded *CCA-tail of a *transfer RNA at the end of the *acceptor arm.

acceptor stem → acceptor arm

ACE method Allows for the chemical solid-phase synthesis of *oligoribonucleotides and is named after the *nucleoside 2'-hydroxyl protection – the bis(acetoxyethyloxy)methyl (ACE) group. First introduced in 1998, the method follows phosphoramidite chemistry (→ phosphoramidite method) for the coupling of appropriately protected nucleoside building blocks and was developed under the aspect that mildly acidic aqueous conditions are ideal for the final 2'-O-deprotection of the synthesized RNA. Therefore, the protection strategy employs two orthogonal *protecting groups – a fluorine ion-labile silyl ether on the 5'-O-position and an acid-labile orthoester on the 2'-O-position. The 3'-O-position is derivatized with a (N,N-diisopropylamino)methoxyphosphinyl residue as the commonly used cyanoethoxyphosphinyl group is unstable in the presence of fluorine reagents required during strand elongation. The structures of the nucleoside phosphoramidite derivatives are depicted in the figure above.

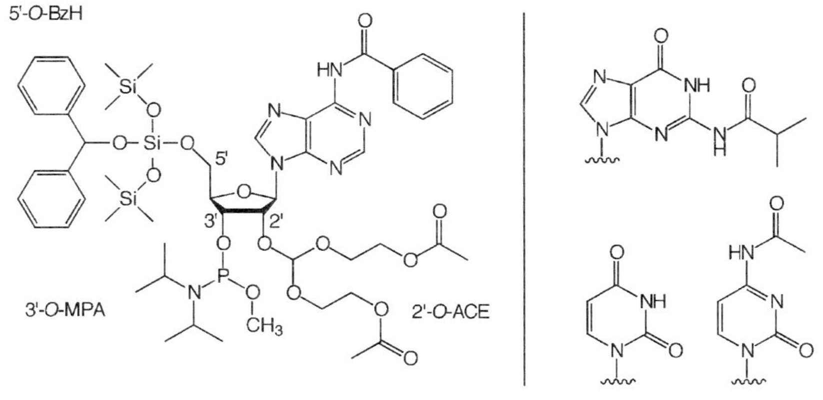

Oligonucleotide assembly proceeds with coupling yields higher than 99% and with coupling speeds as fast as 2 nucleotides per minute. After the assembly, the phosphate methyl protecting groups are removed with disodium 2-carbamoyl-2-cyanoethylene-1,1-dithiolate trihydrate. This step may limit the kind of modifications that can be incorporated into oligoribonucleotides by the ACE method compared to the alternative *TOM or *TBDMS RNA synthesis methods where this treatment is not required. Then, basic conditions (40% aqueous MeNH$_2$) cause release from the solid support, along with the removal of the acyl protecting groups and, importantly, of the acetyl groups on the 2'-orthoesters. The resulting 2'-O-bis(2-hydroxyethyloxy)methyl orthoesters are 10 times more acid labile then prior to the removal of the acetyl groups. Consequently, very mild acidic conditions (pH 3.8, 30 min, 60°C) followed by a short treatment at pH 8.7 to effect hydrolysis of any intermediate 2'-O-formyl groups are sufficient for the final deprotection step.

The access to the 2'-O-bis(2-hydroxyethyloxy)methyl oligoribonucleotides is a major strength of the ACE method as these precursor RNAs are water soluble, can be analyzed by *HPLC and further purified if necessary. Of further significance is that the 2'-O-bis(2-hydroxyethyloxy)methyl protection interrupts *secondary structures, renders the RNA resistant to *nucleases and other forms of degradation, and therefore facilitates handling and long-term storage.

The 5'-O-silyl ethers and 2'-O-orthoesters have been structurally optimized to provide crude oligoribonucleotides of high sequence integrity, excellent purity and biological activity. Using commercial synthesizers (→DNA synthesizer), the ACE method enables routine preparation of oligoribonucleotides up to 60 bases in length, regardless of sequence or secondary

structure. The ACE method produces milligram to kilogram quantities of more than 90% pure RNA and has been successfully commercialized (Dharmacon Inc.).
RONALD MICURA

acidity of nucleosides and nucleotides The acid–base behavior of *nucleotides is a very significant physical characteristic. The pK_a value estimates the charge and the tautomeric structure (→ tautomeric bases), the ability to donate or accept hydrogen bonds, and therefore the possibility for base pairing, and lastly also the structure of nucleic acids. All bases are uncharged in the physiological range (pH 5–9). The 2′,3′-diol in *ribose loses one proton only at pH higher then 12. In nucleotides, the remaining free hydroxyl group is deprotonated only above pH 15. The phosphate group loses one proton at pH 1 and the second proton at pH 7. Due to the amine–imine tautomerism, the character of the C–NH$_2$ bond has approximately 45% double-bond character. Therefore, the ring nitrogens are more readily protonated than the exocyclic amino groups. Due to the negative charge on the phosphorous residue, in nucleotides

pK_a values for bases in nucleosides and nucleotides

Base (site of protonation)		Nucleoside	3'-Nucleotide	5'-Nucleotide
Adenine	N1	3.52	3.70	3.88
Cytosine	N3	4.17	4.43	4.56
Guanine	N7	3.30	(3.5)	(3.6)
Guanine	N1	9.42	9.84	10.00
Thymine	N3	9.93	–	10.47
Uracil	N3	9.38	9.96	10.06

Data correspond to loss of a proton for $pK_a > 9$ and capture of a proton for $pK_a < 5$; calculated to 20 °C.

the ring nitrogens are more basic ($\Delta pk_a = +0.4$) and the ring N–Hs are less acidic ($\Delta pk_a = +0.6$) compared with the corresponding nucleosides. For dissociation constants, see table. BETTINA APPEL

activator *DNA-binding protein that increases the rate of *transcription by enhancing the interaction between *RNA polymerase and the *promoter.

active center The terms active center and *active site are often used as synonyms. However, active center may be a description for active site. For example, the active center of *group I introns is composed of elements P1, P3, P4, P6, P7, J4/5 and J8/7, but only a subset of *functional groups within these structural elements form the active site. The term *catalytic core, employed in the same context to describe the minimal architectural elements required for *catalysis, even goes beyond active center, as it additionally includes RNA elements that buttress the active center (e.g. comprising P4/5/6, P3/7/8/9 and P1/2/10 in group I introns). ROLAND K. HARTMANN, MARIO MÖRL, DAGMAR K. WILLKOMM

active conformation Folding state that predisposes a *catalytic RNA to immediate entry into its functional cycle. For reactions in trans, the RNA requires complexation with the substrate to achieve its active, catalytically competent conformation (→induced-fit model). As an example, the *RNase P RNA-binding module for the *CCA-tail of *precursor tRNAs becomes structured only upon substrate binding, which further entails a rearrangement of metal ions.

Three major factors contribute to the active conformation of RNA catalysts: peripheral structural elements of the RNA involved in *long-range tertiary contacts, metal ions and protein cofactors. In the case of the *hammerhead, it has been discovered only recently that natural variants form tertiary contacts between loop extensions in the two arms of their Y-shaped structure. These contacts seem to restrict the *RNA fold in a state that is equal or close to the catalytically active conformation. The important consequence is a substantially lower Mg^{2+} requirement for activity. To stabilize the *catalytic core and thus adopt an *active conformation, large *ribozymes utilize peripheral structural elements which engage in long-range tertiary interactions. In some *self-splicing introns, long-range interactions are reinforced by protein cofactors. RNase P RNA needs a small basic protein to stabilize local structure in the catalytic core in order to create high-affinity *binding sites for a set of Mg^{2+} ions relevant to substrate binding and catalysis. The protein thereby lowers the Mg^{2+} requirement of the enzyme.

Ribozymes, as protein enzymes, traverse different conformational states during their functional cycle. For example, a large conformational change takes place when the guanosine cofactor of group I introns is replaced with the 3′-terminal omega G after the first *transesterification step. *X-ray analyses of the *hepatitis delta virus ribozyme in pre- and post-cleavage states have revealed a substantial conformational change that occurs after cleavage. This conformational switch can be described as a structural collapse of the product active site, during which the catalytic metal ion is expelled and the catalytic residue C75 is displaced such that the reverse reaction (→ligation) becomes unfavorable.

It is often unknown which fraction of a pool of ribozyme molecules populates a catalytically active conformation. Native polyacrylamide gels may give a first clue about different conformational states and whether the ribozyme population is heterogeneous in conformation.
ROLAND K. HARTMANN, MARIO MÖRL, DAGMAR K. WILLKOMM

active site Term employed for structural elements and functional groups of a *catalytic RNA, which (a) directly interact with or position the reactive *phosphodiester or phosphate, (b) are directly involved in chemistry or (c) that bind *ligands involved in the chemical step, such as catalytic metal ions. The active site may be more solvent exposed, as in the *hammerhead ribozyme, or deeply buried in the folded structure, as in the *hepatitis delta virus (HDV) ribozyme. Different modes of active site construction are found among *ribozymes, e.g. in the HDV ribozyme the active site is formed by a *nested double pseudoknot structure in which five helical segments form two parallel stacks, joined side-by-side by several strand crossovers. In the *hairpin ribozyme, the active site is created by docking of two irregular helices in a side-by-side manner, with the two helical stacks radiating from a perfectly base paired *four-way junction. *Group I introns fold into compact globular structures by assembly of three main domains (P4/5/6, P3/7/8/9 and P1/2/10), with large segments of the universally conserved *catalytic core buried from solvent. The *group I intron active site is created by juxtaposition of elements P1, P3, P4, P6, P7, J4/5 and J8/7.
ROLAND K. HARTMANN, MARIO MÖRL, DAGMAR K. WILLKOMM

active site chemistry The natural *small nucleolytic ribozymes [*hammerhead, *Varkud satellite, *hepatitis delta virus (HDV) and *hairpin ribozyme] catalyze site-specific *phosphodiester hydrolysis with 5′-hydroxyls and 2′,3′-cyclic phosphates as cleavage products. They activate the 2′-hydroxyl at the scissile phosphodiester as the nucleophile. In contrast, the large natural ribozymes (*group I intron, group II intron *RNase P RNA) catalyze phosphodiester hydrolysis or phosphoryl-transfer reactions (→transesterification) that yield 5′-phosphates and 3′-hydroxyls as products. A common salient feature of their catalytic strategies is to prevent the 2′-hydroxyl at the reactive phosphodiester from acting as a nucleophile.

In the hammerhead example, a divalent metal ion is thought to form an *inner-sphere contact with the pro-Rp non-bridging oxygen at the scissile phosphodiester. Linear dependence of the log of the reaction rate on pH has been interpreted to indicate that a hydroxide coordinated to the same metal ion deprotonates the 2'-hydroxyl for nucleophilic attack. In the HDV genomic ribozyme, the N3 function of the catalytic key residue C75 acts as a general base and deprotonates the attacking 2'-hydroxyl. A hydrated metal ion bound nearby then serves as general acid by protonating the 5'-oxyanion leaving group. In the case of the hairpin ribozyme, there is no evidence for a catalytic role of divalent metal ions. Here, the N1 (possibly in its protonated form) and the 6-amino function of residue A38 provide electrostatic stabilization of the *transition state.

The large ribozymes utilize metal-ion-assisted catalysis, with a specific role for two or more magnesium ions, similar to protein phosphoryltransferases. Metal ions activate the nucleophile, stabilize the transition state or leaving group by direct coordination to non-bridging and bridging phosphate oxygens or facilitate proton transfer to the leaving oxygen.
ROLAND K. HARTMANN, MARIO MÖRL, DAGMAR K. WILLKOMM

acyclovir Analog of *guanine in which the nitrogen atom at position 9 is substituted by a hydroxy ethyl methyl ether. It is an antiviral drug, which is primarily used for treating herpes simplex virus. The viral thymidine kinase converts acyclovir selectively into a monophosphate and the monophosphorylated acyclovir is phosphorylated into the active triphosphate form (acyclovir-GTP) by the host cellular kinase.
BETTINA APPEL

adaptor hypothesis →transfer RNA, →wobble hypothesis

adenine One of the four naturally occurring *nucleobases in DNA and RNA.

adenine arabinoside →arabinonucleosides

adenine riboswitch Highly similar in its structure and function to the guanine riboswitch; therefore, both classes are usually referred to as *purine riboswitches.

adenine xyloside →xylonucleosides

adenosine Natural building block of RNA (→nucleosides).

adenosine diphosphate (ADP) →adenosine phosphates, →nucleoside phosphates

adenosine monophosphate (AMP) →adenosine phosphates, →nucleoside phosphates

adenosine phosphates Beyond their function as building blocks for nucleic acids, adenosine phosphates constitute the major form of energy storage in the cell. *ATP is an intermediate product of all cellular processes that produce chemical energy to be stored. Its biosynthesis occurs by phosphorylation of *ADP with the transferred phosphate coming from either high-energy phosphates or creatinine or other *nucleoside triphosphates (→nucleoside phosphates). Energy that is stored in ATP is used for the synthesis of macromolecules as well as for activation of different compounds and metabolites. Furthermore, in a number of cellular processes, adenosine phosphates serve the role as metabolic regulators, e.g. in glycolysis or in the Krebs cycle. In living organisms, *AMP, *ADP and *ATP appear in equilibrium with physiological concentrations of ADP and ATP of about 10^{-3} mol L^{-1}.
SABINE MÜLLER

adenosine triphosphate (ATP)
→adenosine phosphates, →nucleoside phosphates

A-DNA Right-handed double-stranded helix stabilized by *Watson–Crick base pairs. One of the several types of *secondary structures that DNA can adopt, depending on environmental conditions such as counterions and relative humidity. The A-form is favored in solutions that are relatively devoid of water. A-DNA has been well characterized in crystal structure. The reagents used to promote *crystallization of DNA tend to dehydrate it and this leads to a tendency for many DNAs to crystallize in the A-form. A-DNA is shorter and has a greater diameter than the *B-DNA. In A-DNA, 11 *nucleotide pairs complete one *helical turn. *Base pairs are tilted by 13–20° from the perpendicular. Moreover, they are shifted towards the outside of the helix, yielding the deep *major, but shallow *minor groove. The helix axis lies in the major groove, bypassing the bases. The key difference between A- and B-DNA is the *sugar puckering mode; the sugar units in A-DNA are all in the standard C3′-*endo* conformation. Individual residues in A-DNA-type fragments display uniform structural features. This contrasts with the B-DNA *double helix, where sequence-dependent structural modulations are observed. If the salt concentration of the environment is raised or the relative humidity lowered, the B–A transformation takes place. B–A transition occurs cooperatively, indicating the energy barrier between C2′-*endo* and C3′-*endo* sugar puckering. The transition from the B-form double helix to the A-form is essential for biological function, as shown by the existence of the A-form in many protein–DNA complexes. If *DNA–RNA hybrids are generated, they belong to the A-family and resist transformation into B-form (→DNA structures).
NINA DOLINNAYA

ADP →adenosine diphosphate, →adenosine phosphates, →nucleoside phosphates

AFM Abbreviation for atomic force microscopy (→scanning force microscopy).

A-form RNA Type of the double-helical structure of RNA if occurring as a *duplex. The *double helix of A-form RNA closely resembles the helical parameters defined for DNA of the A-family (→A-DNA).
SABINE MÜLLER

algorithmic self-assembly Generalization of crystal growth mechanisms that considers how a mixture of several molecular species can self-assemble to create complex forms. Molecules are modeled as Wang tiles, which are geometric shapes such as squares whose sides are labeled to indicate specific binding interactions. Given a seed structure, tiles may be added if they match on sufficiently many sides. Large, complex structures can be grown using a small number of tile types according to such rules. Wang tiles can, in principle, be created as properly designed proteins or other molecules, but to date experimental demonstrations have used DNA Wang tiles. Algorithmic crystals have been created that contain Sierpinski triangle patterns, binary counting and direct copying of information. Important questions include how to reduce error rates and how to nucleate growth on seed structures. →DNA nanotechnology.
ERIK WINFREE

alignment Arrangement of primary sequences of DNA, RNA or protein aiming at the identification of similarity regions as a putative consequence of functional, evolutionary or structural constraints. It is an integral part in many studies of molecular biology and systematics to reveal homologous regions. There are two main types of alignment algorithms: local and global. Local algorithms like BLAST try to align only parts of sequences often avoiding gaps, whereas global algorithms like CLUSTAL try to align entire sequences and explicitly handle gaps. MAURO SANTOS

alkaline hydrolysis One remarkable difference between DNA and RNA is that unlike RNA, DNA is not degraded under alkaline conditions. This is due to the fact that RNA possesses a hydroxyl group at its 2' carbon which can be activated by OH^- to trigger a nucleophilic attack on the phosphorous, which in turn leads to a *transesterification and ultimately breaks the RNA's *sugar–phosphate backbone. ULI HAHN

alkaline phosphatase Enzyme that cleaves phosphoric acid monoesters under alkaline conditions.

alkylating agent →DNA alkylating agent

alkyltransferase Enzyme supporting the transfer of an alkyl group (→methyltransferase).

allele One of two or more alternative forms of a *gene located at the corresponding site (locus) on *homologous chromosomes.

allosteric activation Induced change of the conformation of a biomolecule upon interaction with low-molecular-weight compounds. Allosteric effects play an important role in the regulation of enzymatic activity. Typically, the binding domain for the *ligand is located in a part of the molecule that is distant from the catalytic part. In the nucleic acids field, a number of *allosteric ribozymes have been constructed. Activity of these *ribozymes is coupled to the presence of a specific ligand that upon binding to the allosteric domain of the ribozyme regulates activity up (positive regulation) or down (negative regulation).

Allosteric activation concerns also the field of gene regulation at the level of *messenger RNA. Binding of metabolites to specific sites in the *5'-untranslated region of certain mRNAs can induce conformational changes that in turn lead to premature *transcription termination or *translation inhibition (→riboswitches). SABINE MÜLLER

allosteric ribozyme Also called *aptazyme, *ribozymes that can be regulated by an allosteric cofactor. An allosteric ribozyme is composed of a RNA motif that binds a specific *ligand with high affinity (→aptamer) and a catalytic RNA motif. Both motifs are connected via a bridge element or communication module that is capable of translating the binding event in the aptamer domain to the catalytic part and thus triggering activity. Allosteric ribozymes can be constructed by rational design, combining a pre-selected *aptamer with a known ribozyme structure. Many examples involve the *hammerhead ribozyme or the *hairpin ribozyme. Alternatively, allosteric ribozymes can be selected from a random *RNA library. In this case, a synthetic RNA library has to be screened for active species in the presence (or absence) of the specific ligand (→allosteric activation, →*in vitro* evolution of nucleic acids, →ribozyme). SABINE MÜLLER

allostery →allosteric activation, →allosteric ribozyme

alternative splicing Pathway to generate different mature *messenger RNA (mRNA) molecules from one *primary transcript. This mechanism was initially discovered in viruses, generally known to depend on very limited amounts of their own genetic material. Today, however, alternative splicing pathways have been identified in most eukaryotes. The possibility to generate more than one functional gene product from the same pre-mRNA is one of the most important features to explain the relatively low complexity of metazoan *genomes, e.g. the surprisingly small number of about 30 000 functional *genes in mammals. In addition to *transcriptional regulation, alternative splicing provides a second important basis for differential *gene expression during cell differentiation and developmental processes such as sex determination. Alternative splicing may not only result in a different selection of *exons from a *pre-mRNA molecule, but may also establish alternative initiation or termination sites for protein biosynthesis, i.e. *start and *stop codons. Although the precise mechanism of alternative splicing remains to be elucidated, it was found that exonic as well as intronic regulatory elements are involved. Those elements may act as either enhancers or silencers of splicing. At the protein level, the large family of *SR (serine/arginine-rich) proteins appears to be involved in splice site selection. In their function, SR proteins seem to be antagonists to the major *heterogeneous nuclear ribonucleoprotein (hnRNP) proteins. Alternative splicing may result in (a) exon skipping, (b) alternative 3'-splice site choice, (c) alternative 5'-splice site choice and (d) *intron retention. Although major class and minor class splice sites are incompatible with each other, a rare intron-within-an-intron architecture has been observed, meaning that a major class intron is flanked by two *atac splice sites and splicing occurs by one or the other pathway, alternatively. BERND-JOACHIM BENECKE

altritol nucleic acid (ANA) RNA analog with a phosphorylated D-altritol backbone. The nucleobase is attached at the 2-(S)-position of the carbohydrate moiety.

amber codon UAG *stop codon which can be decoded by a specific amber *suppressor transfer RNA. The artificial name "amber" (Am) was used to designate a mutation in a laboratory strain which turned out to be a *nonsense mutation. The name commemorates the work of one of the participating graduate students, Harris Bernstein (German: Bernstein = English: amber). STEFAN VÖRTLER

amber mutation Special case of a *nonsense mutation.

amino acid binding site Reaction center in *amioacyl-tRNA synthetases (aaRS) which is responsible for the specific binding and activation of one out of the 20 canonical proteinogenic amino acids. Pauling pointed out as early as 1954 that it is impossible to differentiate between structural isomers (Val and Ile) or amino acids which differ in as little as a methyl group (Thr and Ser or Cys) just on binding affinity. The energetics are far too similar, but aaRS must do exactly that. However, nature utilizes not one, but several reactions to achieve high selectivity by coupling them sequentially. (a) A double-sieve mechanism excludes larger or sterically non-fitting amino acids from binding, while smaller ones are disfavored by suboptimal binding energies. (b) An additional proofreading site tests either amino acid adenylates or already aminoacylated *transfer RNA (tRNA) and hydrolyses off any wrongly attached amino acid. These pre- and post-transfer mechanisms (equations 1 and 2, respectively) are known to operate at various degrees in different aaRS. A classic example is the differentiation of Val and Ile by IleRS:

Pre-transfer:

$$\text{Val-AMP} \cdot \text{IleRS} + \text{tRNA}^{\text{Ile}}$$
$$\rightarrow \text{Val} + \text{AMP} + \text{IleRS} + \text{tRNA}^{\text{Ile}} \quad (1)$$

Post-transfer:

$$\text{Val-tRNA}^{\text{Ile}} \cdot \text{IleRS}$$
$$\rightarrow \text{Val} + \text{tRNA}^{\text{Ile}} + \text{IleRS} \quad (2)$$

Considerable biotechnological interest concerns the activation and incorporation of unnatural amino acids into proteins by *genetic code expansion. To avoid any interference with normal *decoding, usually one of the three *stop codons will be reassigned to decode a *suppressor tRNA. The *ribosome during decoding does not control the nature of the tRNA with respect to the attached amino acid ($\rightarrow$ aminoacyl-tRNA synthetase, $\rightarrow$ translation). Major efforts were therefore concentrated to (a) engineer or evolve aaRS to recognize and activate novel amino acids, which is by nature very difficult as precise amino acid activation is a prime requisite of this enzyme class, and (b) develop novel tRNA which do not interfere with the natural aminoacylation systems of a given organism. Thus, the presence of any additional tRNA will not matter as long as it is not a substrate to endogenous aaRS. It works if yeast or archeal aaRS/tRNA pairs are used in a prokaryotic system as they mutually do not recognize the heterologous molecules. Several such orthogonal aminoacylation systems have been established by large-scale protein as well as cellular engineering techniques. The alternative strategy of chemical aminoacylation and supplementing translation systems with prepared aminoacyl-suppressor tRNA works, but suffers from the instability of the aminoacyl bond, limiting the efficiency as well as yield.

A direct amino acid-binding ability of RNA has been long speculated about and searched for with respect to the *RNA world, the *adaptor hypothesis and the origin of the *genetic code. Artificial *aptamers and *ribozymes with aminoacylation function posses such abilities, as well as natural *riboswitches located in messenger RNA to regulate metabolism. Statistical analysis of known RNA-binding motifs with respect to the *codon assignment of the corresponding amino acids is an active field of investigation and could give clues to a stereochemical origin of the *genetic code: a particular *codon was assigned to a particular amino acid because it participated best in binding it. STEFAN VÖRTLER

aminoacylation The activation of amino acids for protein synthesis by esterifying the carboxy-terminus to the 2'- or 3'-hydroxyl of the terminal *ribose of a *transfer RNA (→aminoacyl-tRNA, →aminoacyl-tRNA synthetase). STEFAN VÖRTLER

aminoacyl-tRNA Product of the *aminoacylation reaction catalyzed by *aminoacyl-tRNA synthetases which activate amino acids for *translation by esterifying them to the terminal *ribose of the corresponding *transfer RNA. The formation of a 2'- or 3'-hydroxylester depends on the class of the corresponding synthetase. Both are of high energy (around 30 kJ mol^{-1}) and reactivity, resulting in short half-lifes. Apart from hydrolysis and aminolysis involving the neighboring α-amino group, a *transesterification equilibrium between the 2'- and 3'-hydroxyls exists as all aminoacyl-tRNA incorporated by the *ribosome are acylated to the 3'-hydroxyl group. Stability of the esters increases dramatically once the α-amino group is involved in a bond, as in *peptidyl-tRNA, or sterically hindered and protected when bound to *elongation factor EF-Tu/EF1α. For this reason protein biosynthesis relies on the rapid binding of aminoacyl-tRNA to a sufficient pool of elongation factors, which are the most abundant proteins in the cytosol (5–10% of total protein).

Activated amino acids are also precursors for porphyrin biosynthesis in plants and photosynthesizing bacteria (glutamate tRNA^{Glu}) or bacterial peptidoglycan synthesis (glycine tRNA^{Gly}). STEFAN VÖRTLER

aminoacyl-tRNA binding site (A-site) Binding site for *aminoacyl-tRNA in the *ribosome (→translation).

aminoglycosides *Antibiotics that are often used to treat meningitis or mucoviscidosis. Aminoglycosides are not assimilated by the gut, so they need to be given intravenously and intramuscularly. The binding of aminoglycosides to the 30S *subunit of bacterial *ribosomes inhibits the *translocation of the *peptidyl-tRNA from the *A-site to the *P-site and may cause miscoding of *messenger RNA. This leads to inhibition of protein synthesis and to the cessation of bacterial growth. BETTINA APPEL

2-aminopurine Fluorescent analog of *adenine. It forms a *base pair with *uracil that is isosteric to a natural AU base pair. Fluorescence of 2-aminopurine is high if the nucleobase is located in a highly flexible region of a molecule, but becomes quenched upon integration of the base analog into a more structured surrounding. Thus, 2-aminopurine has been used to follow structural dynamics of nucleic acids, in particular strand *association and *dissociation as well as tertiary folding (→fluorescence spectroscopy of nucleic acids). SABINE MÜLLER

A-minor motif One of the most interesting *tertiary structure motifs in RNA. It is, for example, found in the *hammerhead ribozyme, the P4–P6 domain of *group I intron and in the *ribosomal RNA. It describes a *nucleotide–nucleotide interaction involving interactions between

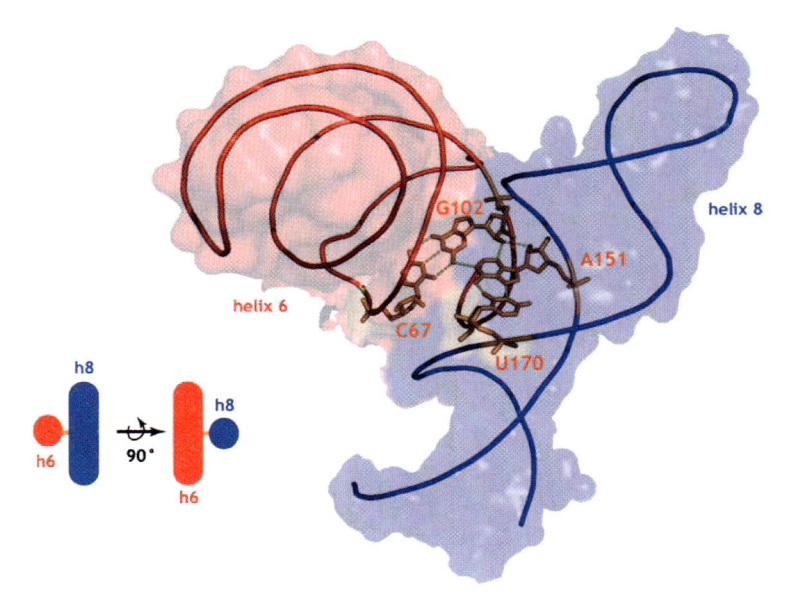

A-minor motif in the 16S ribosomal RNA. Structural representation of how a type II A-minor motif facilitates helix packing of the helices h8 (blue) and h6 (red); the nucleotides involved in the tertiary interaction are shown as sticks in orange, nucleotide A151 forms a type II A-minor motif by interacting with the base pair of G102–C67; A151 itself is additionally involved in a Hoogsteen type of interaction with U170.

*nucleobases and the *ribose moieties of the *nucleotide.

The motif is defined as a single-stranded *adenine that interacts via *hydrogen bonds and *van der Waals contacts in the *minor groove of a given *base pair (either AU or GC). The A-minor motif is classified into two types. Type I describes the interaction of the adenosine with both nucleotides involved in the base pair, whereas type II represents just the interaction with one out of the two base-pair-forming nucleotides. BORIS FÜRTIG, HARALD SCHWALBE

AMP →adenosine monophosphate, →adenosine phosphates, →nucleoside phosphates

amplicon DNA products synthesized using *amplification techniques, such an amplified segment of a *gene or DNA is known as an amplicon. It could also be applied to a cloned, amplified DNA sequence, but is commonly used to describe sequences derived from the *polymerase chain reaction or the *ligase chain reaction. DAVID LOAKES

amplification Production of multiple copies of a DNA sequence starting with one or a few copies. The process of amplification can occur *in vivo by the production of many copies of a *plasmid in a bacterial cell or in culture (→cloning) or *in vitro, which is generally carried out through the *polymerase chain reaction (PCR). Amplification by most *polymerases (DNA and RNA) is usually linear (i.e. a single copy is produced during each round of replication), but occurs logarithmically during the PCR amplification of the target sequence. DAVID LOAKES

ANA →altritol nucleic acid

anchored primer *Genomic DNA and *messenger RNA (mRNA) contain significant regions of repeat sequences or polypurine/polypyrimidine regions, e.g. *poly(A) tails. Such regions are particularly difficult to accurately reproduce or sequence. Thus, an oligo(dT) primer on a poly(A) *template would have many *binding sites resulting in *primer slippage and extension products of varying length and *nucleotide composition as a direct result of mispriming. A method to fix a primer to such a sequence is the use of an anchored primer, which usually contains two specific *nucleotides at the 3'-end that are complementary to the *template sequence. For example, to anchor a primer to a poly(A) sequence which terminates with the nucleotide sequence CG the primer would be designed with GC at the 3'-end of the oligo(dT) region. It anchors the oligo(dT) portion of the primer directly at the junction between the poly(A) tail and the end of the *transcript, and anchors the oligo(dT) primer only to the subset of transcripts that are complementary at the dinucleotide position.

Anchored primers are of particular use in the differential display *polymerase chain reaction (DD-PCR), which is a method for identification of differentially expressed genes by comparative display of arbitrarily amplified *cDNA subsets. The essence of the differential display method is to use, for *reverse transcription, an anchored oligo(dT) primer that anneals to the beginning of a subpopulation of the poly(A) tails of mRNAs. The anchored oligo(dT) primers consist of 11–12 thymidines and two additional 3'-nucleotides to provide specificity. DAVID LOAKES

annealing of oligonucleotides Pairing of single-stranded RNA or DNA to a *complementary sequence by hydrogen

	R^1	R^2
Idarubicin	H	acetyl
Daunomycin	OCH_3	acetyl
Doxorubicin	OCH_3	2-hydroxyacetyl

Epirubicin

anthracyclines

bonding, forming a double-stranded *oligonucleotide. It is mainly used to describe the process of *primer binding to a DNA strand during the *polymerase chain reaction. DENISE STROHBACH

anthracyclines Class of antibiotics used as chemotherapeutic agents to treat a wide range of cancers (e.g. leukemias, lung cancer). They inhibit *DNA replication due to the *intercalation with nucleolar DNA forming a highly stabilized complex. The aromatic D-ring interacts via the *major groove of DNA and the A-ring inserts into the *minor groove. The amino-sugar can make further *hydrogen bonds with the DNA, increased by the presence of water. BETTINA APPEL

antibiotics Compounds that hamper the growth of microorganisms. Several antibiotics interact with DNA or RNA and thus inhibit essential steps of *nucleic acid metabolism (→anthracyclines, →ribosome interaction with antibiotics). SABINE MÜLLER

anticodon Single-stranded 3-nucleotide stretch (positions 34, 35 and 36) at the end of *anticodon arm of a L-shaped *transfer RNA. Rapid formation of a short duplex stretch with the corresponding *complementary *messenger RNA *codon in the ribosomal *A-site is the prerequisite for *decoding of the *genetic code during *translation. STEFAN VÖRTLER

anticodon arm One branch of the L-shaped three-dimensional structure of a *transfer RNA, which carries the *anticodon at its end ready to interact with *messenger RNA at the 30S *ribosomal subunit. It is formed by coaxial stacking of the double-stranded *TΨC and anticodon helices of a tRNA cloverleaf. The *acceptor arm forms the second branch. STEFAN VÖRTLER

anticodon loop Single-stranded loop region at the end of the *anticodon arm of a *transfer RNA containing the 3-nucleotide *anticodon. The loop contains 7 *nucleotides at positions 32–38 in the fringe being stacked to the double-helical stem and an invariant U33 inducing a *uridine turn in the *phosphate backbone. All of this leads to a pre-structured, accessible orientation of the anticodon bases ready to base pair with the mRNA. STEFAN VÖRTLER

***anti* conformation** The *anti* conformation corresponds to a *glycosidic torsion angle κ in the range $180 \pm 90°$. Due to steric hindrance, nucleosides and nucleotides are usually found in the *anti* conformation with $\kappa = -135 \pm 45°$. →glycosidic torsion angle. KLAUS WEISZ

antiparallel orientation Mutual orientation of two chains in *double-stranded DNAs of the B-, A- and Z-families. In the case of antiparallel orientation of two DNA strands, their 5',3'-*phosphodiester linkages run in opposite directions. NINA DOLINNAYA

antisense drug *antisense oligonucleotide used as a therapeutic agent in the treatment of various types of diseases, e.g. cancer or viral infections. The first antisense drug, vitravene, has been approved by the US Food and Drug Administration for the treatment of cytomegalovirus retinitis. IRENE DRUDE

antisense gene Sequence that is expressed as a RNA complementary to the *messenger RNA of a given *gene. Antisense genes efficiently knockdown the expression of endogenous genes and substantiate the *RNA interference knockdown strategies. MAURO SANTOS

antisense oligonucleotides (ASONs) DNA oligomers of 15–20 nucleotides in length that bind to a complementary target RNA by *Watson–Crick base pairing are referred to as ASONs. Their potential to inhibit *gene expression in a sequence-specific manner has been explored since the late 1970s. ASONs hybridize to their *target RNA and induce a cellular endonuclease, *RNase H, which recognizes the unusual *DNA–RNA hybrid and cleaves the RNA moiety of the *heteroduplex. As a consequence, the protein encoded by the targeted RNA is not synthesized. A second mechanism by which ASONs inhibit *translation is a steric blockade of the *ribosome.

```
5'————————CCTAACCGTCATGACATG————————3'  target RNA
              | | | | | | | | | | | | | | | | | |
           3' GGATTGGCAGTACTGTAC 5'   Antisense Oligonucleotide
```

Phosphorothioate DNA   **2'-O-methyl RNA**   **Locked Nucleic Acids (LNA)**

Furthermore, ASONs can be directed against *splice sites to alter the splicing pattern of a *pre-mRNA.

Since natural DNA *oligonucleotides are rapidly degraded in biological fluids, ASONs have to be protected against nucleolytic degradation by the introduction of *modified nucleotides. The most widely used building blocks for ASONs are phosphorothioates in which one of the non-bridging oxygens is replaced by a sulfur atom. *Phosphorothioate oligonucleotides are stable and activate RNase H, but they have a comparably low target affinity and toxic side-effects have been observed due to unspecific binding to certain proteins. Therefore, a second generation of modified nucleotides has been developed containing alkyl groups at the 2'-position of the ribose (e.g. 2'-O-methyl RNA). In recent years hundreds of new modifications have been introduced to further improve the properties of ASONs (e.g. *locked nucleic acids). Chemical modifications of the *ribose usually result in a loss of the ability of ASONs to activate RNase H. To retain this activation, gapmers can be employed which contain modified nucleotides at both ends of the oligonucleotide for protection against exonucleases and DNA monomers or phosphorothioates in the center to induce RNase H cleavage of the target RNA. Another major challenge for *in vivo* applications of ASONs is their efficient delivery to the cells in the targeted tissue.

JENS KURRECK

antisense RNA Cellular and viral *gene expression and the *transcription of *non-coding RNA give rise to transcribed RNAs that are essential for controlled biological processes in living cells. Among the commonly used *antisense strategies, antisense RNA is one out of various chemical forms of *nucleic acids. Those include long-chain nucleic acids that are usually RNA as well as short *antisense oligonucleotides both of which are complementary in nucleotide sequence with a given natural *transcript of above origin. Thus, antisense RNA represents one form of *antisense drugs that may be transcribed *in vitro* by *RNA polymerases such as the phage *T7 RNA polymerase prior to cellular delivery or it may be transcribed intracellularly in living cells. In the latter case, *antisense strands are transcribed by cellular polymerase from recombinant *antisense genes that have to be introduced into target cells and tissues by *gene transfer using strategies that are developed and exploited in the field of *gene therapy. Antisense genes may be under the control of constitutive *promoters giving rise to stable expression of antisense RNA. In the case of viral target RNA this concept has been termed "intracellular immunization". For purposes of correcting aberrant cellular gene expression in a transient fashion one also uses *inducible promoters. This method of transient target suppression by antisense RNA is commonly used in the fields of gene function analysis and target validation. GEORG SCZAKIEL

antisense strand Non-coding strand of *double-stranded DNA that serves as the *template for *messenger RNA synthesis during *transcription (→antisense gene).

antisense strategy *Messenger RNAs (mRNAs) transfer *genetic information (in sense orientation) from the *genomic DNA to the place of protein synthesis. *Oligonucleotides with a complementary sequence to a certain mRNA can be employed to bind to the targeted mRNA and prevent its *translation. These oligonucleotides can be viewed as being orientated in the antisense direction. Originally, *antisense oligonucleotides have been utilized for this purpose, but in a broader sense subsequently developed methods like *ribozyme technologies and *RNA interference can be considered to be antisense strategies as well. The term (post-transcriptional) *gene silencing is also frequently used to depict these applications.

Antisense agents hybridize to their target RNA by *Watson–Crick-type base pairing. Since a sequence of at least 16 nucleotides is statistically unique in the human genome, binding of oligonucleotides of this length can be expected to be highly specific. Antisense strategies therefore allow precise targeting even in cases in which low-molecular-weight compounds are not sufficiently specific due to structural similarities of closely related proteins.

Antisense strategies are widely employed techniques for functional genomics studies. Although the complete sequence of the human genome is now available, the precise function of many *genes and their encoded proteins is not yet known. Antisense strategies allow reverse genetic approaches to elucidate the role of genes by specifically inhibiting their expression. The resulting loss-of-function *phenotype can be analyzed to draw conclusions about the function of the lacking protein. Antisense strategies can be seen as a complementary approach to the generation of knock-out animals. They are, however, more straightforward, faster, cheaper and can even be employed in cases where the lack of the gene under investigation is lethal during embryonic development.

Furthermore, antisense strategies are promising approaches to treat diseases caused by the expression of deleterious genes. *Antisense oligonucleotides, *ribozymes and *small interfering RNAs have thus been tested in clinical trials to treat a broad range of diseases, including cancer, viral infections and inflammatory diseases. JENS KURRECK

antitermination Mechanism of *transcriptional control in bacteria in which *termination is prevented at a specific terminator site, allowing *RNA polymerase to express *downstream *genes. Transcription of *non-coding genes, such as *ribosomal RNA and *transfer RNA, which are mostly highly structured, constitutes a problem for the contiguous synthesis of RNAs. Here, RNA polymerase needs additional proteins, called antitermination factors (Nus proteins in phage λ or the ribosomal protein S10 in *Escherichia coli*) which prevent RNA polymerase from pausing. BEATRIX SÜß

A-platform →AA-platform

AP-site →abasic site

aptamers (Latin: *aptus* = "to fit") Aptamers are macromolecules composed of DNA,

RNA or modified nucleic acids (→ modified DNA, → modified RNA) that bind tightly to a specific molecular target. Their properties are defined by their *nucleotide sequence and most aptamers are in a size range of 15–60 nucleotides. The chain of nucleotides forms intramolecular interactions that fold the molecule into a complex three-dimensional shape, which allows it to bind tightly against the surface of its target molecule. *Induced fit and adaptive binding play important roles in aptamer–target interaction. As a high diversity of molecular shapes exists within the universe of all possible nucleotide sequences, aptamers may be obtained for a wide array of molecular targets, including most proteins and many small molecules.

Aptamers are typically isolated by the *SELEX process (→ aptamer selection). Aptamers can distinguish between closely related, but non-identical molecules, or between different functional or conformational states of the same macromolecule. In addition to high specificity, aptamers have very high affinities to their targets, typically in the picomolar to low nanomolar range. Although aptamers are chemically rather stable, they must be chemically modified for *in vivo* applications to reduce their sensitivity to enzymatic degradation or to improve their pharmacokinetics.

Aptamers hold tremendous potential as targeted molecular therapeutics. Furthermore, they are used for target validation, drug (small-molecule) screening, sensor development (→ reporter ribozyme, → RNA sensor) and affinity separations.

In nature, aptamer domains have been discovered as components of *riboswitches.

The term aptamer is sometimes also used for high-affinity peptide ligands (peptide aptamers). ANDRES JÄSCHKE

aptamer selection Procedure for the selection of *aptamers and a variant of the *SELEX process. The selection procedure allows the isolation of aptamers on the basis of binding between a target and *nucleic acid molecules, and relies on standard molecular biology techniques. Steps of the selection cycle are (a) library preparation, (b) selection, (c) amplification and (d) aptamer isolation. In step (a), a large library (or *sequence pool) is synthesized. Each molecule in the library (typical complexity of up to 10^{15} different compounds) contains a unique nucleotide sequence that can, in principle, adopt a unique three-dimensional shape. Only very few of these molecules – the aptamers – present a shape that is complementary to the target molecule. The selection step is designed to find those molecules with the greatest affinity for the target of interest. The library is incubated with the target and the library members will either associate with the target or remain free in solution. Several methods exist to physically separate the aptamer target complexes from the unbound molecules in the mixture. The most common variants of the selection step are the use of immobilized or biotinylated targets, filter binding of RNA–protein complexes, or chromatographic/electrophoretic isolation. The unbound molecules are discarded, and the target-bound aptamers released and copied enzymatically to generate a new library of molecules that is enriched for those species that can bind the target. This enriched library is used to initiate a new cycle of selection and binding. Typically, after 5–15 cycles, the library is reduced from the 10^{15} unique sequences to a small number of tight-binding molecules. Individual members of this final pool are then isolated, their sequence determined, and their properties with respect to binding affinity and specificity are measured and compared.

Aptamers have been isolated from RNA, DNA and modified nucleic acid libraries. Several variations of the selection technique exist. Aptamers with high affinity can be found for virtually any target, from small organic molecules to peptides, proteins, cell surfaces and tissue sections. Aptamers can be selected either manually or in an automated fashion. ANDRES JÄSCHKE

aptazyme *Ribozyme, the activity of which is controlled by an external *ligand (→allosteric ribozyme). In general, aptazymes consist of two domains – the *aptamer domain that binds specifically to a ligand (also called effector) and the ribozyme domain that catalyzes the cleavage reaction depending on this binding event. In this way the cleavage activity of a ribozyme can be manipulated in both directions: the binding event results either in structural stabilization or destabilization of the ribozyme domain and causes in up- or down-regulation of ribozyme activity.

*Oligonucleotides and *nucleotides, amino acids and proteins, small organic compounds, and even metal ions have been found to be potential ligands. A variety of nucleic acid-based *biosensors use the principle of allosterically switchable ribozymes. DENISE STROHBACH

apurinic site Special case of an *abasic site.

apyrimidinic site Special case of an *abasic site.

1-β-D-arabinofuranosyladenine
→arabinonucleosides

1-β-D-arabinofuranosylcytosine
→arabinonucleosides

1-β-D-arabinofuranosylguanine
→arabinonucleosides

1-β-D-arabinofuranosylthymine
→arabinonucleosides

1-β-D-arabinofuranosyluracil
→arabino-nucleosides

arabinonucleosides Structural analogs of *ribonucleosides in which *ribose is replaced with *arabinose.

arabinose Pentose monosaccharide ($C_5H_{10}O_5$; 150,13 g mol^{-1}) that exists in two enantiomeric conformations: D(−)-arabinose and L(+)-arabinose.

arabinose operon (*ara*) Controls the utilization of L-arabinose in *Escherichia coli*. The three structural genes *araB*, *araA* and *araD* form the *araBAD* *operon, and encode enzymes that convert *arabinose to xylulose-5-phosphate, which can be metabolized via the pentose phosphate pathway.

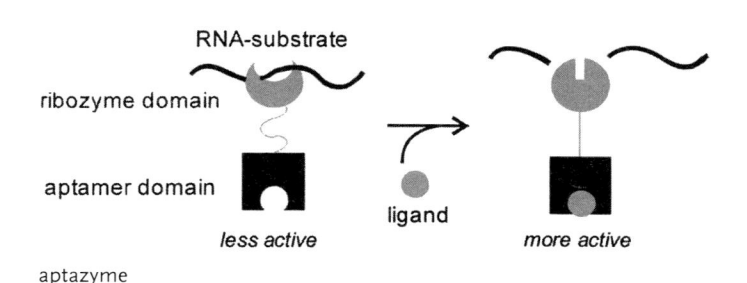

The pBAD promoter is regulated by the arabinose-binding protein AraC. The *araC* *gene is divergently transcribed from the *araBAD* operon.

Depending on the intracellular arabinose concentration, AraC acts either as a *repressor or *activator. In the absence of arabinose, an AraC dimer binds simultaneously to two sites positioned 63 (araI$_1$) and 279 (araO$_2$) residues *upstream of the transcriptional start site of *araBAD*. The intervening DNA is looped and access of the *RNA polymerase to the pBAD promoter is blocked. AraC also is an autoregulator and prevents *transcription from its own *promoter by this mechanism.

In the presence of arabinose, binding of the sugar introduces a conformational change in AraC, converting it into an activator protein that binds to araI$_1$ (–63) and araI$_2$ (–43). Transcription of the *araBAD* operon requires additional binding of the cAMP–CRP (CAP) complex adjacent to AraC (at –92) to ensure that arabinose is used only if the better carbon source glucose is absent.

A series of expression vectors has been constructed on the basis of the pBAD promoter. *Plasmids containing the *araC* gene and the pBAD promoter followed by a *multiple cloning site can be used to clone a gene of interest. Expression of the plasmid-encoded gene in *E. coli* can be tightly controlled by the addition of glucose (repression) or arabinose (induction). FRANZ NARBERHAUS

arabinosides →arabinonucleosides

Argonaute Core component of *RISC, the effector complex of the *RNA interference pathway. In flies, two Argonaute homologs are present. Ago2 is associated with *small interfering RNAs in RISC and responsible for *messenger RNA (mRNA) cleavage, whereas Ago1 is associated with *micro-RNAs and responsible for inhibition of *translation of the targeted mRNA. NICOLAS PIGANEAU

artificial chromosome Linear DNA vector designed to be introduced in eukaryotic cells, and which is stably maintained in the nucleus by means of a *centromere and two *telomere-like termini. It is originally designed as a circular *shuttle vector that can be maintained and amplified in *Escherichia coli*. The vector is linearized using specific *restriction sites, and the region of interest or *insert (stretching from 100kb to 1Mb) is subsequently cloned and transfected into eukaryotic cells, where it is maintained as a nuclear *episome. GEMMA MARFANY, ROSER GONZÀLEZ-DUARTE

artificial nuclease Molecule that imitates the mechanism of enzymes that cleave RNA. These *nucleases can be metal complexes based on Cu^{2+}, Zn^{2+} and lanthanide ions. Moreover, constructs carrying imidazole moieties that are thought to simulate the two histidines in the *active site of *RNase A have been described. Artificial nucleases can also be *oligonucleotides themselves. An example for the latter agents are oligonucleotide-based artificial nucleases, which are *oligonucleotides (in many cases modified to prevent them from being degraded) carrying the cleaving agent covalently linked. →antisense strategy, →artificial ribozyme, →nuclease. JÖRN WOLF

artificial ribozyme *Ribozymes not known to exist in nature. While the currently known natural ribozymes catalyze only a narrow spectrum of chemical reactions (*phosphodiester hydrolysis, formation and *transesterification, and peptide bond formation), *in vitro* selection has successfully been used to isolate *catalysts for various other chemical reactions, such as aminoacylation, *RNA polymerization, N-*glycosidic bond formation and cleavage, *pyrophosphate bond formation and cleavage, amide bond formation, N-alkylation, S-alkylation, porphyrin metalation, biphenyl isomerization, Diels–Alder reactions, Michael or Aldol reactions, and oxidations and reductions.

For the isolation of artificial ribozymes, two different variations of the *SELEX cycle have been developed. The first is the selection of *aptamers using *transition-state analogs as targets and the subsequent screening of the selected aptamers for *catalysis of the reaction that proceeds via the respective transition state. While such aptamers were frequently generated, only few of them displayed catalytic activity and rate acceleration was typically low. The second, more successful strategy is direct selection, which initially allowed only the isolation of RNA-modifying ribozymes. A *RNA library is mixed and incubated with a substance that might attach itself covalently to the RNA. RNA molecules that accelerate such a tagging reaction are then physically separated from the unmodified members of the RNA library, enzymatically amplified, and subjected to further rounds of selection and amplification. By tethering a second substrate to the RNA pool members, it has become possible to expand the scope of RNA catalysis to reactions between two small organic molecules, like the Diels–Alder and Aldol reactions.

While most artificial ribozymes are *single-turnover, self-modifying RNAs, some of them act as true enzymes with *multiple turnover and obey the laws of enzyme kinetics (→Michaelis–Menten kinetics, →Michaelis–Menten model). The catalytic repertoire of ribozymes can be enhanced by incorporating *modified nucleotides that contain additional functional groups. The catalytic versatility of artificial ribozymes provides support for the *RNA world hypothesis. ANDRES JÄSCHKE

A-site →aminoacyl-tRNA binding site

association For *nucleic acids the term association mostly refers to formation of *secondary or *tertiary structures, e.g. formation of a duplex upon *hybridization of two singe nucleic acid strands, or interaction of secondary structure elements such as loop–loop interactions or domain docking. Furthermore, any molecule binding to a nucleic acid associates to it.

A-stem *Acceptor stem–loop of a *transfer RNA

asymmetric base pair Any base interaction that does not have the regular (symmetric) *Watson–Crick geometry.

atac splicing Splicing of *introns marked by non-canonical splice site *consensus sequences. These "rare" introns have been found in a number of metazoan *genes. The splice site sequences mostly consist of AT (5′-splice site) and AC (3′-splice site) nucleotides, therefore "atac" introns. Since the U12-*small nuclear RNP (snRNP), instead of the U2-snRNP, is involved in removal of those rare introns, they have been designated as *U12-type introns. Together with three other rare *U-snRNAs (U11, U4atac and U6atac), U12-snRNA is a constituent of the *minor spliceosome. In vertebrates, the frequency of occurrence of the U12-type introns is about 0.2% relative to the major GT–AG introns, now often designated as U2-type introns. BERND-JOACHIM BENECKE

AT content The amount of adenosine–thymidine base pairs in double-stranded nucleic acids (→GC content).

atomic force microscopy →scanning force microscopy

ATP →adenosine triphosphate, →adenosine phosphates, →nucleoside phosphates

attenuation Regulation of a wide variety of bacterial *operons by controlling *termination of *transcription at a site located before the first structural *gene. Classically, attenuation occurs when the transcribed RNA *upstream of an operon has the ability to fold into two mutually exclusive *RNA-fold structures, one which is termed an antiterminator and the other a terminator. If the terminator hairpin loop is allowed to fold, transcription is ultimately halted. Alternatively, if the antiterminator structure folds, the terminator is precluded from folding and transcription of the operon proceeds. The mechanisms that alternate between these two RNA folds (terminators and antiterminators) are quite diverse and can be mediated by direct RNA–ligand interaction (→riboswitches), by uncharged tRNA (→transcriptional attenuation) or stalled ribosomes (→translational attenuation). Regulation by *antitermination can be differentiated from attenuation by the fact that alteration of the transcription complex (rather than alternate RNA structures) decreases the efficiency of *downstream terminators (→translational control). BEATRIX SÜß

attenuator →attenuation

autosome *Chromosome that appears independent of the sex of a certain eukaryotic organism.

5-azacytidine (5-AC) *Cytidine analog that is used for studying *DNA methylation. If incorporated into DNA, 5-AC changes the methylation pattern. Due to the nitrogen atom at position 5 of the cytosine residue, enzymatic methyl transfer to this position is not possible. This inhibition of methyltransferases leads to undermethylation of the respective DNA and *gene activation, because many genes are turned off by cytosine methylation, particularly in *promoter sequences. SABINE MÜLLER

azidothymidine (Zidovudine) Analog of *thymidine in which the hydroxyl group at C3′ is replaced by an azido group. Azidothymidine is an antiretroviral drug that inhibits the activity of *reverse transcriptase. The azido group increases the lipophilic character of azidothymidine, such that cell membranes can be crossed easily by diffusion. The effective form of azidothymidine is its 5′-triphosphate, which can be synthesized by cellular enzymes. BETTINA APPEL

B

B Abbreviation for *base (→nucleobase).

backbone →sugar–phosphate backbone

band shift assay Method for detecting *DNA-binding proteins [also called gel shift, gel retardation or electrophoretic mobility shift assay (EMSA)]. A band shift is observed when a protein forms a complex with a DNA fragment, because complexes of protein and DNA migrate through a non-denaturing polyacrylamide gel more slowly than free DNA fragments or double-stranded oligonucleotides. An advantage of studying DNA–protein interactions by an electrophoretic assay is the ability to resolve complexes of different stoichiometry or *conformation. Another major advantage for many applications is that the source of the DNA-binding protein may be a crude nuclear or whole-cell extract rather than a purified preparation. Gel shift assays can be used qualitatively to identify sequence-specific DNA-binding proteins (such as *transcription factors) in crude lysates and, in conjunction with *mutagenesis, to identify the important binding sequences within a given gene's *upstream regulatory region. Band shift assays can also be utilized quantitatively to measure thermodynamic and kinetic parameters. The ability to resolve protein–DNA complexes depends largely upon the stability of the complex during the brief time it is migrating into the gel. Sequence-specific interactions are transient and are stabilized by the relatively low ionic strength of the *electrophoresis buffer used. Upon entry into the gel, protein complexes are quickly resolved from free DNA, in effect freezing the equilibrium between bound and free DNA. In the gel, the complex may be stabilized by caging effects of the gel matrix, meaning that if the complex dissociates, its localized concentration remains high, promoting prompt re-association. Therefore, even labile complexes can often be resolved by this method. TOBIAS RESTLE

band shifting Observation of certain nucleic acid species migrating as a band in a gel during *gel electrophoresis (→band shift assay). Classically, the binding of *DNA-binding proteins has been the majority of cases in which band shifts were observed originally. This includes many examples of the binding of *transcription factors and *repressors to their cognate *binding site within *double-stranded DNA, being part of control regions for *gene expression. In addition to *DNA–protein interactions, other interactions of macromolecules with motifs contained within nucleic acids can also be visualized by band shifting. These include interactions between antibodies and protein–nucleic acid complexes, which is called *supershift, as well as interactions between proteins and single-stranded DNA or single- and double-stranded RNA. The observation of band shifting requires a certain minimal *binding affinity of the binding partners under the given experimental conditions. At the experimental level, band shifting does not provide very strong experimental evidence for the existence of protein–nucleic acid interactions, it has to be interpreted with great care and it needs a number of appropriate controls for binding specificity. GEORG SCZAKIEL

bar gene Confers resistance to the herbicide bialaphos and was originally cloned from *Streptomyces hygroscopicus*. It encodes a phosphinothricin acetyl transferase enzyme. MAURO SANTOS

base →nucleobase

base analog →modified base, →rare base

base composition Ratio that indicates the number of adenine and thymine bases compared to the number of guanine and cytosine bases in a DNA double strand (→nucleotide sequence). In taxonomy it is used to classify bacterial species. VALESKA DOMBOS

base edge →edge-to-edge interaction of nucleobases

base excision repair Mutation prevention process intercepting pre-mutagenic *DNA lesions before mutation fixation by *replication occurs. Base excision repair proceeds, in the simplest case, in four steps. (a) Removal of the chemically damaged *nucleobase by a specialized *DNA glycosylase. (b) Strand cleavage at the 5′-side of the resulting *abasic site by endonucleolytic incision brought about by an AP-endonuclease with formation of a 3′-hydroxyl terminus. (c) Extension of that *primer end by at least 1 nucleotide with removal of the 5′-phosphorylated, base-free sugar residue from the neighboring 5′-terminus creating a 5′-phosphate end (potentially followed by a short patch of *nick translation). (d) Sealing of the strand by *DNA ligase. The original information is restored in step (c) since the *DNA polymerase recruits the undamaged nucleotide opposite the pre-mutagenic lesion as the *template. Several alternatives are available for step (b). Certain AP-endonucleases cleave other *phosphodiester bonds next to an AP-site (note that there are four alternatives, two each at the 5′- and the 3′-side of the base-free sugar moiety). Alternatively, strand cleavage at the abasic site can be brought about by β-elimination (as sketched under *abasic site); the corresponding catalytic activity (AP-lyase) is integral part of some *DNA glycosylases. Additional exonucleolytic trimming of the primer end is required in all cases in which step (b) does not directly produce a 3′-hydroxyl terminus. Examples of DNA lesions typically processed by base excision repair are DNA-U (resulting from hydrolytic deamination of C), 3-me-A, 8-hydroxy-G (7-H,8-oxo-G, "8-oxo-G", "OG") and *pyrimidine residues having a saturated 5,6 C–C bond. HANS-JOACHIM FRITZ

base flipping DNA base flipping was first observed in the crystal structure of the *DNA cytosine-C5 methyltransferase (MTase) from *Heamophilus haemolyticus* (M.*Hha*I) in complex with a short duplex *oligodeoxyribonucleotide and the cofactor product S-adenosyl-L-homocysteine. In this complex the target cytosine including the 2′-deoxyribose and both adjacent phosphates are rotated out of the DNA by about 180° (thus base flipping is also called nucleotide flipping) and placed into the *active site next to the cofactor. This energetically unfavorable DNA conformation is stabilized by multiple specific and unspecific contacts of the enzyme to the deformed DNA and the opened space in the DNA is filled by a glutamine residue (Q237) of M.*Hha*I which forms hydrogen bonds with the orphaned guanine.

In addition, base flipping was observed in the protein–DNA cocrystal structures of the DNA cytosine-C5 MTase M.*Hae*III and the DNA adenine-N6 MTases M.*Taq*I and Dam from bacteriophage T4.

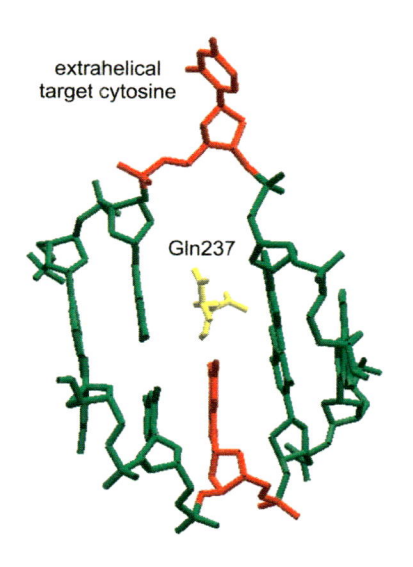

Furthermore, several biochemical assays have been developed to detect base flipping by placing *modified nucleobases at the target position of DNA MTases. These include tighter binding to DNA *mismatches, high *photo-crosslinking yields of 5-idouracil, high reactivity of thymine towards permanganate and a strongly increased *2-aminopurine fluorescence. The *fluorescence increase of 2-aminopurine has also been used to analyze the kinetics of base flipping induced by DNA MTases. They first bind to their DNA *recognition sequences and then flip out their target nucleobases with rate constants (half-lives in the millisecond timescale) which are much higher than their *turnover numbers. From the wealth of available biochemical data it is now concluded that all DNA MTases use a base flipping mechanism.

However, base flipping is not restricted to DNA MTases and has also been observed in many protein–DNA cocrystal structures of DNA *base excision repair *glycosylases which recognize damaged nucleobases in DNA. Examples include pyrimidine dimer DNA glycosylase T4 (endonuclease V), *uracil DNA glycosylase (UDG), 3-methyladenine DNA glycosylase (AAG), mismatch-specific uracil DNA glycosylase (MUG), 3-methyladenine DNA glycolsylase II (AlkA), 8-oxoguanine DNA glycosylase (hOGG1), endonuclease VIII (Nei) and formamidopyrimidine DNA glycosylase (Fpg; also called MutM). Furthermore, crystal structures of T4 phage β-glucosyltransferase and DNA photolyase, which repairs cyclobutane pyrimidine dimers in a light-driven process, demonstrate that they bind their targets in an extrahelical position. These numerous examples demonstrate that base flipping is an important mechanism for many DNA-modifying enzymes, like DNA MTases and DNA glycosylases, and allows these enzymes to gain excess to their target nucleobases which are otherwise hidden in the DNA base stack.

In addition to enzymes, small synthetic molecules can bind to DNA *mismatches and induce base flipping. The bisacridine macrocycle BisA is thought to intercalate into DNA, sandwich one of the neighboring base pairs and flip one of the mismatched nucleobase out of the DNA helix. Another example is binding of two naphthyridine-8-azaquinolone conjugates to DNA containing CAG repeats (A/A mismatch). The 8-azaquinolone moieties recognize the mismatched adenine residues and the naphthyridine moieties bind to the neighboring guanine residues leading to extrusion of the two cytosine residues from the DNA helix. ELMAR WEINHOLD

base modification →rare nucleotides

base pair Hydrogen-bonded pair of *nucleic acid bases. *Hydrogen bonds involving the amino and carbonyl groups of *purine and *pyrimidine bases are the important mode of interaction between bases. If it is supposed that a stable base pair should contain two hydrogen bonds as a minimum, four nucleic acid bases could form 28 homo and hetero pairs. The most important hydrogen-bonding patterns are those defined by Watson and Crick in 1953 in which A bonds specifically to T (or U) and G bonds to C (→Watson–Crick base pair). These two types of base pairs predominate in *double-stranded DNA and RNA. This specific pairing of bases permits the duplication of *genetic information by the synthesis of nucleic acid strands that are complementary to existing strands. Hydrogen bonds between bases permit a complementary *association of two and occasionally three or four strands of nucleic acid.

→edge-to-edge interaction of nucleobases. NINA DOLINNAYA

base pairing rules →base pair, →Watson–Crick base pair

base pair roll Orientation of the *base pair as a whole (the best mean plane through the purine and pyrimidine) about its long axis. If two successive base pairs along the helix are rolled in opposite directions, they open up an angle between them towards either the *major or the *minor groove. The roll angle from one base to the next is defined as positive if the angle between them opens to toward the minor groove. There is no way to measure local base roll angles from fiber diffraction data, but single-crystal studies show that base roll is one of the important helix parameters. Since in *B-DNA the base pairs are nearly perpendicular to the helix axis, there need to be no systematic mean roll angle from one base pair to the next. In the A-form, the DNA ladder is wrapped around the helix axis in a way that requires an accordion like opening of base pairs toward the minor groove, and the mean roll angle in *A-DNA is $6°$.
NINA DOLINNAYA

base pair tilt Inclination of *nucleobases with respect to the axis. In *B-DNA the tilt is close to zero (the *base pairs are stacked nearly perpendicular to the helix axis) and the axis runs through the center of each base pairs. In A-DNA the base pairs are tilted between 13 and $19°$ from perpendicular. NINA DOLINNAYA

base pair twist Rotation between one *base pair and its neighbor. Helices are described as right-handed if they follow the rule: they show clockwise rotation when viewed along their axes and moving from front to rear. As a result of its handedness, a helix has chirality. NINA DOLINNAYA

base replacement → point mutation

base sequence → nucleotide sequence

base stacking One of two main forces that stabilize the nucleic acid *double helix. Stacking interactions represent the important mode of interactions between two or more heterocyclic bases which are positioned with the planes of their aromatic rings parallel. The stacking involves a combination of *van der Waals, *dipole–dipole and *hydrophobic interactions between the bases or base pairs. These interactions help to minimize the contact of bases with water. The intensity of base stacking decreases in the following order: purine–purine>pyrimidine–purine>pyrimidine–pyrimidine, with stacking interaction of bases (→base pairs) being sequence dependent. NINA DOLINNAYA

base triple A single *nucleobase exhibits three main edges that are prone for interactions (→edge-to-edge interaction of nucleobases). The three edges comprise the Watson–Crick edge (defined by the functional groups at atom positions R: 2, 1, 6 or Y: 2, 3, 4), the Hoogsteen edge (R: 6, 5, 7, 8 or Y: 4, 5) and the shallow groove edge (R: 2, 3, 4, 2′ or Y: 1, 2, 2′). Additionally, the pairing diversity is increased by the possibility to form *cis* and *trans* base pairs, where the bases are at the same or the opposite side of a line median to the hydrogen bonds. This structural plasticity of the base-pairing behavior is exploited in RNA–RNA recognition or tertiary interactions.

Arrangements of base triples are common features in *RNA structures. Examples are the base triples G1091–C1100•G1071 and C1092–G1099•C1072 that bend a 9-nucleotide *loop in the L11 binding region of the 23S RNA. The same tertiary motifs are found in the U2-*small nuclear RNA and in the loop region 715

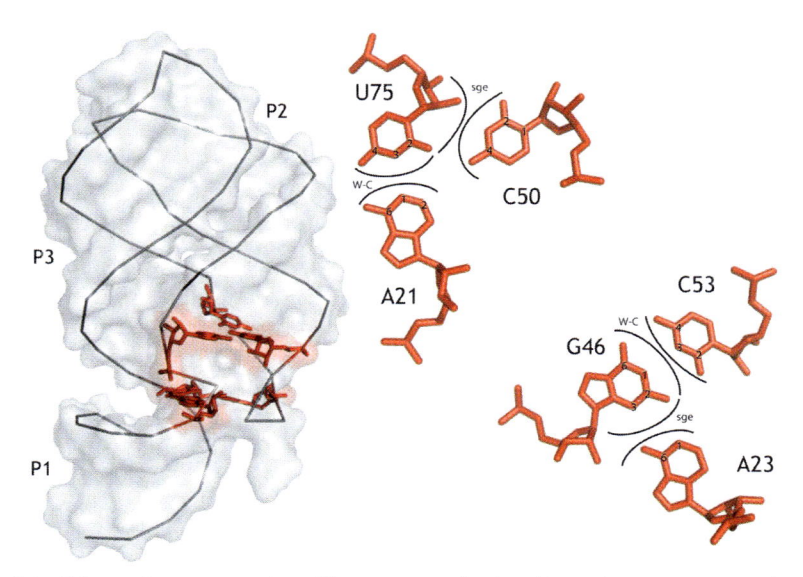

Base triple: Schematic representation of base triples found in the structure of the riboswitch RNA (left panel, PDB code: 1U8D), where they facilitate tertiary interactions. Two base triples are displayed in stick representation showing that these interactions also take place at the Watson–Crick edge (W-C) as well as on the shallow groove edge (sge).

of the 23S RNA. Frequently, base triples are found in RNA *pseudoknots, e.g. in the ScYLV RNA facilitating the fold back of the knotted strands, and in *riboswitch RNA. BORIS FÜRTIG, HARALD SCHWALBE

base triplet Set of three consecutive *nucleotides in a nucleic acid molecule (→codon, →anticodon).

B chromosomes Small and disposable supernumerary *chromosomes found in the nuclei of many eukaryote organisms that have no homologous member in the normal (A) chromosome set. B chromosomes are considered to be parasitic elements that are transmitted by non-Mendelian rules and accumulate in the progeny of their carries. MAURO SANTOS

B-DNA Most stable *double-helix structure for random sequence DNA molecules under physiological conditions (referred to as a *Watson–Crick helix). B-DNA and associated *C- and *D-DNA belong to the B-family. The *base pairs are located on the helix axis and stacked nearly perpendicular to the axis. *Major and *minor grooves are nearly equal in size. B-DNA shows around 10.5 base pairs per turn in aqueous solutions, with a spacing of 0.34nm from one base pair to the next. Double-stranded B-DNA displays systematic sequence-dependent modulations with changes in *helical twist and roll angle, and alterations in *sugar–phosphate torsion angles. The predominant furanose ring puckering mode is C2'-*endo*. B-form helix is the main biologically relevant *secondary structure of DNA. NINA DOLINNAYA

Beer–Lambert law (Beer's law) Describes the direct linearity of *absorbance and concentration of a light-absorbing species:

$$A = a(\lambda) \cdot L \cdot c$$

where A is the measured *absorbance, $a(\lambda)$ is the wavelength-dependent *absorption coefficient, L is the path length and c is the analyte concentration. The Beer–Lambert law is often applied to determine the concentration of nucleic acid solutions photometrically. Working in concentration units of molarity requires a *molar absorption coefficient ε:

$$A = \varepsilon \cdot L \cdot c$$

The Beer–Lambert law has its limitations due to chemical and instrumental factors, one being high concentrations. At concentrations higher than 0.01molL^{-1}, molecules in the analyte come in close proximity, which causes electrostatic interactions and deviations in *absorptivity. Stray light or scattering of light disturbs measurements as well as fluorescence or phosphorescence of the sample or shifts in chemical equilibria as a function of concentration. Additionally, high analyte concentration can alter the refractive index and therefore cause non-linearity. VALESKA DOMBOS

Beer's law →Beer–Lambert law

bending of DNA Specific DNA–protein interaction in which a protein bends a DNA *double helix. Measurements of DNA bending are essential for understanding the mechanism of DNA recognition. MATTHÄUS JANCZYK

bidirectional promoter *Promoter sequence that is arranged head to head and is used in both directions with *transcription start points being less than 1000 nucleotides away from each other. *Genes are situated on both sides of bidirectional promoters and can be transcribed in opposite directions. In several cases, the expression of the gene pair is coordinated and the

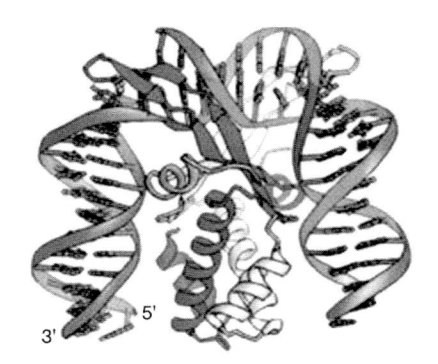

bending of DNA

*transcription products are biologically related. In humans, at least 10% of all *genes are arranged bidirectionally. →promoter.
JÖRN WOLF

bidirectional replication Found for the circular *chromosomes in *Escherichia coli*, *Bacillus subtilis* and other bacteria. *Replication of DNA begins at the *origin of replication in both directions, forming two *replication forks. Replication is terminated when the two forks meet together.
ANDREAS MARX, KARL-HEINZ JUNG

binary vector Genetic construction able to propagate in a bacterial host cell (*Agrobaterium*), designed to transfer *genes to plant cells with the aid of genetic factors located in a second resident plasmid.
GEMMA MARFANY, ROSER GONZÀLEZ-DUARTE

binding affinity Strength (→binding energy) of non-covalent chemical binding between *macromolecules as measured by the *dissociation constant of the complex.

binding energy Reflects the relative stability of a non-covalent macromolecular complex. In other words, it is the energy required to disassemble a whole into separate parts. A bound system has a lower potential energy than its constituent parts – this is what keeps the system together. The usual convention is that this corresponds to a *positive* binding energy. The driving force depends on the strength of the hydrogen bonding of the *macromolecules to each other, compared with the combined strength of their separate hydrogen bonding to water molecules; the stability of the salt bridges between them, compared with the tendency of the individual ions to be solvated by water; the dispersion energies, compared with those in water; and also the hydrophobic bonding. This applies to all molecular recognition phenomena in solution: protein–protein, protein–*ligand, protein–nucleic acid and nucleic acid–nucleic acid interactions.
TOBIAS RESTLE

binding pocket As opposed to a *binding site, a binding pocked usually represents a defined cavity on a protein or a nucleic acid. According to the key-and-lock hypothesis only certain molecules will snugly fit into this cavity to trigger a biological response, others not. This molecular recognition has three main aspects: shape, electrostatic and hydrophobic complementarity. A different scenario is described through the induced-fit mechanism (→induced-fit model). Here, initial binding of a ligand triggers a conformational change of the protein or nucleic acid creating a binding pocket. TOBIAS RESTLE

binding site Defined region on a protein or nucleic acid to which specific other molecules – in this context usually termed *ligands – attach and interact non-covalently. Such sites usually exhibit specificity, which is a measure of the types of ligands that will bind, and affinity, which is a measure of the strength of the macromolecular interaction. TOBIAS RESTLE

biomolecule Compound which naturally occurs in living organisms. A wide range of biomolecules are known and classified into four groups: small molecules (e.g. carbohydrates, disaccharides, vitamins, hormones, lipids, phospholipids), monomers (e.g. *nucleotides, phosphates, monosaccharides, amino acids), polymers (→biopolymers) (e.g. DNA, RNA, polysaccharides, peptides, proteins) and macromolecules (e.g. prions). BETTINA APPEL

biopolymer Composed of similar biomolecule subunits (→nucleotides, amino acids or saccharides) that are linked together, often by polycondensation, to build DNA, RNA, proteins, peptides and polysaccharides. Biopolymers can be produced by biological systems (i.e. microorganisms) or synthetically from biological starting materials. Some biopolymers are used as an alternative to traditional petroleum-based plastics (e.g. a blend of starch-based polymers and polyethylene), which allows for improved environmental properties. Others are developed as materials for storage and information transfer. →DNA nanotechnology. BETTINA APPEL

biotin Also known as vitamin H or vitamin B_7, biotin binds very tightly to the protein avidin (or streptavidin), forming one of the strongest known non-covalent interactions between a protein and a small *ligand. The biotin–avidin interaction is extensively utilized in affinity chromatography, where molecules containing a biotin label are isolated from a solution by incubation with a solid support containing streptavidin. Biotin can be readily incorporated into *nucleic acids (→oligonucleotide labeling). SNORRI TH. SIGURDSSON

biotin labeling Incorporation of *biotin into DNA and RNA (→oligonucleotide labeling).

biotinylation →biotin labeling.

bistable RNA RNA with two alternative *secondary structures that coexist in a monomolecular thermodynamic equilibrium.

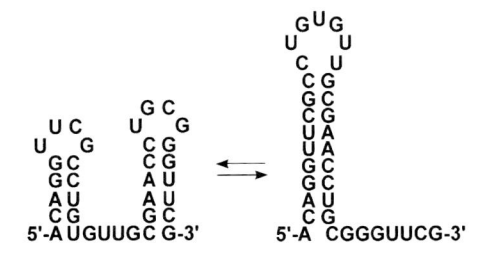

blotting Transfer of DNA, RNA or proteins onto a membrane (e.g. nitrocellulose, polyvinylidenefluoride or nylon) by pressure, capillary power or by an electric field.

In molecular biology and genetics, the blotting technique is used for the specific recognition and quantification of a desired DNA or RNA sequence by *hybridization with a DNA or RNA probe of known *sequence or for the recognition of specific proteins by binding with an antibody targeting the desired protein. In many instances the mixture of molecules to be analyzed is separated by *gel electrophoresis, followed by blotting onto a membrane. Direct transfer of samples onto a membrane is also possible. Visualization of a DNA or RNA sequence is obtained by a labeled DNA or RNA probe (→labeling) with *complementary sequence. Washing of the membrane with a solution containing the labeled DNA or RNA probe results in recognition and binding of the probe to the complementary target sequence via *Watson–Crick base pairing.

Unbound probes can be eliminated by iterative washing steps with an appropriate washing buffer. *Oligonucleotides to be analyzed can be visualized in dependence on the specific *labeling of the DNA or RNA probe. Detection can be carried out directly, if the probe is radioactively labeled (→radioactive labeling) or marked by a *fluorescent dye, or by antibodies, if the probe is labeled with a molecule (digoxygenin, *biotin) that is recognized by a specific antibody. Antibodies used to detect a specific molecule are mostly coupled with a reporter enzyme, which in the case of a positive binding event initiates either a colored or a chemiluminescent reaction on the membrane.

Depending on the type of molecule to be analyzed the blotting technique is distinguished into *Northern blot (RNA), *Southern blot (DNA), Western blot (protein), *colony blot or *plaque blot.
IRENE DRUDE

blunt end ligation Enzymatic fusion of *blunt-ended restriction fragments using *DNA ligase (→cohesive end ligation, →ligation).

blunt ends Flush ends of a linear *double-stranded DNA molecule produced by *restriction endonucleases (e.g. PvuII) or by enzymatic changes of *cohesive ends. Single-stranded cohesive DNA ends are blunted by *exonucleases or by *fill-in reactions. 5'-Single-stranded *overhangs are filled using the 5'–3' polymerase activity of a *T4 DNA polymerase or the *Klenow fragment. The 3'-single-stranded overhangs are blunted using the 3'–5' exonuclease activity of the *T4 DNA polymerase. DNA polymerases (e.g. Pwo DNA polymerase) with 3'–5' exonuclease (proofreading) activity generate blunt-ended *polymerase chain reaction products. ANNEGRET WILDE

borano nucleic acid Artificial *nucleic acid structure in which one of the non-bridging phosphate oxygen atoms is substituted with borane. It was introduced by Shaw et al., who synthesized DNA with substituted oxygen atoms. In a later study, this synthetic approach was also extended to *ribonucleoside α-boranotriphosphates. *Solid-phase synthesis of the DNA derivative was achieved using the *H-phosphonate approach. Also, incorporation of the borane group in DNA and RNA is possible, since borane substituted *nucleoside triphosphates and *dNTPs are recognized as substrates by the respective *polymerases. Apart from being isoelectric to natural nucleic acids, conformational features of DNA like torsion angles, *sugar puckering, etc., change only slightly upon introduction of the borano group.

$R = H, OH$
$B_1, B_2, B = A, C, G, T$ or U

Their potential to base pair with high specificity and affinity along with their *nuclease resistance make them extremely well-suited for therapeutic applications such as *RNA interference. JÖRN WOLF

branched DNA DNA structure with discontinuities which interrupt the classical double-helical structure and result in *branch sites. In many cases, these are formed by more than two interacting single DNA strands. *In vivo*, the single strands are held together by classical *Watson–Crick base pairs, often supported by protein components, forming *helical junctions at their *branch points. Branched DNA is very abundant in the organism, e.g. during *DNA replication (→replication fork), or during meiosis in eukaryotes, when *homologous chromosomes are juxtaposed and exchange genetic material by DNA strand cleavage and successive crossover.

In vitro, there are many possibilities to construct branched DNA. Several structures carrying branched DNA motifs have been synthesized. Those carry three to eight helical arms at fixed branch points. These arms overlap partially by Watson–Crick base pairs. As a result of their partial overlap, they have unpaired regions at the ends that are not involved in Watson–Crick pairing. These *sticky ends can pair with arms reaching out from other junctions. Thus, regular geometric structures such as cubes, polyhedra or other topological figures made of branched DNA can be formed.

Furthermore, nucleic acid interactions other than pure interactions can be used to form synthetic branched DNA *in vitro*, e.g. junctions can be formed by 5′-biotinylation of DNA at the 5′-end and successive reactions with streptavidin, forming a four-way branch site with *biotin at its center. JÖRN WOLF

branched nucleic acid →branched DNA, →branched RNA

branched RNA Any RNA in which double-helical regions alternate with unusual structures such as single-stranded regions, *bulges, *loops and *junctions. Many *ribozymes such as the *hammerhead ribozyme or *hairpin ribozyme contain branches that are necessary for their catalytic function. In these ribozymes, junctions are formed by non-covalent interactions. Other branched RNA structures in which a modified nucleotide accounts for the *branch site are formed during *splicing. The branched RNAs that are formed in this process are called *lariats. These can also be synthesized *in vitro*, e.g. by means of *DNAzymes. JÖRN WOLF

branch migration The migration of a *branch point in nucleic acid structures is very common. It is seen during the process of *crossing over, where the branch point of a *Holliday junction slides along the DNA duplexes. This process forms DNA *heteroduplexes consisting of one strand of each parental DNA duplex. The distance of the migration start and end defines the length of this heteroduplex. It also appears during the process of *DNA replication, where the parent DNA duplex branches out into two *double-stranded DNA daughter duplexes. There are also many cases reported of branch migration occurring in *RNA structures, especially during processes like folding of *ribosomal RNA or *splicing. SLAWOMIR GWIAZDA

branch point →branch site

branch site Site where a regular nucleic acid structure is split. Branch sites appear in DNA, e.g. during *replication or *recombination, as well as in RNA during *splicing. In the latter case, the branch site marks a position 18–40 *nucleotides *upstream of the 3′-splice site of a *pre-mRNA. During splicing, an intermediate with branched circular configuration is formed (→lariat). SLAWOMIR GWIAZDA

breathing →DNA breathing

breathing bubble →DNA bubble

5-bromouracil (5-BrU) Nucleobase analog that replaces thymine in DNA and thus can induce DNA *mutations. It exists in two tautomers (→tautomeric bases) with different pairing properties. While the keto form base pairs with *adenine, the enol form base pairs with *guanosine. This characteristic feature leads to *point mutations. Experimentally, 5-BrU often is incorporated into nucleic acids for crystallization to improve diffraction and to lower anisotropy (→crystallization of nucleic acids). SABINE MÜLLER

bulge Occurs in double-stranded *nucleic acid molecules exhibiting unpaired *nucleotides that form a bulge on only one strand. Whereas a single *pyrimidine bulge is mostly extrahelical, single *purine bulges have a strong tendency to stack in the *double helix (→base stacking). This finding is in line with the *stacking properties of single *nucleobases. The strength of *stacking for the single nucleobases decreases in the following order $G>A>C>>U$. Bulges with more than a single unpaired nucleotide bend the helix. The exact bend angle is a function of the exact sequence as well as of the effect of divalent ions such as Mg^{2+} that compensate repulsion forces between the negatively charged *backbone phosphates. The TAR element (→TAR RNA) of HIV represents one of the most prominent RNA structures containing a bulge. Here the bulge consists of 2 or 3 nucleotides depending on the HIV type. It plays a critical role in the essential interaction with a single arginine residue of the TAT-protein. In the free form, the bulge nucleotides of TAR are stacked between the two helices and have no *tertiary interactions with stem nucleotides. This conformation changes in the complexed state of TAT-TAR in which the bulge nucleotides are no longer stacked. A *base triple is formed between a bulge U and an AU *base pair in the upper stem of TAR to stabilize the interaction of the TAT arginine residue with a GC pair and the *phosphodiester backbone along the *major groove. BORIS FÜRTIG, HARALD SCHWALBE

C

C →cytosine

CAAT-box Part of a *conserved sequence located around 75–80 bp *upstream to the initial position of eukaryotic *transcription. The CAAT-box is recognized by many *transcription factors. MAURO SANTOS

caged nucleoside/nucleotide Biologically active *nucleosides or *nucleotides that are temporarily inactivated, e.g. by photo-labile *protecting groups. Under radiation of a certain wavelength, depending on the introduced protecting group, the *protecting group is irreversibly cleaved and the molecule regains full biological activity. The term "caged" may be confusing and therefore these molecules are sometimes also referred to as light-activated or masked nucleotides. The first reported caged nucleotide was an *adenosine triphosphate which was "caged" by a *ortho*-nitrobenzyl group.

Caged molecules, in general, are used to investigate the biology of receptors, ion channels, neuronal activation sites or the mode of action of enzymes, peptides and nucleic acids (extensively reviewed in G. Mayer et al., *Angew Chem Int Ed* **2006**, 45, 4900 ff.). Caged nucleotides could be used to control *translation or *gene regulation. Researchers could already show that caged *messenger RNA transfected into cells stayed inactive until exposure of the cells to light partly restored translational activity. VALESKA DOMBOS

calf thymus DNA DNA isolated from calf thymus and used as *carrier DNA for ethanol *precipitation of nucleic acids or during *blotting of nucleic acids to reduce *hybridization background signals caused by non-specific binding of *labeled probes to the membrane. STÉPHANIE VAULÉON

calorimetry Measures the heat of chemical reactions or physical changes as well as heat capacities. Differential scanning calorimetry (DSC) and isothermal titration calorimetry (ITC) are the most common forms of calorimetry in use for biochemical applications.

DSC measures the heat capacity of a sample as a function of temperature. In a differential scanning calorimeter, heat flow into a sample cell is measured differentially against the flow into a reference cell with an equal volume of solvent or buffer. The temperature of the calorimeter is raised linearly with time, i.e. the heating rate is kept constant. In case of a thermal transition, e.g. *denaturation of double- to single-stranded DNA, excess heat is absorbed by the sample giving rise to a peak in the DSC curve. From the integral of this peak the enthalpy of *melting and from its onset the *melting temperature can be determined.

In ITC the heat of reaction as it is taken up or released during a biomolecular interaction is monitored. If *ligand is titrated to a receptor molecule, the measured heat gives a direct readout of the extent of interaction at any given concentration. This allows not only the direct determination of the observed change in molar enthalpy, but also of the binding or *dissociation constant. Consequently, ITC allows for a full thermodynamic description of the interaction at a given temperature. KLAUS WEISZ

cAMP →cyclic adenosine-3′,5′-monophosphate

cancerogen →carcinogen

canonical sequence Also called a *consensus sequence. It can either be a sequence that is found very frequently at a defined position of a DNA strand or it can refer to a characteristic nucleotide sequence in a functional part of a *gene which occurs in the same context in other genes, even in other organisms. Well-known examples are the *TATA-box, the *CAAT-box and the *Shine–Dalgarno sequence.
MATTHÄUS JANCZYK

CAP-binding protein A 24-kDa protein with binding affinity for the CAP structure at the 5′-end of *messenger RNA (mRNA) that is thought to assist together with other initiation factors for protein *translation in binding the mRNA to the 40S *ribosomal subunit. GEORG SCZAKIEL

capillary electrophoresis *Electrophoresis technique in free solution to separate charged *macromolecules, usually proteins or nucleic acids. Differently charged molecules are separated via silica capillaries 25–100 μm in diameter and 20–100 cm in length by electro-osmotic flow. This technique only needs small amounts of sample and can be very sensitive (10^{-18} mol or below), depending on the mode of sample detection (ultraviolet detection, fluorescence detection, etc.). It is very well-suited for analytical purposes, e.g. *DNA sequencing. VALESKA DOMBOS

capping Masking of unreacted functional groups at a solid support to prevent them from reaction in the next synthesis cycle (→oligonucleotide synthesis).

cap structure Very early *post-transcriptional modification of the 5′-end of the growing eukaryotic *pre-mRNA by a nuclear guanylyltransferase. This modification results in the addition of an extra *GMP residue to the start nucleotide (+1) which is always a *purine. In this reaction group, one phosphate of the start nucleotide is removed and the GMP residue coupled to the RNA chain via an unusual 5′–5′ triphosphate bridge resulting in a m^7GpppN sequence element. Subsequently the cap structure serves as substrate for several methylation reactions. Invariantly, in a first step, a methyl group is added to position N7 of the extra GMP resulting in a 7-methylguanosine (m^7G) *nucleotide. Typically, the original start nucleotide is then methylated at the 2′-hydroxyl, resulting in a 2′-O-methyl group. Further methylation may occur at position N6, in the case that the start nucleotide is an adenine. In some species, 2′-O-methylation is also found at the third nucleotide of a capped mRNA. Capping of eukaryotic mRNA is essential for efficient initiation of *translation. Other cap structures are found in *U-snRNA and *small nucleolar RNA.
BERND-JOACHIM BENECKE, TINA PERSSON

carcinogen Chemical substance, virus, hormone or radiation which causes cancer. Cancer is characterized by uncontrolled division of cells and may spread through the bloodstream and lymphatic system to other parts of the body. Two different types of carcinogens are distinguished: genotoxic cap structure

carcinogens, which directly react with DNA (e.g. ionizing radiation), and epigenetic carcinogens, which do not damage the DNA itself, but cause alterations that predispose to cancer (e.g. arsenic). The International Agency for Research on Cancer (IARC) classified agents, mixtures and exposures into five different categories (→oncogene, →mutagen, →mutation). BETTINA APPEL

carrier DNA Most eukaryotic cells need excess *nucleic acid or carrier DNA to be transfected or transformed (yeast). The carrier DNA is used as a filling DNA to carry over the relevant DNA to the nucleus. Herring sperm DNA, *calf thymus DNA, total genomic yeast DNA or empty DNA *vectors are usually used as carrier DNA. GEMMA MARFANY, ROSER GONZÀLEZ-DUARTE

carrier RNA Sometimes RNA (mostly yeast *transfer RNA), is preferred as the carrier *nucleic acid instead of DNA in some *transfections or *transformations (→carrier DNA). GEMMA MARFANY, ROSER GONZÀLEZ-DUARTE

IARC categories of carcinogens

Category	Definition	Examples
1	carcinogenic to humans	aflatoxine, arsenic, benzene, dioxin, formaldehyde, tobacco smoking
2A	probably carcinogenic to humans	acrylamide, benzo[a]pyrene, epichlorhydrin, nitrogen mustard, ultraviolet radiation
2B	possibly carcinogenic to humans	acetaldehyde, chloroform, dichlormethane, phenolphthalein, styrene, vinyl acetate
3	not classifiable as to its carcinogenicity	aciclovir, aniline, azobenzene, dimethylformamide, eosin, hepatitis D virus, polystyrene, pyridine
4	probably not carcinogenic to humans	caprolactam

$\Delta G^{\ddagger}_{uncat}$ activation energy of uncatalyzed reaction
$\Delta G^{\ddagger}_{cat}$ activation energy of catalyzed reaction
ΔG_B free binding energy of the transition-state-complex
$\Delta G°$ standard free energy of the reaction

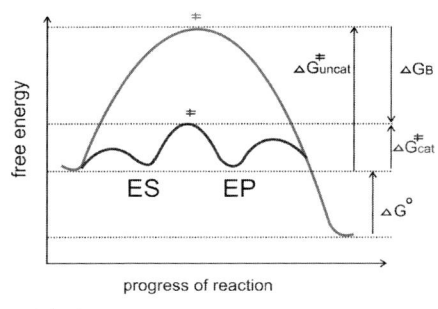

catalyst

catalysis Chemical or biochemical process mediated by a *catalyst.

catalyst Participant of a chemical or biochemical reaction that is itself not consumed by the overall reaction. Catalysts generally react with one or more reactants to form a chemical intermediate that subsequently reacts to form the final reaction product. By stabilizing the transition state of the reaction, the activation energy is reduced and the reaction rate is increased.

Biochemical catalysts are *enzymes. In addition to proteins, RNA (→ribozyme) or even DNA (→DNAzyme) can also fulfill these requirements. In a first step, enzyme E and substrate S form a complex ES that is converted into the transition-state complex $EX^{\ddagger}$. After formation of the enzyme–product complex EP, the product P dissociates. The enzyme is unchanged and ready to catalyze another event.

$$E + S \leftrightarrow ES \leftrightarrow EX^{\ddagger} \leftrightarrow EP \leftrightarrow E + P$$

To lower the activation barrier of a biochemical process, enzymes use different mechanistic strategies such as *precise substrate orientation, preferential *transition state binding, *electrostatic catalysis and *general acid–base catalysis. All these interactions stabilize the transition state, thus the free binding energy of the transition state complex ΔG_B compensates partly the activation energy $\Delta G^{\ddagger}_{uncat}$ of the uncatalyzed reaction. Not all of these strategies are necessarily used by an enzyme.

The equilibrium constant is not changed, since enzymes accelerate the rates of the forward and reverse reaction equally. Also, the standard free energy $\Delta G°$ is not altered. In most cases ribozymes act repeatedly following a true *Michaelis–Menten kinetic that describes the behavior of real enzymes. DENISE STROHBACH

catalytic center →catalytic core

catalytic constants →Michaelis–Menten equation

catalytic core Domain of an enzyme (e.g. *ribozyme) that contains the *active site. The nucleotide sequence of this region is highly conserved, since a complex network of specific interactions of these *nucleotides provides a basis for *catalysis. DENISE STROHBACH

catalytic molecular beacons (CMBs) Modular constructs combining *deoxyribozymes (or *ribozymes) and stem–loops (→stem–loop structure) used for *molecular beacons. If the stem blocks substrate access, the binding of *complementary *oligonucleotides to the *loop region opens up the stem, allowing the substrate to bind and the catalytic reaction to proceed. CMBs report the presence of oligonucleotides by cleavage of fluorogenic substrates. CMBs are also the basic computing element, sometimes called YES, sensor or detector gate. MILAN STOJANOVIC, DARKO STEFANOVIC

catalytic RNA →ribozyme

CCA-tail Conserved single-stranded 3'-terminus of a *transfer RNA forming the end of the *acceptor arm. The *ribose of the terminal A is acylated to the amino acid, thereby activating it for translation. Base pairing between *ribosomal RNA and the two C insures proper placement of this reactive site, explaining the strict conservation and mechanistic importance of the CCA-tail. Special enzymes insure the integrity of the CCA-tail during the tRNA life cycle or even introduce it to newly synthesized eukaryotic tRNA, as it is not encoded in the *pre-tRNA *transcripts. STEFAN VÖRTLER

***cdc* genes** Encode proteins (cyclin-dependent kinases) that regulate the cell division cycle in eukaryotes.

C-DNA *Right-handed *double helix belonging to the B-family and identified by the X-ray fiber diffraction method. This form is produced if the lithium salt of natural DNA is maintained at rather low relative humidity, 44–66%, or if sodium ions are used as counterions at high salt content and at humidity conditions intermediate between those required to produce *A- and *B-DNA; low d(G + C) content facilitates B–C transition. C-DNA resembles B-DNA, with conformational parameters of the *nucleotide residues changed only slightly: 9–9.3 base pairs per turn of helix with *base pairs are tilted by −8°. The helix axis is displaced toward the *minor groove, the *major groove being a little more shallow and the minor groove a little more deep. Synthetic DNAs with defined sequence can also transform into C-type structure. NINA DOLINNAYA

cDNA →copy DNA

cDNA library Usually generated *in vitro* by *reverse transcription of all the *messenger RNAs derived from a cell, a specific tissue or organism. If such a collection of *DNA sequences contains the entire genomic information of an organism, it is called a *gene library. As this collection of *copy DNAs (cDNAs) only contains protein-coding DNA, it can be used, after *cloning into an appropriate organism, to express proteins. mRNA is usually reverse transcribed to cDNA because cDNA is more stable, easier to work with and can be stored for a longer period of time.

For selection of nucleic acids with specific functions or for screening of new *aptamers, *ligands or nucleic acid effectors, cDNA libraries can be also synthesized via nucleic acid solid-phase synthesis or by *hybridization of two partially complementary rationally designed DNA *primers and subsequent *fill-in reaction with a *DNA polymerase. The DNA sequences are then either entirely randomized or only at certain nucleotide positions. →reverse transcription polymerase chain reaction. VALESKA DOMBOS

CDP Abbreviation for cytidine diphosphate (→nucleoside phosphates).

CD spectroscopy →circular dichroism spectroscopy

CeNA →cyclohexenyl nucleic acid

central dogma of molecular biology The working machinery of present-day organisms relies mainly on proteins (the

proteome) and a few, but crucial, *ribonucleoprotein complexes like the *ribosome and the *spliceosome. All blueprints are maintained in the *genetic information made of DNA (→genome). To build and adapt the amount of the needed *macromolecules, DNA has to be transcribed into structural and functional *transfer RNA, *ribosomal RNA, *spliceosomal RNA, etc., as well as the intermediate *messenger RNA, which serves as a *template for ribosomal protein synthesis (→translation). This flow of the genetic information was suggested to occur only in one direction, which formed the central dogma of molecular biology. Discovery of *reverse transcriptases allowing viruses to copy the information contained in RNA back into DNA modified this view. Still, "reverse translatases" are not known and actually not needed. Evolution occurs by *mutations in the genetic material, which form mutant proteins. If these offer advantages, the cell will survive and maintain the altered information in its *genome.
STEFAN VÖRTLER

centromere Constricted region in eukaryotic *chromosomes that contains a complex system of fibers called the kinetochore that attaches to spindle microtubules during nuclear division and becomes duplicated when chromosomes divide. In most eukaryotes the bulk of the *heterochromatin is found in the regions surrounding the centromeres, where it is thought to have an important centromere function. MAURO SANTOS

cGMP →cyclic guanosine-3′,5′-monophosphate

Chargaff's rule In a *DNA double helix, the amount of *adenine is always equal to *thymine and the amount of *cytosine is equal to *guanine (→GC–AT rule). A *pyrimidine base in one chain pairs always with a *purine base in a second chain. Thus, guanosine always pairs with cytosine, thymidine with adenosine. The specificity of base pairing is determined by the nature of *hydrogen bonds. There are two bonds between A and T, and three between G and C. As all bases have to fit into the double helix, a small base (T, C) makes hydrogen bonds with a large one (G, A). CT pairs would be too small, GA pairs too large for isosteric base pairing. MATTHÄUS JANCZYK

charged tRNA *Transfer RNA with esterified amino acid on its *CCA-tail (→aminoacyl-tRNA synthetase).

charge transfer in DNA Since the first consideration of the regularly stacked B-form DNA as a pathway for charge transfer processes which was published over 40 years ago, the possibility of DNA as a conducting *biopolymer for potential applications in nanotechnology, *biosensors and molecular diagnostics was discussed in a highly controversial scientific dispute. Remarkably, DNA was considered to be a molecular wire, on the one hand, or semiconductor or insulator, on the other hand. Such controversial results were supported by both experimental and theoretical work. Central questions of all theoretical and experimental investigations were the efficiencies and the rates of charge transfer processes through the DNA, the basic mechanisms, and the possibility of a distance limit. The interest in this subject has grown enormously in the scientific community over the last two decades. Apart from the biological relevance in *DNA damage, DNA was considered as a unique and exciting medium for physical chemists and physicists. The access to synthetic oligonucleotide substrates as the molecular basis for the investigation of DNA-mediated charge transfer processes was provided by automated phosphoramidite chemistry (→oligonucleotide synthesis) which was further improved by bioorganic chemists.

Most of the past experiments were performed according to the following strategy. (a) *Labeling of the DNA with redox-active probes through *intercalation and/or covalent linkages. (b) Photochemical or electrochemical charge injection into the *DNA double helix. (c) Detection of the charge transfer processes by steady-state or time-resolved spectroscopy methods, by electrochemical assays, or by chemical or biochemical analysis of the charge-transfer-related DNA products. It is now possible to obtain a pretty clear picture about charge transfer phenomena in DNA and the initially extreme controversy has been solved by the description of different mechanistic aspects, mainly the superexchange which occurs over short distances (3–4 bp) and the hopping mechanism (→electron hopping) over longer distances (at least 200 Å).

In principal, DNA-mediated charge transfer can be categorized as either oxidative hole transfer or reductive electron transfer processes. The description of hole transfer is misleading since it also describes an electron transfer, but in the opposite direction. Hence, both processes are electron migration reactions, but with different orbital control, either the highest occupied molecular orbital (oxidative) or lowest unoccupied molecular orbital (reductive). Thus, this categorization is not just a formalism about the different direction of the electron. The oxidative type of DNA-mediated charge transport processes represents the primary step in the formation of oxidative damage to the DNA that results in *mutagenesis, apoptosis or cancer. Hence, the biologically motivated research was focused on the photochemically induced oxidation of DNA and, furthermore, on the mobility of the created positive radical charge, which can be considered as oxidative hole transport. On the other hand, the mobility of an excess electron in the DNA base stack can be described as reductive electron transport and recent results strongly support the idea that this type of charge transport has a high potential for the development of electrochemical *DNA chips, e.g. for the detection of *single-base mutations and the application of electronic nanodevices based on DNA or DNA-inspired architectures.

A few measurements of the electronic conductivity of DNA have been performed. It is important to point out that in many cases the DNA *secondary structure that is crucial for a fast and efficient electron transport remained undetermined due to the absence of water in the experiments. Nevertheless, it was shown that it is possible to transport electrons along DNA molecules, but the conductivity is rather poor. Taken together with the results from the photoinduced electron transport experiments it becomes clear that electron transport through longer single DNA molecules (larger than 10nm) requires more research efforts. Hence, in order to use DNA as a molecular wire it is necessary to develop DNA-inspired materials which contain the typical DNA structural features, but exhibit improved electron transport capabilities.

HANS-ACHIM WAGENKNECHT

"C–H" edge →edge-to-edge interaction of nucleobases

chemical and enzymatic structure mapping of RNAs A technique to obtain insights on the structure of RNAs of any size. The folded RNA is subjected to chemical or enzymatic reactions that modify RNA *bases, phosphates or *riboses and/or induce strand scission at specific RNA sites. The probes are used under limited conditions where less than one cleavage or

The most commonly used nucleases to probe RNA secondary structure

Enzymes	Single or double stranded	Sequence requirements
RNase T1	single	3' of guanines
RNase T2	single	none – with a preference for adenine
Nuclease S1	single	none
RNase V1	double	none

modification occurs per RNA molecule with a statistical distribution. *Nucleases used in RNA structure analysis exhibit single- or double-stranded specificity and several of them exhibit sequence specificity.

Several classes of chemicals have been described. (a) Base-specific reagents have been largely used to define *RNA secondary structure models. The combination of dimethylsulfate (DMS), 1-cyclohexyl-3-(2-morpholinoethyl)carbodiimide metho-p-toluene sulfonate (CMCT) and β-ethoxy-α-ketobutyraldehyde (kethoxal) maps the accessibility of the four *nucleobases at one of their Watson–Crick positions (→edge-to-edge interaction of nucleobases). These reagents will modify the target nucleotides that are not involved in *Watson–Crick base pairings. Position N7 of *purines can be modified by diethyl pyrocarbonate (DEPC) for *adenines and by DMS or nickel complex for *guanines. Modifications will occur on purines that are not involved in Hoogsteen or reverse Hoogsteen interactions (→edge-to-edge interaction of nucleobases). Nickel complex and DEPC are very sensitive to stacking of the base rings and thus N7 of purines within a helix are never reactive except if the deep groove of the helix is widened. (b) Divalent metal ions are required for *RNA folding by neutralizing and bringing the negatively charged *phosphoribose backbone. Under particular conditions, binding of the divalent ion can promote cleavages in RNA. As an example, Pb^{2+} ions cleave RNA within single-stranded regions, *loops and *bulges. (c) Another class of chemical probes, encompassing ethylnitrosourea (ENU) and hydroxyl radicals, attacks the ribose–phosphate backbone. ENU is an alkylating reagent that ethylates phosphates. The resulting ethyl *phosphotriesters are unstable and can be easily cleaved by a mild alkaline treatment. Hydroxyl radicals are generated by reaction of Fe(II)–EDTA complex with hydrogen peroxide, and attack hydrogens of positions C1' and C4' of the *ribose. In one single experiment, it is possible to gain information on the inside and outside regions of large and highly structured RNAs. Hydroxyl radicals can also be produced by potassium peroxonitrite via transiently formed peroxonitrous acid. Radiolysis of water with a synchrotron X-ray beam allows sufficient production of hydroxyl radicals in the millisecond range. This time-resolved small-angle X-ray scattering (SAXS) is an approach to detect transient RNA–RNA interaction and to measure the fast global shape changes of large RNAs. Probing the structure using chemicals can also be performed under different experimental conditions (i.e. by varying the temperature or the concentration of divalent ions).

The identification of the cleavages or of the modifications can be done by two different methodologies depending on the length of the RNA molecule and on the nature

Chemical and enzymatic structure mapping of RNAs

Chemicals	Specificity
Base-specific reagents	
DMS	C(N3), A(N1), G(N7)
CMCT	U(N3), G(N1)
Kethoxal	G(N1, N2)
DEPC	A(N7)
NiCR	G(N7)
Divalent ions	
Pb^{2+}	specific divalent ion-binding site single-stranded regions
Ribose phosphate-specific reagents	
ENU	phosphates
Fe–EDTA, potassium peroxonitrite	riboses

of the modification. The first approach, which uses end-labeled RNA (→end labeling, →oligonucleotide labeling), only detects scissions and is limited to RNA containing less than 300 nucleotides. The second approach, which uses *primer extension, detects stops of *reverse transcription at *modified nucleotides or cleavages. This method can be applied to RNA of any size, and can be used with all enzymes and chemicals. PASCALE ROMBY, PIERRE FECHTER

chemical interference mapping Defines a set of *nucleotides of nucleic acids that cause a loss of function when they are modified at a specific position by a chemical probe. The approach is applicable to any RNA that has an assayable function that can be used to distinguish active and inactive variants. RNA functions that are amenable to chemical interference include *catalysis, folding, protein or *ligand binding and the ability to act as a reaction substrate. PASCALE ROMBY, PIERRE FECHTER

chemical labeling Chemical approach for incorporation of labels such as *fluorescent dyes into DNA and RNA (→oligonucleotide labeling).

chemical ligation Chemically induced *oligonucleotide coupling reaction accelerated simply by bringing reactants into proximity using nucleic acid *hybridization. Nucleic acid templated synthesis plays a central role in fundamental biological processes, including the *replication of *genetic information, and the *transcription and the *translation of RNA into proteins. In earliest attempts to apply chemical ligation to non-biological reactants, DNA or RNA hybridization was used to form the *phosphodiester bonds or other structural mimics of the *nucleic acid backbone: *pyrophosphate, *phosphoramidate, diphosphoryldisulfide, phosphothioester and phosphoselenoester links. The template architecture used to support these reactions most frequently places the site of reaction at the center of a nicked DNA duplex. In other cases, the polymerization or concatemerization of mono(oligo)nucleotides takes place. Water-soluble condensing reagents or pre-activated functionalized oligonucleotides were used to mediate

these aqueous-compatible reactions. The products form efficiently when the template and reagents sequences are complementary. The precision with which the reactive groups are aligned into a DNA-like *conformation could influence the increase in their effective molarity upon DNA *hybridization. It is conceivable that only those reactions that proceed through *transition states consistent with the *conformation of duplex DNA may be suitable for chemical ligation. Nick-sealing ability was shown to be sensitive even to sequence-dependent modulations in the DNA local structure. Oligonucleotide analogs have also served as templates for chemical ligation reactions: non-natural nucleic acids containing a hexose sugar, *pyranosyl-RNA. In addition to analogs of the *sugar–phosphate backbone, products that mimic the structure of stacked nucleobases have also been generated. Nucleic acid templated synthesis can maintain sequence-specific control even when the structures of reactants and products are unrelated to that of *nucleic acids. C–C bond-forming reactions and organometallic coupling were explored in DNA-templated format. The specific architecture in which a DNA template anneals to *complementary DNA reagents could profoundly affect the nature of the reaction. In end-of-helix systems, the ability of chemical ligation to direct a collection of chemical reactions without requiring the precise alignment of reactive groups into DNA-like conformation was demonstrated. The rate constants of product formation were shown to not significantly change when the distance between hybridized reactive groups on the ends of *DNA duplex (e.g. amine and carboxylate) was varied from 1 to 30 bases. A growing understanding of the simple but powerful principles underlying DNA-templated synthesis has rapidly expanded its synthetic capabilities, and has also led to emerging chemical and biological applications, including the assembly of complex biological structures, sequence-specific DNA modifications and chemical targeting of *biomolecules.
NINA DOLINNAYA

chemical mutagenesis Technique that induces *mutation by application of chemical *mutagens. Chemical mutagens possess different potency to enter living cells: their toxicity, their interaction with DNA and the probability to correct the induced chemical change in DNA by a repair system (→ DNA repair). Examples widely used are ethylmethylsulfonate and 2-hydroxy-4-methoxy-benzophenone. BETTINA APPEL

chemical sequencing   → Maxam–Gilbert sequencing

chemical synthesis of DNA and RNA → oligonucleotide synthesis

chimera Individual assembled from genetically different types of cells. In the context of molecular biology, it refers to a construct which carries DNA sequences from different sources. MATTHÄUS JANCZYK

chimera DNA DNA that has been recombined with a *gene from a different organism. *Plasmids that are used for *transfection of genes from an external source into a cell carry chimera DNA. MATTHÄUS JANCZYK

chi sequence The *hotspot of the *homologous recombination sequence 5′-GCTGGTGG-3′ in the *Escherichia coli* *genome recognized by the RecBCD enzyme. The 4 639 221-bp complete nucleotide sequence of *E. coli* reveals 1009 chi sites, or 1 per 4.6kb. It seems that the distribution of chi is enriched within *open

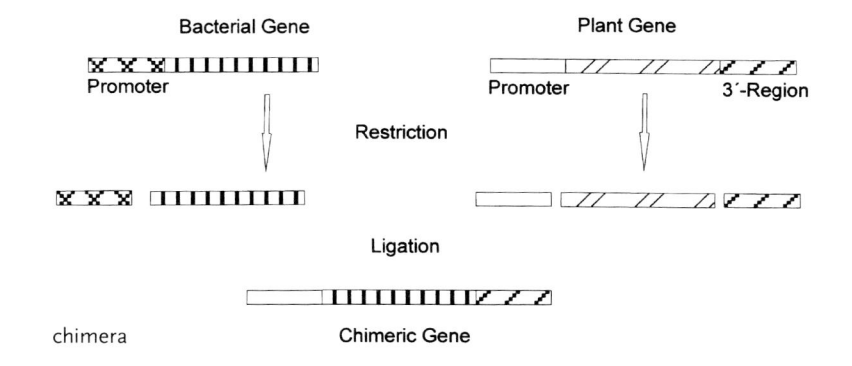

```
        Bacterial Gene              Plant Gene
    ▨▨▨ ▯▯▯▯▯▯▯▯▯▯        ▭ ▨ ▨ ▨▨▨
    Promoter                     Promoter         3'-Region
                ⇓    Restriction    ⇓
    ▨▨▨   ▯▯▯▯▯▯▯▯         ▭    ▨ ▨ ▨ ▨▨▨
                      Ligation
                   ▭▯▯▯▯▯▯▯▯▨▨▨
    chimera              Chimeric Gene
```

reading frames. Chi is the acronym of crossing-over hot-spot instigator. MAURO SANTOS

chi structure Four-branched intermediate DNA structure observed during *homologous recombination arising from the physical exchange between two duplex DNAs. The ultramicroscopy image observed between *plasmids undergoing homologous recombination reproduces the shape of the Greek letter χ, and hence the name. The quadruplex helix at the core is named the *Holliday junction, after the name of the author of the first genetic model of homologous recombination. GEMMA MARFANY, ROSER GONZÀLEZ-DUARTE

chloroplast DNA *Circular DNA found in the photosynthetic organelle of plants and algae. Chloroplast DNA derived from the *genome of the cyanobacterial ancestor and typically encodes 110–120 *genes. Some algae have retained a larger chloroplast genome with more than 200 genes. Chloroplasts of parasitic plants do not contain chlorophyll. These organelles are called plastids and their DNA *plastid DNA. ANNEGRET WILDE

chromatid One of a pair of duplicated *chromosomes produced by *replication.

chromatin Complex of DNA and protein in the nucleus that stains with basic dyes.

chromosomal DNA Large *macromolecule of DNA that carries all or part of the *genetic information in all living organisms and many viruses.

chromosomal RNA Large *macromolecule of RNA that constitutes the genetic material of RNA viruses. The *nucleic acid is usually *single-stranded RNA, but can occasionally be *double-stranded RNA. Human pathogenic RNA viruses include influenza and hepatitis C virus. MAURO SANTOS

chromosome (Greek: $chroma$ = color and $soma$ = body) Structural unit of the genetic material consisting of a single, *double-stranded DNA (dsDNA) molecule and associated proteins. It contains many *genes, regulatory elements and other intervening nucleotide sequences (→intervening sequence). Each member of a species normally possesses the same number of chromosomes in its somatic cells, characteristic of the species. In prokaryotes, a single, circular dsDNA molecule constitutes the bulk of the genetic material. In the chromosomes of eukaryotes, the uncondensed DNA exists in a quasi-ordered structure inside the nucleus, where it wraps around *histones, and where this composite material is called *chromatin. During mitosis (cell division), the chromosomes are condensed and called metaphasic chromosomes. This is the only natural context in which individual chromosomes are visible with an optical microscope. Prokaryotes do not possess histones or nuclei. In its relaxed state (→relaxed DNA), the DNA can be accessed for *transcription, regulation and *replication. BEATRIX SÜß

chromosome walking Method for identifying and sequencing long stretches of DNA. The principle is to first clone a genomic fragment that serves as a starting point for the walk. The *genomic library is then screened with the chosen *clone as probe to identify adjacent clones to the left or to the right. The selected adjacent clone serves as a new probe to reiterate the process. MAURO SANTOS

cIMP Abbreviation for cyclic inosine monophosphate (→inosine phosphates).

circular dichroism (CD) spectroscopy Measures differences in the absorption of left-handed polarized light versus right-handed polarized light which arise due to chiral molecules, structural asymmetry or an asymmetric environment. One source of chirality in *nucleosides is the *ribose sugar to which the heterocyclic base is attached. This effect is, however, less significant than the chirality originating from nucleic acid helicity. CD spectroscopy is extremely sensitive to changes in mutual position of bases and is therefore uniquely suited to study conformational isomerizations of *nucleic acids in solution like *helix–coil or B–Z transitions, formation of *triplexes or guanine *tetraplexes. KLAUS WEISZ

circular DNA Form of a *DNA double helix in which the helix is covalently closed into a circle. Circular DNA is found in *plasmids, mitochondria, chloroplasts and viral *genomes. A feature of closed circular DNA is that its topological state cannot be altered by any conformational rearrangement, but requires the action of a *topoisomerase. →supercoiled DNA. MATTHÄUS JANCZYK

circularization of DNA Process where linear DNA strands are transferred into a circular molecule (→circular DNA). For efficient circularization, DNA fragments with *sticky ends are required to be covalently joined by DNA ligase. MATTHÄUS JANCZYK

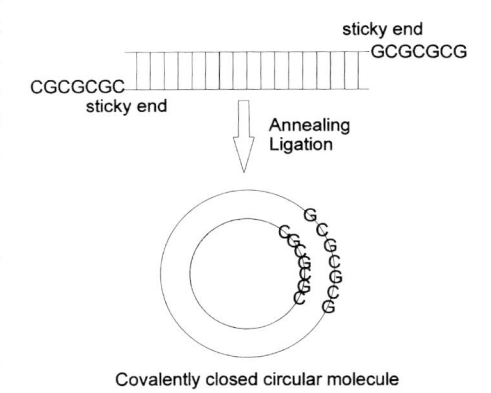

cis-**acting element** Sequence that controls *gene expression and is physically linked (at a distance of up to several kilobases) to the regulatory target.

cis-**active** Genetic element located close to a *gene that stimulates (enhancer) or suppresses the *transcription of eukaryotic genes. It is also used in the context of *RNA catalysis to describe an intramolecular reaction. MAURO SANTOS, SABINE MÜLLER

cis-**active sequence** →signal sequence

cisplatin Diamminedichloroplatinum(II). Antitumor agent, licensed in 1979. Cisplatin interacts with DNA by *ligand exchange, particularly by binding to N7 of *guanine and *adenine. This leads to inter- as well as intrastrand *crosslinks and as a result to cell apoptosis. SABINE MÜLLER

cis-splicing Intramolecular splicing reaction that ligates *exons from the very same *pre-mRNA molecule.

cistron Genetic unit that encodes a single polypeptide.

clone Copy of a *DNA sequence, a *gene or a whole *genome that is identical to its ancestor (→cloning).

clone analysis Assessment of *clones. Depending on the context, the positively isolated clones from library screenings are phage plaques or *plasmid clones.

cloning General process by which *amplification of the number of identical copies of a particular DNA is achieved. It usually involves: (a) *ligation of the DNA of interest into a chosen *plasmid vector using specific *restriction enzyme sites, (b) transformation in *Escherichia coli*, (c) amplification of the *clone by bacterial growth to obtain sufficient amounts of plasmid, and (d) assessment of the cloning process by restriction and/or *sequencing. The term cloning is also used concerning the generation of organisms from a single somatic cell by mitotic divisions. The term is generally used in the context of "therapeutic cloning" or "reproductive cloning". In these two cases, it refers to the introduction of a somatic cell nucleus from the donor into an enucleated oocyte to generate a new embryo from which genetically identical cells/tissues (therapeutic cloning) or organisms (reproductive cloning) are obtained. GEMMA MARFANY, ROSER GONZÀLEZ-DUARTE

cloning vector DNA molecule (→plasmid, →cosmid) used in recombinant DNA experiments as a carrier of *foreign DNA. Cloning vectors usually contain a selectable marker, e.g. antibiotic *resistance genes, unique *recognition sites for *restriction enzymes to insert DNA fragments and replicate autonomously in appropriate host cells. The first synthetic plasmid cloning vector pBR322 consists of DNA fragments from several naturally occurring plasmids. ANNEGRET WILDE

closed reading frame As opposed to an *open reading frame, a reading frame that contains *stop codons, thus preventing the *translation of following *nucleotides into protein. MAURO SANTOS

cloverleaf structure *Secondary structure representation of a *transfer RNA (tRNA) showing four double-stranded base paired regions in a *cruciform with the *CCA-tail as well as three *loops (*DHU, TC and *anticodon) at each end. It was initially described as one out of three possible *secondary structures by Holley after sequencing the first tRNAs. It not only turned out to be correct, but also to be universal, even for unconventional organelle tRNAs missing the entire *DHU arm. STEFAN VÖRTLER

CMP Abbreviation for cytidine monophosphate (→nucleoside phosphates).

CNA →cyclohexane nucleic acid

coaxial stack Structural motif, where two helical arms stack upon each other along their helical axes (→base stacking).

code In genetics, the correspondence between one *triplet (→codon) in DNA or RNA and the amino acid in the polypeptide. As a consequence of the degeneracy of the *genetic code many codons are redundant and there has been much speculation about why nature picked the one we observe. Conjectures range from the frozen accident hypothesis to the highly robust nature of the actual code against *point mutations (→genetic code). MAURO SANTOS

coding capacity Actual and potential protein-coding *exons in a *genome.

coding density Fraction of *base pairs in *codons, related to the number of *genes in a *chromosome or a *genome.

coding region For a *gene it is the fraction of DNA that is transcribed into *messenger RNA and translated into protein.

coding sequence *Nucleotide sequence in DNA that codes for a *messenger RNA molecule and thus defines the amino acid sequence of a protein. Conventionally, all other sequences are "non-coding". MAURO SANTOS

coding strand DNA strand that has the same *nucleotide sequence as the transcribed *messenger RNA and contains the sequence of *codons (nucleotide triplets) which interact with the *anticodons of the *transfer RNAs during *translation leading to the primary sequence of the resulting proteins. The coding strand is identical to the *sense strand. ANDREAS MARX, KARL-HEINZ JUNG

codogenic strand Complementary to the *coding strand (sense strand) and therefore identical to the *antisense strand.

codon Linear 3-nucleotide stretch in a *messenger RNA coding for a particular amino acid. Due to the permutation of the four building blocks, the *nucleotides A, C, G and U/T, $4^3 = 64$ different variations exist and form the universal *genetic code, used with a few exceptions by all forms of life (→start codon, →stop codon, →reading frame, →wobble hypothesis, →transfer RNA). STEFAN VÖRTLER

codon–anticodon interaction →ribosome

codon usage Frequency with which one *codon occurs in a particular or on average in all *genes of an organism. For those amino acids which are represented in the *genetic code by only one codon (Met and Trp) there are no alternatives and the codon usage is 100%. All others are encoded by more than one codon: two for Asn, Asp, Cys, Gln, Glu, His, Lys, Phe, Tyr; three for Ile; four for Ala, Gly, Pro, Thr, Val; and six for Arg, Leu, Ser. The codons are, however, not used equivalently, resulting in a typical distribution characteristic for the gene and/or the organism. The reasons are not only *translational control, but also the optimization of the corresponding isoaccepting *transfer RNA (tRNA) pools as it is an energy expenditure to maintain for all codons similar aminoacylated tRNA amounts. *Wobble base pairing and different concentrations of isoaccepting tRNA (major and *minor tRNA) are the consequence. Variation in codon usage between species is of particular importance for the expression of heterologous genes, as in the case of eukaryotic proteins in a prokaryotic host like *Escherichia coli*. Supplementing plasmid-encoded tRNA genes to raise the concentration of minor tRNAs or exchange of isoaccepting codons to better adapt to the codon usage of the host without any change in the protein sequence (a so-called artificial gene) are two strategies to overcome often observed expression problems. STEFAN VÖRTLER

coenzyme B_{12} riboswitch The coenzyme B_{12} (adenosylcobalamine, AdoCbl) *riboswitch is widespread in both Gram-positive and -negative bacteria, where it generally controls expression of cobalamine transport or biosynthesis *genes. The phylogenetically conserved *aptamer is one of the largest and most complex metabolite-binding RNA sequences comprising up to 12 *stem–loop substructures,

AdoCbl riboswitch aptamer consensus sequence

Adenosylcobalamin (Coenzyme B₁₂)

which coincides with AdoCbl being one of the largest metabolic compounds known to be bound by *riboswitches.

AdoCbl recognition by the aptamer is proposed to be modular, where domain P1–P7 seems to bind the cobalamine part and domain P8–P11 the adenosyl portion. This could explain how the aptamer selectively binds AdoCbl over similar metabolic compounds like cyano- or methylcobalamine that differ in the ligand at the central cobalt ion, as well as, for instance, 13-epi- AdoCbl that has only changed stereo configuration in C13 of the cobinamide ring system. RÜDIGER WELZ

coexpression Concomitant expression of two or more *genes. It usually refers to transformed or transfected cells that contain different expression constructs.

cognate codon *Nucleotide triplet of *messenger RNA that is recognized and specifically bound by the *anticodon of *transfer RNA (→ribosome).

cognate sequence Sequence motif for which DNA- or RNA-binding proteins display a preference.

cohesive end ligation Enzymatic fusion of *sticky ended restriction fragments using *DNA ligase (→blunt end ligation, →ligation).

cohesive ends Short overhanging single-stranded ends of a linear *double-stranded DNA molecule usually produced by *restriction endonucleases (e.g. *Hind*III). Complementary single-stranded ends are often used to join DNA fragments (→ligation). Naturally occurring cohesive ends are the *cos* (cohesive site) sites of the phage λ DNA. The λ DNA contains single-stranded overhangs of 12 nucleotides at both ends of the molecule that are complementary to each other. After infection of the host cell the phage DNA molecules are circularized using the cohesive ends. ANNEGRET WILDE

cold shock Microorganisms often encounter harmful temperature shifts, and have developed numerous strategies for responding and adapting to *heat shock and cold shock conditions. Low temperatures decrease membrane fluidity and result in secondary RNA structures that reduce *transcription and *translation efficiency. As protective measure, cells mount a cold shock response that alters the membrane lipid composition and produces *RNA chaperones. FRANZ NARBERHAUS

colony blot Method for identification of bacteria colonies carrying chimeric *vectors by *in situ* *hybridization with a labeled *DNA probe.

Recombinant bacteria *clones growing on an agar plate are formed onto a filter and grown on the filter that has been placed on the surface of an agar plate. A reference plate is obtained by stamping the filter onto a second agar plate, which can be stored at 4°C. After lyses of the colonies on the filter their DNA is denatured *in situ* and fixed on the filter. The membrane is then treated with a labeled DNA or RNA probe with a *complementary sequence to the DNA sequence to be analyzed, allowing the hybridization of the probe with the target sequence. DNA sequences to be analyzed can be visualized based on the labeled properties of the DNA or RNA probe (autoradiographically, by a *fluorescent dye or by chemiluminescence). Based on positive signals, colonies containing a specific DNA sequence that is being searched for can be identified. IRENE DRUDE

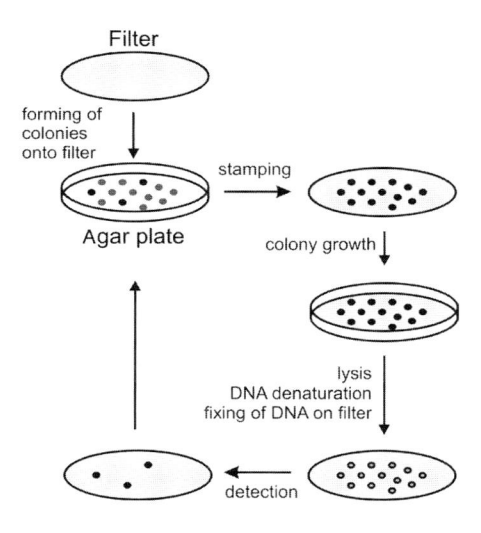

colony hybridization →colony blot

communication module Structural motif in nucleic acids that bridges two functionally distinct structural units and allosterically transfers information about events in one module to the other. SABINE MÜLLER

competence Ability of certain bacteria (*Neisseria gonorrhoeae*, *Bacillus subtilis*, *Streptococcus pneumoniae*, etc.) to absorb specific DNA from the environment. This ability can be natural or induced. →signal sequence. JÖRN WOLF

competitive PCR →competitive polymerase chain reaction

competitive polymerase chain reaction (PCR) Method for quantify DNA using the *real-time polymerase chain reaction. It is based on a coamplification of the target DNA with a homologous or heterologous competitor DNA standard which competes with the sample *template DNA for the same set of PCR *primers. The internal control needs to be distinguished from target DNA, e.g. it will be of a different size or have a different *restriction fragments pattern. It is also necessary to know the concentration of the control in order to determine the quantity of final product. The *amplification products can be separated on an agarose gel and the fluorescence of these products is proportional to the amount of amplified DNA.

Competitive PCR is different from real-time PCR in that it is an endpoint method in which quantification is based on the amount of amplified material obtained at the last amplification cycle. DAVID LOAKES

complementary base Correspondent base to a given nucleobase to form a *Watson–Crick base pair. The shape and chemical structure of the bases allow for *hydrogen bonds to form efficiently only between A and T (U in RNA) or G and C, thus there is only one complementary base for any of the bases found in DNA and in RNA. DENISE STROHBACH

complementary DNA Product of *reverse transcription, derived from a RNA (usually messenger RNA) template (→copy DNA).

complementary nucleotide Nucleotide that can form a specific *hydrogen bond interaction with a given "template" nucleotide according to *Watson–Crick base-pairing rules. SABINE MÜLLER

complementary oligonucleotide Oligonucleotide that forms contiguous *Watson–Crick base pairs with an oligonucleotide of opposite *strand polarity.

complementary RNA (1) Accurate product of *transcription, derived from a DNA template or (2) RNA molecule, derived from a sense RNA through the activity of an *RNA-dependent RNA polymerase.
DENISE STROHBACH

complementary strand DNA strand with a sequence of nucleotides according to *Watson–Crick base pairing with the *template strand, i.e. AT and GC, respectively.

concatemer DNA or RNA molecule which consists of identical monomeric units which are linearly repeated in the same orientation (e.g. λ phage multimers).

concatenate Structure formed by the chain-like interlocking of *circular DNA molecules.

configuration Steric constitution of molecules without regard to the different *conformation of atoms. The configuration of molecules describes their permanent geometry. All stereoisomers except of conformers are configuration isomers.
BETTINA APPEL

conformation The conformation of a molecule is its structure. Molecules with the same connection of atoms are conformation isomers. They only differ in rotation around single bonds. In polymers, different *secondary structures and *tertiary structures are called conformers, e.g. *A-, *B- and *Z-DNA are conformers. In *nucleosides, the base can exist in two different orientations about the N-*glycosidic bond: *anti and *syn conformation. The conformation of an individual *nucleotide in DNA and RNA molecule can be described by eight torsions along the *backbone, sugar and base. →DNA structures, →RNA structures. BETTINA APPEL

conjugate The verb "conjugate" is used frequently in the area of nucleic acids chemistry, where it usually refers to connecting a *nucleic acid, *nucleotide or *nucleoside to an organic molecule with a covalent bond. SNORRI TH. SIGURDSSON

conjugation Process of connecting or joining two molecules together to form a *conjugate.

consensus motif →consensus sequence, →consensus structure

consensus sequence Primary nucleotide or amino acid sequence that is conserved among different nucleic acids or peptides, assumed to be relevant for function (→consensus structure). MAURO SANTOS

consensus structure Structural motif that is conserved among different nucleic acids or peptides, assumed to be relevant for function (→consensus sequence, →RNA secondary structure prediction). MAURO SANTOS

conserved nucleotide Nucleotide within DNA or RNA sequence that is essential to function.

conserved regions In one or more linear sequences of *nucleotides or amino acids, conserved regions are stretches of high similarity often related to an important cellular function. Thus, highly conserved proteins are often required for basic cellular function, stability or reproduction. MAURO SANTOS

conserved sequence Mainly refers to a sequence of *nucleotides or amino acids that is similar or even identical in several organisms. Sequence similarities provide evidence for structural and functional conservation, as well as for evolutionary relationships between the sequences. MAURO SANTOS

constitutive genes *Genes that are continually transcribed.

control genes *Genes that switch on and off the expression of other genes.

copy DNA (cDNA) A cDNA sequence is obtained from *messenger RNA (mRNA) by the action of an *RNA-dependent DNA polymerase, such as a *reverse transcriptase. The resultant cDNA sequence is complementary to the *template mRNA. cDNA is used to study *gene expression because it is more stable than RNA and is more amenable to recombinant cloning techniques. The single-stranded copy is also often used as a probe to identify complementary sequences in DNA fragments or genes of interest. A cDNA library represents the entire expressed DNA in a cell (→complementary DNA). DAVID LOAKES

core →catalytic core

core DNA The 146-bp DNA contained on a *core particle of *chromatin.

core particle In the nucleosome, consists of 146-bp DNA that wraps four identical pairs of *histones H2A, H2B, H3 and H4 collectively known as the histone octamer. The crystal structure of the nucleosome core particle at a resolution of 2.8 Å was determined in 1997. MAURO SANTOS

corepressor Peptide or nucleic acid molecule that enhances, additively or synergistically, the repression mediated by another molecule.

core sequence *Conserved sequence, relevant for function, which corresponds to the *catalytic center of a protein or is embedded in a *nucleic acid motif.

cosmid Genetically engineered *vector containing phage and *plasmid genetic elements which accepts inserts of larger length (average of 45 kb) than phage (15 kb) or plasmid (1–10 kb) derived vectors. The *replication origin and antibiotic resistance gene come from plasmids. The *cos* sites required for packaging come from phages. Cosmids are packaged into capsids and are able to infect host cells, thus efficiently injecting the *vector DNA which is then amplified and maintained as a giant plasmid in the host cell. GEMMA MARFANY, ROSER GONZÀLEZ-DUARTE

CoTC ribozyme Catalytic RNA structure identified in the human β-globin *gene *messenger RNA (mRNA). *Transcriptional termination for the human β-globin gene was found to occur by cotranscriptional cleavage (CoTC). The primary cleavage event downstream of the mRNA *poly(A) site involves RNA self-cleavage activity and plays a functional role in efficient termination *in vivo*. SABINE MÜLLER

cotransduction Simultaneous transfer of several *genes from a host cell to another, by means of virus (eukaryotic) or phages (bacteria).

cotransfection Introduction of two or more DNA fragments to the same prokaryotic or eukaryotic host cell, generally used when concomitant expression of two or more *genes within the same cell is needed. GEMMA MARFANY, ROSER GONZÀLEZ-DUARTE

CPG Abbreviation for controlled pore glass (→ solid support).

CpG islands Regions in the backbone of DNA with a large number of *cytosine–*guanine linked by *phosphodiester bonds. The "p" in the notation is used to distinguish CpG from the *G + C content, which is the proportion of Watson and Crick GC base pairs in the DNA molecule or *genome sequence (→ GC content). Many *genes in vertebrate have CpG islands near their *promoter regions that are supposed to be relevant to *gene expression. MAURO SANTOS

CpG oligonucleotides → CpG-rich oligonucleotides

CpG-rich oligonucleotides DNA *oligonucleotides containing *CpG islands are known to activate vertebrate innate immune cells and B cells. This induction can be explained by the fact that unmethylated CpG dinucleotides are more frequently found in viral and bacterial DNA than in mammalian *genomes, suggesting that they are a marker for the immune system to signal infection. The level of immune stimulation depends on the sequences flanking the CpG motive. Oligonucleotides containing CpG dinucleotides are recognized by Toll-like receptor 9 and induce secretion of various cytokines, including interleukins, tumor necrosis factor and interferon. Most of the *antisense oligonucleotides currently being tested in clinical trials contain CpG sequences and it has been speculated that their major activity *in vivo* is immune stimulation rather than specific inhibition of *gene expression by a true antisense mechanism. This unspecific mode of action can be prevented by C5-methylation of the *cytosine residue. The stimulatory potential of CpG-rich oligonucleotides has, however, also been explored to activate the immune system with the purpose to treat viral infections, inflammatory diseases and cancer. Several promising clinical trials are currently ongoing. JENS KURRECK

crossing over Process during meiosis in which there is exchange of genetic material between *homologous chromosomes and which is responsible for genetic *recombination. MAURO SANTOS

crosslink Crosswise connecting part (atom or group) that connects one polymer chain to another, inter- or intramolecularly (→crosslinking, →crosslinker, →crosslinking of DNA, →ultraviolet crosslinking, →ultraviolet laser crosslinking). TATIANA S. ORETSKAYA, VALERIY G. METELEV

crosslinker Compound that could crosslink *biopolymers. Crosslinkers can be either homobifunctional or heterobifunctional. A homobifunctional crosslinker has two identical reactive groups and often is used in one-step reaction. Heterobifunctional crosslinkers possess two different reactive groups that allow for sequential (two-stage) conjugations, helping to minimize undesirable polymerization or self-conjugation (→crosslinking, →crosslinker, →crosslinking of DNA, →ultraviolet crosslinking, →ultraviolet laser crosslinking). TATIANA S. ORETSKAYA, VALERIY G. METELEV

crosslinking Establishment of chemical links between the molecular chains in polymers. Crosslinking can be accomplished by chemical reagents, ultraviolet light and electron bombardment. Crosslinking of proteins to nucleic acids is based on either the use of non-specific binding reagents (e.g. formaldehyde) or selective introduction of reactive groups into protein or nucleic acid reagents. *Crosslinkers contain reactive groups specific to primary amines, sulfhydryl groups, etc., on nucleic acids, their derivatives, proteins or other molecules. Crosslinkers have been used to assist in determination of near-neighbor relationships, three-dimensional structures of proteins and molecular associations in cell membranes. They also are useful for solid-phase immobilization, hapten–carrier protein conjugation, and preparing antibody–enzyme conjugates, immunotoxins and other labeled protein reagents. Other uses include modification of nucleic acids, drugs and solid surfaces. In most cases, crosslinking is irreversible. In some cases, though, if the crosslink bonds are sufficiently different, chemically, from the bonds forming the polymers, the process can be reversed. Nucleic acid reagents containing reactive groups react selectively with spatially close amino acid residues of the protein without any external factors. The introduction of a reactive group into a preset position of the *oligonucleotide chain allows the probing of point contacts between different regions of *biomolecules, whereas variations of the chemical nature of the reactive group enable selective binding of a definite type of amino acid residues within the nucleic acid–protein complex. Reactive groups can be introduced into any position of the nucleic acid, i.e. the heterocyclic *base, the *sugar–phosphate backbone or the ends of the oligonucleotide chain. A recent development combines the use of protein engineering and site-specific DNA modification techniques to introduce disulfide crosslinks into protein–DNA interfaces. In disulfide trapping, a Cys residue is introduced by *site-directed mutagenesis into the DNA-binding surface of a protein and an alkanethiol tether is incorporated at a nearby site in the protein-binding surface of the DNA through synthetic chemistry. Application of this technology requires some knowledge about likely contact points in the protein and DNA.

Disulfide bonds formation can be carried out under conditions of thiol/disulfide equilibration. The crosslinked product can be obtained in yields greater than 80% (→crosslinking, →crosslinking of DNA, →ultraviolet crosslinking, →ultraviolet laser crosslinking). TATIANA S. ORETSKAYA, VALERIY G. METELEV

crosslinking of DNA Occurs when various exogenous or endogenous agents react with two different positions in the DNA. This can either occur in the same strand (intrastrand crosslinking) or in the opposite strands of the DNA (interstrand crosslinking). Agents that crosslink DNA can be divided into two categories. (a) *Exogenous crosslinkers*: alkylating agents such as 1,3-bis(2-chloroethyl)-1-nitrosourea and nitrogen mustard which are used in chemotherapy can crosslink with DNA at the N7 of *guanine on the opposite strands forming interstrand crosslink; and *cisplatin and its derivative forms DNA crosslinks as monoadducts, interstrand crosslinks, intrastrand crosslinks or DNA–protein crosslinks. (b) *Endogenous crosslinkers*. (i) Nitrous acid formed in the stomach dietary source nitrites. It induces formation of interstrand DNA crosslink at amino group of exocyclic N2 of guanine at the CG sequences. (ii) Reactive chemicals such as malondialdehyde which are formed endogenously as the product of lipid peroxidation. Psoralens are natural compounds (furocoumarins) present in plants. These compounds are activated in the presence of ultraviolet irradiation. They form covalent adducts with *pyrimidines. Psoralens can form two types of monoadducts and one diadduct (an interstrand crosslink) reacting with *thymine. The crosslinking reaction by psoralens targets TA sequences intercalating in DNA and linking one base of the DNA with the one below it. Psoralen adducts causes replication arrest, and are used in the treatment of psoriasis and vitiligo. DNA–protein crosslinking aldehydes such as acrolein and crotonaldehyde found in tobacco smoke or automotive exhaust can form DNA interstrand crosslinks in DNA. TATIANA S. ORETSKAYA, VALERIY G. METELEV

cross-reactive arrays of aptameric sensors
Given a domain of objects (e.g. steroids), a *sensor is a device which measures a specific feature of those objects. Often the given domain of objects can be partitioned into several classes of interest. If each of these classes has a distinct value for a single given feature, then a single sensor measuring that feature can be used to determine the class of an object. However, often classes cannot be distinguished using only a single feature. A cross-reactive array of sensors is a collection of sensors which each measure a different, independent feature of objects taken from a given domain. By measuring many independent features, the task of classifying objects becomes much easier as there is more information available to describe the differences between classes. A single non-specific sensor can usually be used to determine concentrations of an analyte only if no other analytes that interact with that sensor are present. In other words, a single readout from a sensor cannot be unambiguously attributed to a certain concentration of that analyte, but could be the result of various concentrations of other analytes or even mixtures of analytes. However, multiple sensors with differential cross-reactivities can be organized in a cross-reactive array, which can then be used to generate a pattern of responses to individual analytes over all sensors in the array. Patterns identified through "training sets" (i.e. "standards") could be

further used to identify ("classify") unknowns into classes such as [*analyte, concentration*]. While *aptamers are renowned for their selectivity, *fluorescent sensors based on hydrophobic pockets within *three-way junctions interact non-specifically with various steroids. Some steroids can be classified on the basis of the response patterns from arrays made of three-way junction variations.
MARK OLAH, MILAN STOJANOVIC, DARKO STEFANOVIC

cruciform DNA structure motif caused by *inverted repeat nucleotide sequences. A double-stranded helix can in such case rearrange itself, forming two base-paired hairpins. The two new hairpin–stems and the two stems of the now discontinued DNA form a *four-way junction. The formation of cruciform motifs is believed to play an important role in various regulatory processes. SLAWOMIR GWIAZDA

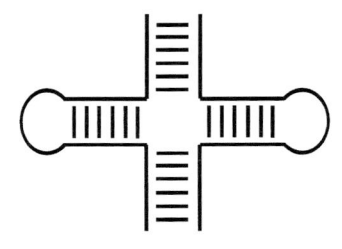

cryptic gene Silent *gene not normally expressed during the life cycle of an individual, but may be reactivated by various genetic mechanisms. It is expected that cryptic genes would accumulate *mutations and become permanently inactivated since they do not normally contribute to fitness, and would thus be rare in the populations. The fact that cryptic genes are commonly observed in microorganisms, however, indicates that there is selection for their retention in microbial populations.
MAURO SANTOS

cryptic splice site Sequence in *pre-mRNA that resembles the authentic *consensus sequences of *splice sites. In case of *mutations affecting the natural splice site of a *gene, the splicing apparatus may select a nearby cryptic splice site for the reaction. The result is an altered polypeptide sequence with divergent functional properties or an entirely non-functional truncated protein. BERND-JOACHIM BENECKE

crystallization of nucleic acids DNA structure is predominantly restricted to right-handed doubl-helical forms ($\rightarrow$ B-DNA, $\rightarrow$ A-DNA), a consequence of the fact that DNA strands always have a perfect complement available in the cell. In special sequence contexts more exotic structures such as a left-handed helix ($\rightarrow$ Z-DNA), triple-stranded arrangements, *guanine quadruplexes or parallel-stranded intercalated structures ($\rightarrow$ i-motifs) can be adopted. Often special structures, such as the *Holliday junction or transitions between different helical forms, are preferentially or exclusively adopted by DNA when bound to proteins. *DNAzymes have been evolved *in vitro* which presumably fold into more intricate three-dimensional structures. While double-helical portions are major structuring elements in RNA as well, additional single-stranded segments allow RNAs to fold into complex three-dimensional arrangements that ultimately determine their functions as already demonstrated for transfer RNA^{Phe} in the 1970s. X-ray crystallography is a powerful method to experimentally determine atomic structures of nucleic acids. It requires growth of single three-dimensional crystals and subsequent structure analysis by X-ray diffraction.

When purified by denaturing means, the nucleic acid of interest has to be subsequently refolded, which usually occurs by heating and slow cooling (annealing) in a suitable buffer. The procedure is robust for short double-stranded nucleic acid fragments. However, care has to be taken to provide longer molecules with every ingredient necessary for proper folding. In many cases, divalent metal ions (typically Mg^{2+}) have to be added. Cases are known in which a RNA molecule can be trapped in a non-functional conformation during refolding. Denaturation at high concentrations of urea and refolding by dialysis is an alternative that can overcome conformational traps in some cases. In any case, a biochemical assay for the function of the target molecule could prove useful to establish proper refolding conditions. Nevertheless, the rather high concentrations of nucleic acids required for crystallization can lead to surprising effects in special cases, e.g. molecules designed to form short hairpins can alternatively pair into self-complementary duplexes with an internal stretch of *non-canonical base pairs. During initial attempts to crystallize RNA *stem–loop structures comprising stable *tetraloops, the alternative, non-Watson–Crick-paired duplexes crystallized instead. For crystallization, the sample is ultimately brought into a buffer that typically comprises a low concentration of the buffering substance (e.g. 10 mM) and the minimum concentrations of ingredients required for biological function. In that way, reservoir solutions added during the crystallization trials can take maximal effect. In the case of RNA, the final sample should be tested for *RNase activity, especially when proteins are included in the crystallization trials.

For the crystallization of nucleic acids, the same basic methods as in protein crystallization are employed. Independent of the molecular weight, a concentration of 5–10 mg mL^{-1} of the target molecule is a suitable starting value to check for crystallizability. Such values translate to around 1–2 mM concentrations for short oligomers or around 0.5 mM for intermediate size *ribozymes. The most widespread experimental set-up for crystallization trials is vapor diffusion with either sitting or hanging drops. A small drop containing the molecule of interest is thereby suspended above a larger reservoir solution containing precipitants and additives. Reservoir and sample are mixed in the crystallization drop in about equal volume ratios, the system is sealed and the chemical activities of the ingredients in the drop and reservoir are allowed to reach equilibrium by adjusting through volatile components via the gas phase. An alternative is microbatch crystallization in which the nucleic acid of interest is combined with precipitant solutions and covered with inert oil that avoids or strongly reduces the rate of evaporation. The trials are incubated at a given temperature without further sealing. Manually set-up vapor diffusion or microbatch experiments consume about 1 μL of sample per trial; robotics can reduce sample consumption by more than an order of magnitude. In the future, chip-based microfluidic systems in microbatch format or using free interface diffusion may further downscale sample consumption.

During the crystallization experiments, the chemical space that could sponsor crystallization is screened. Since complete systematic coverage of this multi-dimensional space is not possible (and as it turns out not necessary), incomplete factorial screens or sparse matrix screens as known from protein crystallization are usually employed. In general, all commercially available screens, primarily sold for protein crystallization, could be used. However, certain ingredients have been found particularly useful for the crystallization of nucleic acids and such substances are highly represented in screens that have been developed specifically for the crystallization of nucleic acids. As for refolding, divalent metal ions often prove useful or necessary since they not only support the intrinsic folds, but can also efficiently mediate crystal packing interactions via negatively charged phosphate groups of the backbone. The same holds true for polyamines such as spermine, spermidine or putrescine or metal-amines such as cobalt or osmium hexammine. The latter substances could be useful phasing vehicles as well. As precipitants, low-molecular weight polyethylene glycols (PEG) and alcohols like 2-methyl-2,3-pentanediol (MPD) have been widely used.

Once crystals have been obtained they have to be prepared for diffraction analysis. Traditionally, macromolecule crystals have been measured in thin-walled special glass capillaries that are sealed with mother liquor on one side to avoid dehydration and loss of crystalline order. Diffraction analysis at cryogenic temperatures dramatically reduces the loss of diffraction power as a consequence of radiation damage and its propagation throughout the crystal. Normally, macromolecular crystals have to be transferred into cryoprotecting solutions before freezing to avoid ice formation. In some instances precipitants such as MPD or low-molecular-weight PEGs provide for cryoprotection intrinsically. In other cases, various substances have to be tested and collections of cryoprotectants are commercially available. Synthetic mother liquors with cryoprotectants sometimes exert an advantageous limited dehydration of the crystals that leads to improved order and diffraction power. MARKUS WAHL

C-strand →coding strand

CTP Abbreviation for cytidine triphosphate (→nucleoside phosphates).

cyclic adenosine-3′,5′-monophosphate
→nucleoside phosphates

cyclic DNA →circular DNA

cyclic guanosine-3′,5′-monophosphate
→nucleoside phosphates

cyclic phosphate Nucleotide in which two neighboring hydroxyl groups form an ester with phosphoric acid. The 3′,5′-cyclic phosphates of *adenosine and *guanosine are of particular biochemical relevance, acting as second messengers in signal transduction processes. The 2′,3′-cyclic phosphates are formed in RNA cleavage reactions by induced nucleophilic attack of the 2′-hydroxyl group onto the neighboring phosphate. →nucleoside phosphates, →transesterification, →RNase, →ribozyme. SABINE MÜLLER

cyclohexane nucleic acid (CNA) Nucleic acid in which the sugar residue is replaced by cyclohexane. CNA is used for studies in *antisense technology.

cyclohexenyl nucleic acid (CeNA) Nucleic acid in which the sugar moiety is replaced by cyclohexene. CeNA is a conformationally flexible oligonucleotide and is used for several binding studies with *nucleic acids and proteins as well as for *antisense technologies. CeNA is able to form a double strand with RNA and exposure to *RNase H triggers *RNA degradation. BETTINA APPEL

cytidine Natural building block of RNA (→nucleosides).

cytidine diphosphate (CDP) →nucleoside phosphates

cytidine monophosphate (CMP) →nucleoside phosphates

cytidine phosphates →nucleoside phosphates

cytidine triphosphate (CTP) →nucleoside phosphates

cytosine →nucleobases

cytosine arabinonucleoside →arabinonucleosides

cytosine ribonucleoside →nucleosides

cytosine xylonucleoside →xylonucleosides

D

dA →2′-deoxyadenosine

dADP Abbreviation for 2′-deoxyadenosine diphosphate (→nucleoside phosphates).

Dam →DNA methylation

dAMP Abbreviation for 2′-deoxyadenosine monophosphate (→nucleoside phosphates).

dATP Abbreviation for 2′-deoxyadenosine triphosphate (→nucleoside phosphates).

daunomycin (Daunorubicin) Anticancer drug belonging to the class of *anthracyclines. It is naturally produced by the bacterium *Streptomyces peucetius*. The daunosamine residue intercalates into the *minor groove of DNA. The highest priority is given to sequences with neighboring GC *base pairs flanked on the 5′-side by an AT base pair. Daunomycin shows interaction to all three base pairs and causes local unwinding, but insignificant falsification of helical *DNA conformation. In this way, it inhibits *DNA replication in rapidly growing cancer cells. Daunomycin is given as intravenous infusion to treat acute lymphoblastic leukemia and acute myeloid leukemia. BETTINA APPEL

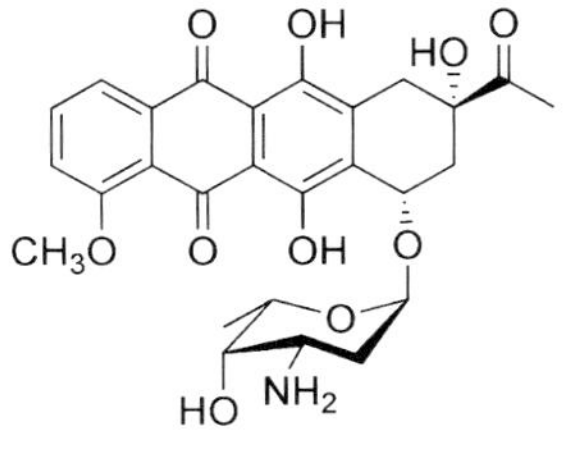

dC →2′-deoxycytidine

dCDP Abbreviation for 2′-deoxycytidine diphosphate (→nucleoside phosphates).

dCMP Abbreviation for 2′-deoxycytidine monophosphate (→nucleoside phosphates).

dCTP Abbreviation for 2′-deoxycytidine triphosphate (→nucleoside phosphates).

ddATP Abbreviation for 2′,3′-dideoxyadenosine triphosphate (→dideoxynucleoside triphosphates, →DNA sequence analysis, →Sanger sequencing).

ddCTP Abbreviation for 2′,3′-dideoxycytidine triphosphate (→dideoxynucleoside triphosphates, →DNA sequence analysis, →Sanger sequencing).

ddGTP Abbreviation for 2′,3′-dideoxyguanosine triphosphate (→dideoxynucleoside triphosphates, →DNA sequence analysis, →Sanger sequencing).

D-DNA Polymorph of DNA, belonging to the B-family. D-DNA is only formed by DNA with alternating AT sequence and by highly modified phage T2 DNA. Compared with *B-DNA, the D-DNA helix is overwound with only 8 bp completing one turn, i.e. the rotation from one *nucleotide to the adjacent one is 45°. The helix axis is positioned in the *minor groove of the *duplex, and the base pairs are tilted –16° toward the helix axis, contrasting the +20° for *A-DNA. D-DNA displaces a very deep, narrow minor groove, a good cavity for trapping water and cations, and a more shallow *major groove with respect to B- and *C-DNA. In contrast to cooperative B–A

Nucleic Acids from A to Z: A Concise Encyclopedia. Edited by Sabine Müller

ISBN: 978-3-527-31211-5

transition, the gradual course of the B–D helical transformation testifies that there is no internal potential barrier to be overcome. NINA DOLINNAYA

ddNTP Abbreviation for 2′,3′-dideoxynucleoside triphosphate (→dideoxynucleoside triphosphates, →DNA sequence analysis, →Sanger sequencing).

ddTTP Abbreviation for 2′,3′-dideoxythymidine triphosphate (→dideoxynucleoside triphosphates, →DNA sequence analysis, →Sanger sequencing).

decoding →ribosome

decoding center One of the two reaction centers of *ribosomes essential for protein synthesis (→translation), with the *peptidyl transferase center being the other.

deletion mutation Mutation caused by the removal of one or more DNA (RNA) residues. Short deletions typically arise during *DNA replication or repair by local *primer/*template slippage, long ones by recombinational events (→insertion mutation, →substitution mutation). HANS-JOACHIM FRITZ

demethylation of DNA →DNA demethylation

denaturation Separation of double-stranded *oligonucleotides into single strands by heating or denaturing agents such as urea or formamide.

2′-deoxyadenosine Natural building block of DNA (→nucleoside).

2′-deoxyadenosine diphosphate
→nucleoside phosphates, →nucleotide

2′-deoxyadenosine monophosphate
→nucleoside phosphates, →nucleotide

2′-deoxyadenosine-5′-triphosphate
→nucleoside phosphates, →nucleotide

2′-deoxycytidine Natural building block of DNA (→nucleoside).

2′-deoxycytidine diphosphate →nucleoside phosphates, →nucleotide

2′-deoxycytidine monophosphate
→nucleoside phosphates, →nucleotide

2′-deoxycytidine triphosphate →nucleoside phosphates, →nucleotide

2′-deoxyguanosine Natural building block of DNA (→nucleoside).

2′-deoxyguanosine diphosphate
→nucleoside phosphates, →nucleotide

2′-deoxyguanosine monophosphate
→nucleoside phosphates, →nucleotide

2′-deoxyguanosine triphosphate
→nucleoside phosphates, →nucleotide

2′-deoxynucleoside →nucleoside

deoxynucleoside diphosphate
→nucleoside diphosphate

deoxynucleoside monophosphate
→nucleoside monophosphate

deoxynucleoside triphosphate
→nucleoside triphosphate

2′-deoxynucleotide Monomeric building block of DNA (→nucleotide).

deoxynucleotidyl transferase
→terminal deoxynucleotidyl transferase

deoxyribonuclease →DNase

deoxyribonucleic acid →DNA

2′-deoxyribonucleoside →nucleoside

2′-deoxyribonucleoside diphosphate
→nucleoside phosphates, →nucleotide

2'-deoxyribonucleoside monophosphate
→nucleoside phosphates, →nucleotide

2'-deoxyribonucleoside triphosphate
→nucleoside phosphates, →nucleotide

2'-deoxyribonucleotide   →nucleotides

2'-deoxyribose Sugar unit in natural nucleosides constituting DNA. Only the D-enantiomer occurs in natural DNA (→nucleosides, →nucleotides).

deoxyribosyltransferase Enzyme that catalyzes cleavage of the N-*glycosidic bond of a *2'-deoxyribonucleoside and subsequent exchange of the cleaved off nucleobase against another nucleobase. Class I N-deoxyribosyltransferases or purine transdeoxyribosyltransferases can only catalyze the exchange of *purine bases. Class II N-deoxyribosyltransferases or nucleoside 2-deoxyribosyltransferases catalyze the purine–purine exchange as well as the purine–pyrimidine or pyrimidine–purine exchange. VALESKA DOMBOS

deoxyribozymes (DNAzymes or DNA enzymes) DNA molecules that possess catalytic activity. Like proteins and RNA, DNA can fold into complex three-dimensional structures featuring *binding pockets for *ligands and *catalytic centers for a broad range of chemical reactions. To date, no naturally occurring deoxyribozymes are known. However, a large number of artificial deoxyribozymes have been created using *in vitro selection from combinatorial *DNA libraries. The first deoxyribozyme was discovered in 1994 and cleaves a specific RNA linkage embedded within a longer *nucleic acid strand. According to the chemistry of the reactions, deoxyribozymes can be assigned to the following groups. (a) Those that cleave RNA. This group constitutes by far the largest and includes molecules that found application as designer restriction enzymes for RNA, including the "8–17 deoxyribozyme" and the "10–23 desoxyribozyme" that can be designed to specifically cleave virtually any mRNA. These DNA enzymes typically require divalent metal ions for activity. (b) Those that covalently modify nucleic acids. Some of these reactions involve changes in the phosphorylation status of a DNA or RNA strand, specifically DNA phosphorylation and adenylation. Other reactions include DNA deglycosylation, *thymine dimer photoreversion and DNA cleavage. The latter is a Cu^{2+}-mediated oxidative cleavage reaction. (c) Those that ligate DNA or RNA, forming either linear or branched connections. (d) Those that catalyze other reactions (porphyrin metalation). In addition to the widespread use as *restriction enzymes, applications of RNA-cleaving deoxyribozymes include *in vivo* targeting of *messenger RNA, mapping of RNA branching points or for the construction of logical computations circuits (→deoxyribozyme- and ribozyme-based logic gates, →deoxyribozyme and ribozyme circuits, →deoxyribozyme-based automata). The catalytic repertoire of deoxyribozymes can be enhanced by incorporating *modified nucleotides that contain additional *functional groups.
ANDRES JÄSCHKE

deoxyribozyme- and ribozyme-based logic gates Molecular logic gates interact with one or more constituents in solution and modify their properties (function) according to a set of rules. Such sets of rules can be interpreted as Boolean calculations. Nucleic acid catalysts act as logic gates if they integrate information about the presence (1) or absence (0) of one or more allosteric modulators, and if they produce (1) or do not produce (0) an output. The rules assigning values of functional outputs

(either 1 or 0) to all possible combinations of inputs can be organized in a truth table. Logic gates analyzing the presence of small molecules and *oligonucleotides have been reported. Examples of logic gates include: (a) sensor gates, e.g. *catalytic molecular beacons, *riboswitches and *aptazymes, (b) NOT gates, which analyze the presence or absence of one input and produce an output if the input is not present, (c) AND gates, which analyze the presence or absence of two inputs and produce an output if both are present, (d) OR gates which analyze the presence or absence of two inputs and produce an output if either one or both are present, (e) ANDNOT gates, which analyze the presence or absence of two inputs and produce an output if the first is present but the second is not, and (6) ANDANDNOT gates (INHIBIT), which analyze the presence or absence of three inputs, and produce an output if the first two are present and the third is not. Examples of logic gates based on *phosphodiesterase and *ligase enzymes have been reported.
MILAN STOJANOVIC, DARKO STEFANOVIC

deoxyribozyme and ribozyme circuits Multiple logic gates can be arranged in parallel (i.e. behaving independently) and in series (output of an upstream gate is cascaded into a downstream gate) into circuits with more complex behavior. Circuits can, in principle, analyze an arbitrary number of inputs and produce an arbitrary number of outputs. Some circuits that have been reported are half- and full-adder, tic-tac-toe playing automata, and *ligase–*phosphodiesterase cascades. Further circuits can be arranged by mixing logic gates with *aptamers.
MILAN STOJANOVIC, DARKO STEFANOVIC

deoxyribozyme-based automata Circuits based on *deoxyribozyme-based logic gates that are designed to respond to a sequence of inputs in a meaningful fashion. Two tic-tac-toe automata MAYA and MAYA-II have been reported. In these automata logic gates are distributed in wells representing the tic-tac-toe board (a 3×3 field). Human moves are encoded as *oligonucleotide inputs that are added to all wells. The addition of inputs triggers a response from the circuit contained in one particular well (and no other well); this response represents the automaton's countermove. MAYA consists of 23 logic gates and plays symmetry-pruned tic-tac-toe, analyzing eight possible inputs and encoding 19 different games. MAYA-II plays the general game and consists of 128 logic gates encoding 76 different games (→ Shapiro–Benenson–Rothemund automata). MILAN STOJANOVIC, DARKO STEFANOVIC

2'-deoxyuridine Uracil linked to 2'-deoxyribose (→ nucleosides).

deoxyuridine mutagenesis → Kunkel mutagenesis

dephosphorylation Process of removing a phosphate group by hydrolysis. A well-known enzyme which catalyzes dephosphorylation is *alkaline phosphatase.

deprotection After *oligonucleotide synthesis, the fully protected oligonucleotide is attached at the *solid support, and has to be released thereof and liberated from *protection groups. For DNA, release and deprotection proceed under basic reaction conditions; for RNA, additional deprotection of the orthogonal 2'-O-protecting groups is required (→ TBDMS method, → TOM method, → ACE method). RONALD MICURA

depurination A reaction by which a *purine residue is lost from DNA (RNA) through hydrolysis of the *glycosidic bond. The result of a depurination reaction is an *abasic site. HANS-JOACHIM FRITZ

dG →2'-deoxyguanosine

DHU arm *Dihydrouridine stem–loop in a cloverleaf *transfer RNA.

DHU loop *Dihydrouridine stem–loop of a *transfer RNA.

dialysis Method for separating large molecules like nucleic acids and proteins from ions and other small molecules. The dissolved molecules are transferred into a semipermeable membrane (e.g. cellulose membrane) with pores of a defined diameter. This so-called dialysis bag is placed in a container equipped with pure water or a different solvent. Smaller molecules that are able to pass through the pores of the membrane will diffuse towards the lower concentration outside the membrane, whereas the larger molecules are retained inside. This technique is usually applied to desalt nucleic acids and proteins. VALESKA DOMBOS

Dicer Eukaryotic *double-stranded RNA-specific endonuclease. Dicer cleavage generates 21- to 26-nucleotide long RNA *duplexes with 2-nucleotide long 3'-protruding ends. Dicer is involved in the maturation of double-stranded RNAs into *small interfering RNAs and of *pre-miRNA into *micro-RNA. →RNA interference. NICOLAS PIGANEAU

2',3'-dideoxyadenosine-5'-triphosphate →dideoxynucleoside triphosphate

2',3'-dideoxycytidine-5'-triphosphate →dideoxynucleoside triphosphate

2',3'-dideoxyguanosine-5'-triphosphate →dideoxynucleoside triphosphate

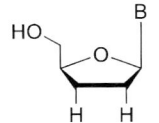

B = Adenine
Cytosine
Guanine
Thymin
Uracil

dideoxy method →DNA sequencing, →Sanger sequencing

2',3'-dideoxynucleoside Nucleoside lacking the 2'- and 3'-hydroxyl function.

2',3'-dideoxynucleoside triphosphate Lacks the 3'-hydroxyl function at the ribose moiety which is necessary for the formation of a *phosphodiester bond with a subsequent *nucleotide. If dideoxynucleoside triphosphates are employed in a *primer-extension reaction, their incorporation by a *polymerase leads to chain termination. Dideoxynucleoside triphosphates are used in *Sanger sequencing reactions; they can be linked to fluorophores (dye terminators) coding for one of the bases A, G, C or T, thereby facilitating the automated reading of sequencing electropherograms. SUSANNE BRAKMANN

2',3'-dideoxyribonucleoside →2',3'-dideoxynucleoside

2',3'-dideoxyribonucleoside-5'-triphosphate →dideoxynucleoside triphosphate

2',3'-dideoxythymidine-5'-triphosphate →dideoxynucleotide triphosphate

Diels–Alderase ribozyme or Diels–Alder ribozyme *Artificial *ribozyme generated by *in vitro selection that catalyzes the formation of C–C bonds by Diels–Alder reaction – a [4 + 2] cycloaddition reaction central to organic chemistry. Two types of Diels–Alderase ribozymes are known. (a) Ribozymes that catalyze the *single-turnover reaction of an aliphatic, butadiene-derived diene with maleimide. These ribozymes contain pyridyl residues appended as "side-chains" and require Cu^{2+} for activity. (b) Unmodified RNA molecules that catalyze the fast *multiple-turnover reaction of anthracene dienes with maleimide dienophiles (panel A). These ribozymes display saturation-type *Michaelis–Menten kinetics and accelerate the reaction with high stereoselectivity. A synthetic mirror-image ribozyme (→ Spiegelmer) composed of unnatural L-RNA has the same catalytic activity but opposite stereoselectivity. This Diels–Alder ribozyme is the first artificial ribozyme for which the three-dimensional structure has been solved by X-ray crystallography (panel B). The RNA adopts a λ-shaped *nested pseudoknot architecture, whose pre-formed wedge-shaped hydrophobic pocket shows precise shape complementarity to the *transition state of the reaction (panels C and D).

*RNA folding and product binding are dictated by extensive stacking and hydrogen bonding, while stereoselection is governed by the shape of the catalytic pocket. *Catalysis is apparently achieved by a combination of proximity, complementarity and electronic effects. Striking structural parallels are observed in the independently evolved catalytic pocket architectures for ribozyme- and antibody-catalyzed Diels–Alder C–C bond-forming reactions, suggesting that RNA, although equipped with a much less varied arsenal of *functional groups than proteins, has independently evolved similar strategies for generating catalytic pockets capable of facilitating C–C bond-forming reactions, with comparable catalytic efficiency and enantioselectivity. A nested pseudoknot architecture is also found in the structure of the *hepatitis delta viroid ribozyme. ANDRES JÄSCHKE

digestion Enzymatic degradation of DNA or RNA into monomer units.

dihydrouridine →pseudouridine

dihydrouridine stem–loop One branch in the *cloverleaf structure of a *transfer RNA containing the frequently appearing *rare nucleoside dihydrouridine (→pseudouridine) in the loop positions 16 and 17. It is an essential region for structure formation as the neighboring loop positions G19 and G20 span up the L-shaped three-dimensional structure by base pairing to the *TΨC loop nucleotides C54 and Ψ55. STEFAN VÖRTLER

dimethoxytrityl group Standard protecting group for the 5′-hydroxyl function of nucleotide building blocks in *oligonucleotide synthesis.

dinucleotide Pair of *mononucleotides linked to each other by a *phosphodiester bond. Biological important dinucleotides are *ADP, *NAD/NADP, *FAD and *CpG units. BETTINA APPEL

directed mutagenesis →site-directed mutagenesis

direct gene transfer Introduction of nude (uncoated) DNA into host cells using physical (electroporation, microinjection, biolistics, etc.) approaches.

discriminator base Base at the 3′-end of *transfer RNAs (tRNA) which is crucial for aminoacyl specificity. It is situated at position 73, lies directly before the *CCA-tail and is important for the recognition by the *aminoacyl-tRNA synthetase to load the correct amino acid onto the respective tRNA. Substitution of the discriminator base by other nucleobases *in vitro* results in different acceptor characteristics (specificity, transfer reaction speed). JÖRN WOLF

displacement loop Ubiquitous structure in mammalian *mitochondrial DNA that is created when the DNA around the *origin of *replication of the heavy strand becomes regionally separated and forms a structure reminiscent of the letter "D". Regions in which displacement loops occur lack protein-coding function. The formation is prompted by a DNA segment that hybridizes to the light strand and displaces the heavy strand. JÖRN WOLF

dissociation Separation of a molecule or a complex structure into two or more fragments. For a reversible equilibrium, the *dissociation constant K_d is the ratio of dissociated to undissociated compound. The deprotonating/protonating ratio of acids in solution is described by the dissociation constant K_a, normally given as the negative logarithm (pK_a) (→acidity of nucleotides and nucleosides). Also the melting of DNA and RNA (conversion of double strands into single strands) is a dissociation. The dissociation constant is an important implement to indicate the affinity of a *ligand to a receptor (e.g. nucleic acid–ligand interaction). BETTINA APPEL

dissociation constant Used in order to represent the equilibrium between a (usually non-covalent) complex and the two constituent species A and B. Since this special *equilibrium constant refers to the dissociation reaction of the complex, the dissociation constant, K_d, is the product of the concentrations of A and B divided by the complex concentration and has the unit moles per liter.

The value of the dissociation constant is determined in titration experiments, where the complex concentration (relative to the maximum possible) is measured in dependence of the concentration of one of the compounds. If the titration experiment is set up with the concentration of the titrant A, c_A, high above the concentration of the other species B the equation for the dissociation constant can be rearranged and approximated by $\Theta = c_A/(K_d + c_A)$, where Θ represents the ratio of complex concentration and total concentration of species B (i.e. fractional binding). Usually, a spectroscopic signal is employed for detection which is proportional to concentrations (e.g. fluorescence or optical density). Therefore, Θ is equal to the ratio of the actual change of the respective signal (due to complex formation) and the maximum signal change. According to this equation saturation is established at concentrations 100-fold or more above the K_d value while half maximum complex concentration is reached at a concentration of species A equal to the K_d value. CHRISTIAN HERRMANN

D-loop *Dihydrouridine stem–loop of a *transfer RNA.

DNA (deoxyribonucleic acid) *Biopolymer composed of *2′-deoxyribonucleotides (→nucleic acid). DNA carries the genetic information of viruses, bacteria and all higher organisms (→genetic code). DNA may occur single- or double-stranded with the 2′-deoxynucleotides connected by 3′,5′-*phosphodiester bonds. In *double-stranded DNA, two *complementary strands are wound around each other in opposite directions (→antiparallel orientation) forming a *right-handed or a *left-handed helix (→DNA structures, →Watson–Crick model). The two strands are held together by specific hydrogen bonding between the nucleobases (→Watson–Crick base pairs) as well as *base-stacking interactions. In different organisms, the relative amount of individual bases varies. However, the content of A always equals the content of T as well as the content of C always equals the content of G (→GC content, →Chargaff's rule). Single-stranded DNA as well as double-stranded DNA may also occur as cyclic molecules (e.g. in bacteria and mitochondria). These cycles may be twisted forming super helices of hypertwisted configuration (→super helix).

While prokaryotic DNA is membrane bound or is contained in *plasmids or *episomes, eukaryotic DNA (about 95%) is localized in the nucleus and is bound to specific proteins (→chromatin, →histone, →DNA–protein interaction, →DNA-binding proteins). Mitochondria and chloroplasts also contain DNA, with *mitochondrial DNA representing only 1–2% of the entire DNA of a cell, *chloroplast DNA about 5%. The transfer of *genetic information from parental to daughter cells occurs by *DNA replication. SABINE MÜLLER

DNA adenine-N6 methyltransferase DNA adenine-N6 methyltransferase (MTase) and DNA cytosine-N4 MTase methylate exocyclic amino groups of *nucleobases and are grouped together as N-DNA MTases. They catalyze the methyl group transfer from the cofactor S-adenosyl-L-methionine (AdoMet, →DNA methyltransferase) to the 6 amino group of *adenine or the 4 amino group of

target adenine out of the DNA helix and into their *active sites (→ base flipping). In the M.*Taq*I active site adenine forms a face-to-face π-stacking interaction with a conserved tyrosine residue (Tyr108) and the 6 amino group forms two weak hydrogen bonds with two carbonyl groups of a conserved asparagine (Asn105) and proline (Pro106) residue (N/D/S**PP**Y/F motif of N-DNA MTases). The amino group is placed in close proximity to the activated methyl group of the cofactor AdoMet and methyl group transfer to the p orbital of the sp^2-hybridized 6 nitrogen atom should lead to a hybridization change to sp^3. The resulting tetrahedral 6-methylammonium group will be stabilized by cation–π interactions with Tyr108 and strong hydrogen bonds to the carbonyl groups of Asn105 and Pro106. Afterwards, deprotonation of the 6-methylammonium group will regenerate sp^2 hybridization at the 6 nitrogen atom leading to N6-methyladenine.
ELMAR WEINHOLD

DNA adenine-N6 methyltransferase: catalytic mechanism of N-DNA MTases using M.*Taq*I as an example.

*cytosine within specific DNA sequences. Structure-guided sequence comparison of N-DNA MTases revealed nine conserved sequence motifs, which correspond to nine motifs defined in C-DNA MTases (→ DNA cytosine-C5 methyltransferases). Depending on the sequential order of the conserved motifs and positioning of a variable domain they can be further subgrouped into α, β and γ N-DNA MTases. Three-dimensional structures in complex with DNA are available for the DNA adenine-N6 MTases M.*Taq*I and Dam from bacteriophage T4, and show that they flip their

DNA alkylating agents Organic chemicals that attach an alkyl group to DNA. Alkylating agents are electrophilic compounds with high affinity to nucleophilic positions in DNA *bases. They are either monofunctional with only one reactive group or polyfunctional with the property to covalently attach more than one alkyl group to different positions of the DNA molecule. Due to their ability to inhibit cell division, some alkylating agents are used as anticancer drugs, provided that they are active under physiological "inner cell" conditions. However, their potency is limited by different toxic side effects in normal tissues. Alkylating agents can attack DNA bases at twelve different sites. Main target of monofunctional alkylating agents are N7 of *guanine and N3 of *adenine. Alkylation at these positions causes destabilization of the N-*glycosidic bond and consequently affords *depurination. Strong methylating agents like dimethylsulfate and methylmethanesulfonate cause only a few *point mutations but show strong clastogenicity, because of the participation of N-alkylated bases in *chromosome aberration. Other predisposed positions for alkylation are positions O6 of guanine and O4 of thymine. They are alkylated, for example, by N-alkylnitrosamine and N-alkylnitrosamide. During *replication, O^6-methylguanine can *base pair with thymine affording a *transition mutation of GC to AT. O^4-methylthymine causes after replication transition of AT into GC. Both types of modified bases show clastogenicity. Other types of alkylating agents are nitrogen mustards (e.g. cyclophosphamide), alkyl sulfonates and triazenes. *DNA damage caused by alkylation is repaired by repair enzymes. BETTINA APPEL

DNA alkylation Process in which alkyl groups are attached to DNA, mainly to the *nucleobases (→DNA alkylating agents).

DNA backbone →sugar–phosphate backbone

DNA barcoding →mitochondrial DNA

DNA bending →bending of DNA

DNA-binding antibiotics *Antibiotics binding to DNA and RNA.

DNA-binding polyamides →minor groove binding polyamides, →DNA-binding proteins.

DNA-binding proteins Any protein that binds to double- or single-stranded DNA. Examples include *transcriptional regulators (e.g. *transcription factors acting as *repressors or *activators), proteins involved in the packaging of DNA within the nucleus (such as *histones), enzymes that cut, modify, transcribe or replicate DNA (e.g. *nucleases, *topoisomerases, *glycosidases, *helicases and *polymerases), or any of many accessory proteins which are involved in these processes (e.g. *single-stranded DNA-binding proteins). The physiological role of DNA-binding proteins is determined by the affinity and specificity of the DNA–protein interaction. These properties depend upon the precise interactions between amino acids in the DNA-binding protein and nucleotides in the DNA molecule. In addition, protein–protein interactions are sometimes required for efficient DNA–protein interactions. Sequence-specific DNA-binding proteins generally interact with the *major groove of *B-DNA. Sequence specificity is achieved through *hydrogen bonding and *van der Waals interactions between protein sidechains and exposed edges of *base pairs or sequence-dependent bendability or deformability of DNA. There are several classes of sequence-specific DNA-binding proteins like *helix–turn–helix proteins, *homeodomains, zinc finger proteins, steroid receptors, leucine zipper proteins, *helix–loop–helix proteins and β-sheet motif-containing proteins. Each motif involves simple secondary structure that is complementary to B-DNA. On the other hand, proteins involved in *transcription or *replication of DNA usually interact with nucleic acids independently of the sequence context. Here the macromolecular interaction occurs via contacts with the *minor groove of *B-DNA. TOBIAS RESTLE

DNA biosynthesis →DNA replication

DNA blotting →blotting, →Southern blot

DNA breathing *DNA double-helix denatures spontaneously even under physiological conditions in a reversible manner.

Eventually formed single-stranded DNA regions called *DNA bubbles are recognized and subsequently bound by *DNA-binding proteins such as *polymerase. The DNA breathing motion depends on the strength of *base pair interactions. BETTINA APPEL

DNA bubble During *transcription of DNA the *double helix is unzipped by enzymes (→DNA helicase, →DNA topoisomerase) resulting in a local single-strand region called a DNA transcription bubble. During replication there are also single-stranded regions caused by *polymerases called replication bubbles. DNA bubbles of 20–30 bp occur in *DNA breathing modes and hence are called DNA *breathing bubbles. BETTINA APPEL

DNA chip Combined semiconductor technology and *gene technology to identify genes or to measure their activity on a very small surface which is organized into many microarrays. Since the 1980s it has been possible to identify up to 100 000 samples simultaneously, even from different specimens. Microarrays are coated with defined *single-strand DNA fragments (covalently bound to the array) which act as probes to bind samples containing a particular DNA or RNA sequence. These samples can be tagged with *fluorescent markers. A high-resolution camera provides data about the position, intensity and wavelength of the signal. The single-strand fragments can be attached via different chemical techniques to the array surface:

- Aldehyde slides produce imines with amino-modified DNA strands.
- Epoxy slides produce amines with amino-modified DNA strands.
- N-hydroxysuccinimide slides produce amides with amino-modified DNA strands.
- *Streptavidin slides bind, *biotinmodified oligonucleotides.

MATTHÄUS JANCZYK

DNA computing (RNA, nucleic acid based) Generic term that describes either DNA being used to solve certain types of mathematical (broadly defined) problems or is used to describe experiments in which mixtures of DNA display programmed behavior. Originally, it referred to a *non-autonomous* procedure in which (a) a large number of DNA strands is generated, such that they encode candidate answers (i.e. solutions) to a given combinatorial problem, (b) a human-guided biochemical procedure is performed on all of these DNA strands simultaneously to either select the strand(s) encoding a veritable answer or to eliminate all strands encoding non-answers, and (c) the selected (or remaining) DNA strands are read out by some biochemical means and are interpreted as the solution for the given problem. This is termed Adleman's computing paradigm (after Leonard Adleman who was the first to demonstrate it) or massively parallel DNA computing. Examples of *autonomous* DNA computing include *algorithmic self-assembly, *Shapiro–Benenson–Rothemund automata and various molecular circuits based on DNA. MILAN STOJANOVIC, DARKO STEFANOVIC

DNA conformation DNA is a dynamic molecule existing in a number of different *conformations (main families: *A-, *B- and *Z-DNA). The equilibrium between conformations is influenced by DNA sequence, ionic strength, surrounding medium as well as by the presence of proteins (e.g. *histones or *DNA-binding proteins) and the extent of topological stress on the DNA molecule. SABINE MÜLLER

DNA crystallization →crystallization of nucleic acids

DNA curvature →bending of DNA

DNA cytosine-C5 methyltransferase DNA cytosine-C5 methyltransferases (MTases) catalyze the methyl group transfer from the cofactor S-adenosyl-L-methionine (AdoMet, →DNA methyltransferase) to the C5 of *cytosine residues within specific DNA sequences and therefore are also called C-DNA MTases. They form a homogenous group of enzymes and share 10 highly conserved amino acid sequence motifs arranged in a linear order within their catalytic domains. Three-dimensional structures in complex with DNA are available for two C-DNA MTases, i.e. M.*Hha*I and M.*Hae*III, and show that they use a *base flipping mechanism to gain access to their target cytosines. In addition, the catalytic mechanism of C-DNA MTases is well understood.

The target cytosine placed in the *active site is attacked at C6 by a conserved cysteine residue and the resulting covalent enzyme–DNA intermediate is stabilized by protonation of N3 (addition reaction). This leads to a strongly nucleophilic character of C5 which attacks the activated methyl group of AdoMet in a S_N2-type reaction (methyl group transfer) and produces the demethylated cofactor S-adenosyl-L-homocysteine (AdoHcy). Subsequently, deprotonation at C5 and leaving of the active site cysteine residue restores the double bond between C5 and C6 (elimination). This catalytic mechanism is analogous to that of thymidylate synthase except that methylation of 2′-deoxyuridine monophosphate is performed with N^5, N^{10}-methylenetetrahydrofolate instead of AdoMet. In addition, this catalytic mechanism involves access to N3 of cytosine, which is involved in *Watson–Crick base pairing in DNA and thus requires opening of the target base pair as achieved by DNA *base flipping. ELMAR WEINHOLD

DNA cytosine-N4 methyltransferase →DNA methyltransferase, →DNA adenine-N6 methyltransferases

DNA damage →DNA repair

DNA cytosine C5 methyltransferase: catalytic mechanism of C-DNA MTases.

DNA demethylation Has been shown to occur *in vivo*. The existence of an enzyme which catalyzes DNA demethylation without excision of the methylated nucleobase is still under discussion. The known demethylation pathways involve a *DNA glycosylase. BETTINA APPEL

DNA-dependent DNA polymerase *DNA polymerase which uses *double-stranded DNA or *single-stranded DNA as a *template for *DNA replication.

DNA-dependent RNA polymerase *RNA polymerase which uses a *DNA template for *RNA polymerization.

DNA double helix → Watson–Crick model, → DNA, → DNA conformation

DNA duplex → double-stranded DNA, → duplex

DNA enzyme → deoxyribozyme

DNA fiber diffraction *X-ray and neutron fiber diffraction method used to determine the structure of biological *macromolecules. DNA forms orientated fibers in which the axes of the molecules line up parallel to each other. The *Watson–Crick model of the *DNA double helix was developed in 1952 based on X-ray fiber diffraction data. BETTINA APPEL

DNA fingerprint *Genotype analysis of several polymorphic markers from a single individual. The specific *allele composition of each individual allows its identification and statistical discrimination from a population of the same species. This term is mostly used in the context of forensic DNA. GEMMA MARFANY, ROSER GONZÀLEZ-DUARTE

DNA-functionalized carbon nanotubes (CNTs) Nanosized molecular tubes formed by graphene sheets wrapped onto themselves with the joined edges. Single-walled CNTs (SWCNTs) are composed of one graphene sheet (1–2 nm diameter), whereas multi-walled CNTs (MWCNTs) are made of several graphene layers with the distances between the walls of around 0.34–0.36 nm (reaching several centimeters in length). CNTs show conductive or semiconductive electronic properties depending on the folding modes of the graphene walls.

DNA-functionalized CNTs are hybrid bionanomaterials offering biological properties of DNA combined with unique electronic, optoelectronic, thermal and mechanical properties of CNTs. DNA can be associated with the CNTs by three different modes: (a) entrapment into the internal channel of CNTs, (b) wrapping around the external walls, and (c) covalent binding to the *functional groups generated at defect sites of the graphene walls or existing at the nanotubes edges. Open-end CNTs provide internal channels that are capable of accommodating DNA chains. DNA transport through a single CNT channel was directly followed by fluorescence microscopy, while molecular dynamics simulations have indicated that DNA can be encapsulated inside CNTs in a water solute environment via an extremely rapid dynamic interaction process, provided that the tube size exceeds a certain critical value. DNA wrapping around the external sides of CNTs walls results in the increased solubility of CNTs in water. Wrapping of SWCNTs by *single-stranded DNA (ssDNA) was found to be sequence dependent. A systematic search of the ss*DNA library selected a sequence $d(GT)_n$ ($n = 10–45$) that self-assembles into a helical structure around individual nanotubes in such a way that the electrostatics of the DNA–SWCNT hybrid depends on tube diameter and electronic properties, enabling separation of SWCNTs with

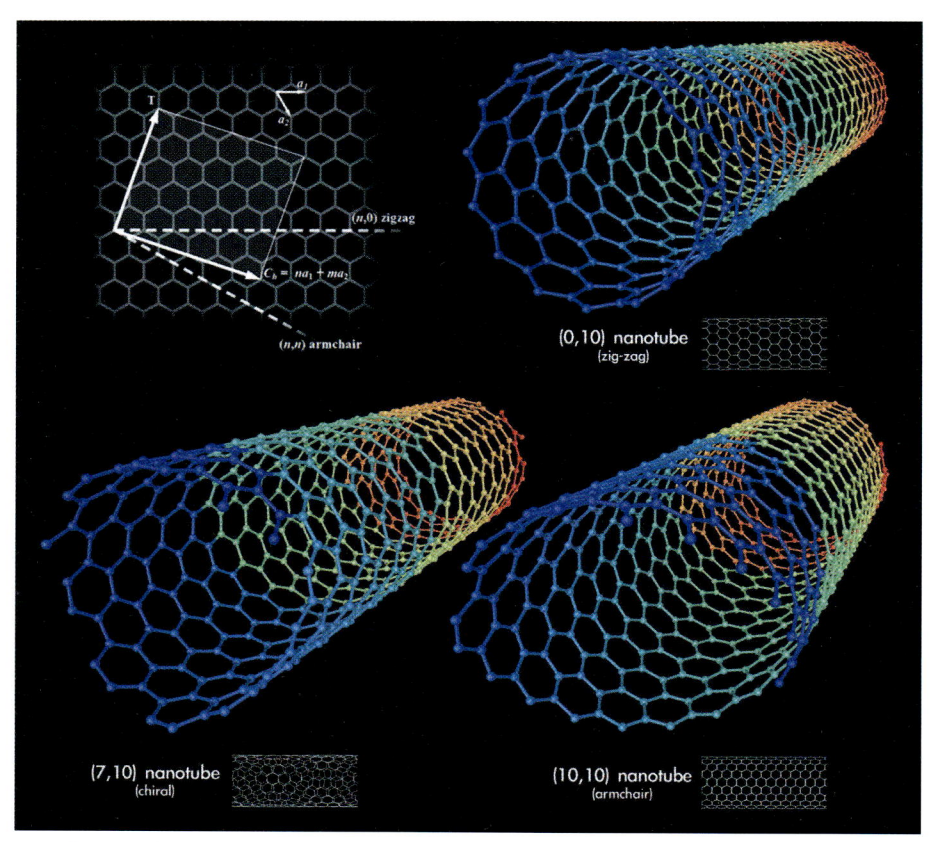

DNA-functionalized carbon nanotubes (CNTs): CNTs structures with the different folding modes of the graphene walls. The (n,m) CNT naming scheme can be thought of as a vector (C_h) in a graphene sheet that describes how to "roll up" it to make the nanotube. T denotes the tube axis, and a_1 and a_2 are the unit vectors of the graphene layer. (From Wikipedia with permission; picture created by M. Strock: http://en.wikipedia.org/wiki/Image:Types_of_Carbon_Nanotubes.png.)

different diameters, walls structures and electronic properties by chromatography. Functional groups for covalent coupling of DNA can be generated specifically at the open ends of CNTs or at defect sites of the graphene walls by their chemical or photochemical etching. Alternatively, polyaromatic anchor groups (e.g. pyrene) can be attached to the external graphene walls via π–π stacking, offering functional groups for the DNA covalent binding. DNA chains of different length were covalently linked to the functional groups generated on CNTs. Short oligonucleotides were first covalently bound to the CNTs and then grafted *in situ* to yield DNA of the controlled sequence and length. CNTs functionalized with DNA are frequently used as heavy labels or high effective-area platforms for DNA sensing. DNA-functionalized CNTs are applied for site-specific placement of conductive or semiconductive CNTs in nanocircuits, nanotransistors and other nanoelectronic elements. The specific placement is controlled by the DNA *hybridization with the *complementary DNA strands organized on solid supports by nanolithographic means. CNTs can operate as carriers that transport and deliver DNA into biological cells, thus offering new techniques for targeted medication in the frame of a new concept of nanomedicine. EVGENY KATZ

DNA-functionalized nanoparticles (NPs)
Hybrid materials composed of various NPs (metallic, semiconductive, magnetic) and DNA demonstrate a unique combination of biological properties of DNA (→hybridization and enzyme-catalyzed reactions: scission, →transcription, →translation, →replication, etc.) and physical properties of NPs (electronic, optoelectronic, magnetic, etc.). Usually DNA chains are functionalized with anchor groups (most frequently thiols) for the strong binding to NPs. Thiol-terminated DNA chains were used to form shell structures around metal (usually Au) and semiconductor (e.g. CdS, CdSe) NPs. The conformation of *single-stranded DNA (ssDNA) attached to NPs through thiol anchor groups depends on the ssDNA length and loading. For low surface coverage, nonspecific wrapping of the ssDNA around NPs was observed. For high surface coverage, short oligonucleotides are oriented perpendicular to the surface in a fully stretched configuration. For high surface coverage and long *oligonucleotides, the inner part of the oligonucleotides, close to the NP surface, is fully stretched in a perpendicular configuration to the surface, whereas the outer part adopts *random coil shape. While *hybridization processes for the bound ssDNA are almost unaffected by NPs cores, some of the enzyme-catalyzed DNA reactions (e.g. polymerization) require the association of enzymes with the DNA chains, thus they might be substantially inhibited by the cores due to the increased steric problems. Hybridization of ssDNA chains results in the controlled aggregation of the DNA-functionalized NPs. This results in shortening of the distances between the metal cores of NPs, yielding changes in their plasmon coupling. The respective optical changes are used to follow the hybridization process by spectral means, thus allowing simple optical DNA sensors ("Northwestern" spot test). NPs are frequently used as label units to detect and amplify signals (optical or electronic) originating from DNA reactions (usually hybridization) occurring in solutions or at surfaces. Fluorescent quantum dots (CdSe) and photocurrent-generating semiconducting NPs (CdS) are used as label units to follow the DNA hybridization by optoelectronic means. Metal NPs (e.g. Ag) and semiconductor NPs (e.g. CdS, ZnS, PbS) are applied as electrochemically active labels in the DNA analysis based on stripping voltammetry. Magnetic NPs (Ni, Fe_3O_4) associated with DNA are used as magneto-readable labels in DNA analysis. DNA-functionalized magnetic NPs also serve as nano-transporting units allowing magneto-controlled *translocation, separation and purification of DNA upon different biosensing procedures. Controlled aggregation of NPs upon DNA hybridization is used to architecture various networks of interconnected NPs, to build complex nanowires, nanocircuitries and nanoelectronic devices. DNA *templates for the controlled positioning of DNA-functionalized NPs can be generated by DNA *polymerization (catalyzed by polymerase *Klenow fragment) or telomerization (catalyzed by *telomerase). The DNA templates allow complex geometry of

the generated nanocircuitry with distance-controlled positioning of NPs. Metal NPs associated with the DNA templates can be chemically enlarged to yield continuous conductive metal nanowires or nanocircuitries of complex geometry. Domains formed by NPs with different electronic properties (conductive, semiconductive) can be generated along the DNA template resulting in nanoelectronic devices with the complex design. EVGENY KATZ

DNA glycosylase Enzyme catalyzing the hydrolytic cleavage of the *glycosidic bond by which a chemically damaged *nucleobase is attached to the DNA *backbone. The reaction is the first step of *base excision repair which reconstitutes the state before the chemical damage occurred. A critical feature of DNA glycosylases is the selectivity with which they have to discriminate their substrate (damaged) nucleobase from a vast excess of ordinary DNA constituents. For this, all DNA glycosylases studied to date employ a mechanism of flipping the damaged nucleotide from the continuous stack in the center of the *DNA double helix to its periphery and burying the base in a tightly tailored pocket on the enzyme surface (→base flipping). The mechanism of rate enhancement is best known for *uracil DNA glycosylase (UDG). In this case, it is caused by a combination of three effects: (a) *general acid–base catalysis, (b) stereoelectronic destabilization of the glycosidic bond brought about by enzyme-induced distortion of the damaged nucleotide and (c) substrate participation in catalysis by stabilizing a partial positive charge that transiently builds up during the reaction at carbon center C1' by juxtaposition of negatively charged *phosphomonoester residues of the DNA backbone. The various *DNA lesions listed under base excision repair are all processed by their own, specialized DNA glycosylases, examples being UDG, AlkA, OggI and EndoIII. HANS-JOACHIM FRITZ

DNA gyrase To enable *replication, the bacterial type II *DNA topoisomerase (gyrase) removes *supercoils and catenanes by cutting both strands of the *DNA double helix. After passage of another DNA molecule it reseals the break. Gyrase is also able to make supercoils after the *replication. ANDREAS MARX, KARL-HEINZ JUNG

DNA helicase Unwinds the two strands of the *DNA double helix by breaking the *hydrogen bonds to form two separated strands.

DNA labeling →oligonucleotide labeling

DNA lesion Site of DNA containing a damaged nucleoside (→base excision repair).

DNA library →sequence pool

DNA ligase Enzyme that in a *double-stranded DNA (dsDNA) forms a *phosphodiester bond between the 5'-end of one *oligonucleotide and the 3'-end of another DNA fragment. DNA ligases require *ATP (in eukaryotic organisms or phage) or *nicotineamide adenine dinucleotide (in case of bacterial DNA ligases) for reaction. Dependent on the specific enzyme, ligation of two dsDNA fragments with *sticky ends or with *blunt ends can occur. *In vivo*, DNA ligases take part in *DNA repair as well as in *DNA *recombination. In molecular biology, DNA ligases are used for creating *recombinant DNAs by combination of DNA fragments. IRENE DRUDE

DNA ligation →ligation

DNA machines →molecular automata, →deoxyribozyme-based automata, →Shapiro–Benenson–Rothemund automata, →DNA nanoarchitectures

DNA methylases →DNA methyltransferase

DNA methylation In addition to *cytosine, *adenine, *guanine and *thymine, the DNA of most organisms contains the methylated *nucleobases C5-methylcytosine, N4-methylcytosine or N6-methyladenine. The methyl group is attached either to the C5 of cytosine or to the exocyclic amino groups of cytosine and adenine, and all three methylated nucleobases are capable of forming regular *Watson–Crick base pairs with opposite guanine or thymine residues in DNA.

C5-methylcytosine

N4-methylcytosine   N6-methyladenine

The methyl groups are introduced enzymatically by *DNA methyltransferases (MTases) after *DNA replication. These enzymes catalyze the methyl group transfer from the ubiquitous cofactor S-adenosyl-L-methionine (AdoMet) (→DNA methyltransferase) to cytosine or adenine residues within their DNA *recognition sequences ranging generally from 2 to 8 bp. Most DNA MTases recognize *palindromic sequences containing two target nucleobases and methylate both target nucleobases in the upper and lower strand leading to fully methylated DNA. After *semiconservative DNA replication the DNA methylation of the newly synthesized strand is erased resulting in *hemimethylated DNA which, after a second round of DNA replication, is transformed into one double strand of unmethylated and one double strand of hemimethylated DNA. Thus, DNA methylation can be lost after several rounds of DNA replication (passive *DNA demethylation). Depending on the presence or absence of *DNA MTases in the cell or the time after DNA replication a certain *DNA sequence can be either in its fully methylated, hemimethylated or unmethylated state. This leads to an increase of the information content of DNA and the methylated nucleobases can be regarded as the fifth, sixth and seventh letters of the genetic alphabet.

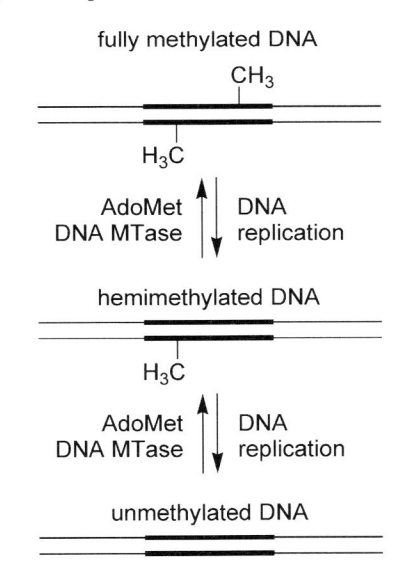

fully methylated DNA

hemimethylated DNA

unmethylated DNA

The biological function of DNA methylation is very diverse. In prokaryotes all three types of methylated nucleobases are found, and DNA methylation is involved in distinction of self and *foreign DNA, direction of DNA *mismatch repair and cell cycle control. The vast majority of DNA MTases are part of *restriction-modification systems consisting of a DNA MTase and a cognate *restriction endonuclease. Both enzymes share the same DNA recognition sequence and DNA methylation protects the host DNA from fragmentation by the endogenous restriction endonuclease. However, invading bacteriophage DNA is generally not methylated within the recognition sequence and cleavage by the restriction endonuclease can readily occur. Thus, restriction-modification systems have been regarded as primitive immune systems. The role of DNA methylation in mismatch repair and cell cycle control is best understood in *Escherichia coli*. This bacterium contains the DNA-adenine methyltransferase *Dam which is not part of a restriction-modification system and methylates the two adenine residues within the palindromic GATC sequence. After DNA replication and before Dam methylation the DNA stays for a few seconds in its hemimethylated form where the parental strand is tagged by methylation while the newly synthesized strand is unmethylated. This gives mismatch repair enzymes the possibility to correct replication errors in favor of the correct parental strand. In addition, *chromosome replication is initiated by Dam methylation. The *origin of replication is inactive in its hemimethylated form and transformation in its active fully methylated form by Dam is retarded for about 20 min by DNA binding competition with the SeqA protein.

In higher eukaryotes DNA methylation is mainly involved in regulation of *gene expression. The DNA of mammals contains in addition to the other four nucleobases only C5-methylcytosine within the CG sequence (→CpG islands). However, not all CG sequences are methylated, and a methylation pattern is established and inherited. This is possible because the DNA MTase Dnmt1 has a high preference for hemimethylated over non-methylated target sequences and after DNA replication hemimethylated CG sequences are fully methylated while non-methylated CG sequences remain non-methylated. In general, methylation of CG sequences in *promotor regions leads to a reduction of gene expression by blocking binding of *transcription factors, recruiting C5-methylcytosine binding proteins that act as *repressors or inducing chromatin condensation via *histone deacetylation. *Gene silencing by DNA methylation plays an important role in cell differentiation, *genomic imprinting and *X-chromosome inactivation. In addition, alterations in the DNA methylation pattern are frequently observed in cancer cells. ELMAR WEINHOLD

DNA methyltransferase (MTase) Catalyzes the nucleophilic attack of either *cytosine or *adenine residues within specific *double-stranded DNA sequences onto the activated methyl group of the cofactor *S*-adenosyl-L-methionine (AdoMet) leading to C5-methylcytosine, N4-methylcytosine or N6-methyladenine residues in DNA (→DNA methylation). Most DNA MTases recognize *palindromic DNA sequences of 2–8 bp containing one target *nucleobase in each DNA strand. The natural substrate of most DNA MTases is *hemimethylated DNA which is formed after one round of *DNA replication. However, most prokaryotic DNA MTases also methylate non-methylated DNA to hemimethylated DNA and subsequently fully methylated DNA.

DNA MTases can be categorized into two classes based on conserved sequence motives and according to the atom that is modified. C-DNA MTases catalyze the methylation of the C5 of cytosine (→DNA cytosine-C5 methyltransferase) and

N-DNA MTases catalyze the methylation of the exocyclic amino group of either cytosine or adenine (→DNA adenine-N6 methyltransferase). Prokaryotic DNA MTases are built of two domains – one larger catalytic domain which binds the cofactor AdoMet and a smaller domain. Both domains form a positively charged cleft where DNA binding occurs. Most interestingly, sequence-specific DNA binding leads to a drastic conformational change of the DNA in which the target *nucleobase is rotated out of the *DNA double helix (→base flipping) and placed near the bound cofactor. Base flipping allows the catalytic machinery of DNA MTases to gain access to their target nucleobases which are normally hidden in the base stack of double-helical DNA.
ELMAR WEINHOLD

DNA microarray →DNA chip

DNA nanoarchitectures Constructs that can be self-assembled from *branched DNA molecules. Their components may be simple branched species or more complex structural motifs. Simple branched DNA *junctions have been produced that contain 3–12 double helices flanking a *branch point. The species can be assembled and/or ligated into DNA stick polyhedra, where the edges are DNA double helices and the vertices correspond to the branch points of the junctions. The first such molecule was a DNA molecule with the connectivity of a cube. Other polyhedra produced to date include a tetrahedron, an octahedron and a truncated octahedron. Branched junctions are somewhat floppy, so only the branching and linking topologies of polyhedra are well defined unless all the faces are triangles. Other individual objects that have been built are topological targets, such as knots and Borromean rings. DNA is an ideal species to use as a topological building block because a half-turn of DNA is equivalent to a node, which is the fundamental topological feature of a knot or a catenane.

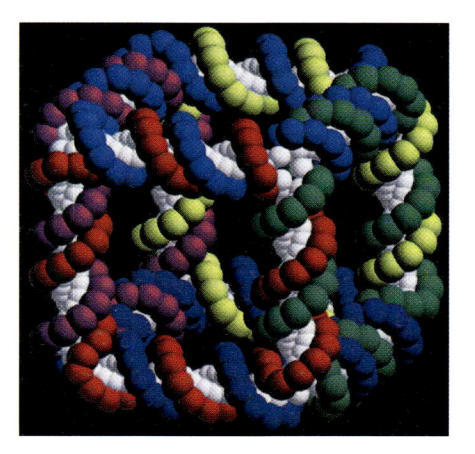

The DNA double-crossover (DX) molecule is another key element in DNA nanoarchitectures. This motif consists of two helices joined twice by strands that connect them, leading to parallel helix axes; the connection points are separated typically by

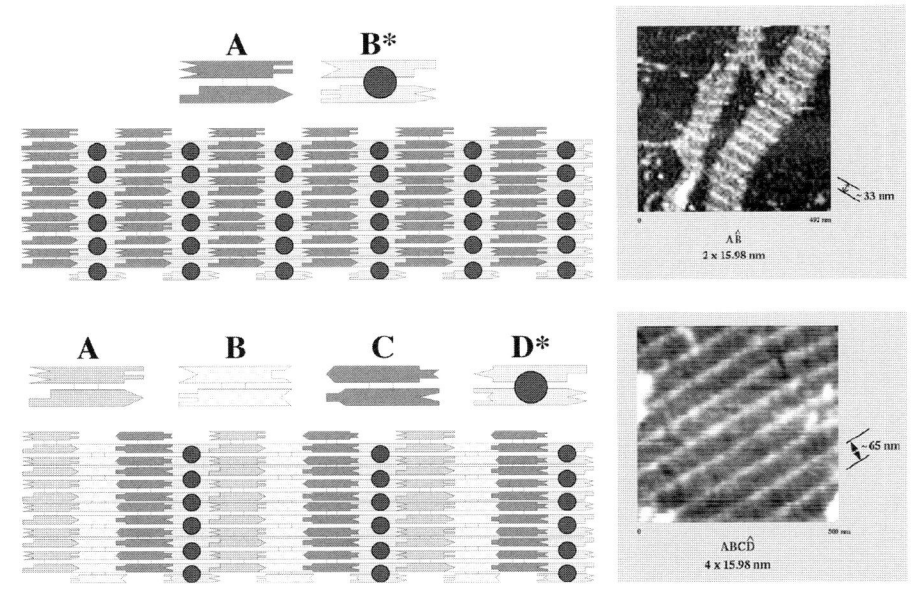

Two-dimensional DNA lattice.

one and two double *helical turns. Each of the connection points is a four-arm junction, so the motif can be described as two four-arm junctions joined twice to each other at adjacent arms. These are robust motifs, usually three to six double helical turns in length and their structures can be reliably predicted. This system can be extended, leading to molecules containing three or more helices joined laterally. Although most often built to be roughly planar motifs, angles can be varied between pairs of helices, using the helicity of DNA, e.g. a six-helix cyclic motif has been reported that approximates a hexagonal tube (→DNA nanotubes). DX molecules and their relatives can be exploited as tiles to produce two-dimensional crystalline arrangements by self-assembly (→DNA self-assembly). An extra motif can be included in these tiles, visible when the crystal is viewed in an atomic force microscope. The accompanying picture shows how arrangements of two 16 × 4 nm tiles produce 32-nm stripes (top) or four tiles produce 64-nm stripes (bottom). In addition to periodic arrangements, aperiodic patterns can also be generated algorithmically.

Single-stranded bacteriophages have been used to produce greatly extended versions of the parallel DNA motif, capable of yielding highly elaborate patterns, in a method called *DNA origami. This is done by using the bacteriophage *genome (several thousand *nucleotides) as a *template to which a large number of "staple strands" are added to fold the genome into a specific shape, including holes in the middle; the addition of strands containing extra domains enable the generation of further features. Smiley faces and a map of the western hemisphere are examples of patterns generated by this method.
NADRIAN C. SEEMAN

DNA nano-objects Discrete structural or topological species that are produced by

*DNA self-assembly (→ DNA nanoarchitectures).

DNA nanotechnology Enterprise of making new nanoscale species from DNA or other nucleic acid-like molecules (RNA, *peptide nucleic acid, etc.), and using those new systems to organize other species. The key reason for using DNA is the predictability of its intermolecular interactions via *Watson–Crick base pairing (→ DNA self-assembly). There are two types of DNA nanotechnology. One of these, compositional DNA nanotechnology, uses DNA as a "smart glue", taking advantage of programmable affinity, without too much concern for local structural features. This type of DNA nanotechnology can produce specific associations, e.g. by joining nanoparticles whose surfaces contain *complementary DNA molecules or perhaps bridging them with a third strand. This approach can produce ordered structures, but on a length scale larger than the DNA molecules involved.

The other type of DNA nanotechnology is structural DNA nanotechnology. In this approach, DNA is used not only to direct intermolecular interactions, but also to program the structures of the components; thus, in addition to the "mortar" of the "smart glue" approach, DNA is used as the "bricks" themselves. This approach leads to stronger topological and structural control, usually producing structures with accuracies down to around a nanometer. In its strongest form, structural DNA nanotechnology produces robust motifs whose intramolecular structures are programmed by their sequences. Strengths of DNA for nanotechnology include the availability of convenient automated synthesis (often commercially available, → oligonucleotide synthesis), convenient modifying enzymes (also commercially available), the stiffness of the DNA (persistence length around 50 nm), the robustness of DNA single strands to heat and its amenability to molecular biology techniques. Other advantages are that DNA boasts an externally readable *code, even when paired, that it has high *functional group density, that it is potentially self-replicable and selectable, and that many derivatives have been prepared. Structural DNA nanotechnology has been used to produce a variety of static nanoarchitectures, such as DNA knots and polyhedral catenanes, as well as periodic and aperiodic DNA crystals (→ DNA nanoarchitectures).

In addition to static structures, structural DNA nanotechnology has led to a variety of DNA-based nanomechanical devices. These devices usually have at least one stage in their machine cycle that is not well structured, but the other stages are. A variety of devices have been produced that rely on changes in a global parameter, such as pH, or the presence of an effector molecule, such as $Co(NH_3)_6^{3+}$, which activates the B → Z transition of DNA, converting conventional right-handed *B-DNA into left-handed *Z-DNA. Devices based on the formation of *G-tetrads or the presence of a particular protein have also been produced.

Such devices certainly work, but they do not take advantage of the greatest strength of DNA, i.e. that its site of action is programmable through sequence. Thus, N two-state devices that are programmed by sequence can lead to 2^N different structural states. A number of sequence-dependent devices have been prepared, including tweezers, a shape-shifting device that rotates one end relative to the other by a half turn, a translation machine and bipedal walkers. These devices may be activated by the addition of DNA strands to the solution or they may respond to the presence of RNA molecules generated intracellularly. A programmable DNA device has been incorporated into a cassette that targets a particular site in a two-dimensional DNA array, which is likely to lead to more sophisticated combinations of devices.

All of these devices require the addition in some fashion of a DNA strand. However, in recent years, progress has been made in producing autonomous DNA devices that can convert *DNAzyme or enzymatic activity into triggers for additional motion, without further intervention by the experimenter; downhill cascades have also been designed that accomplish the same purpose. NADRIAN C. SEEMAN

DNA nanotubes Self-assembled one-dimensional DNA structures containing a central cavity. Two types of DNA nanotubes have been reported. In one type, DNA double-crossover (DX) molecules (→DNA nanoarchitectures) self-assemble to produce a curved surface because of the inherent twist designed into their sticky-ended connections. In a second type of DNA nanotube, the tube has been designed deliberately. For example, a six-helix bundle motif has been produced with 120° angles between successive DX segments, yielding a hexagonal cross-section. Placing complementary *sticky ends at either end of each helix results in the self-assembly of multi-micron tube-like units. Other tubes can be designed, and there is no restriction that the tube cross-section be convex. NADRIAN C. SEEMAN

DNA nanowires One-dimensional DNA structures. These may be formed from the *G-tetrad motif or from clusters of DNA helices, such as a three-helix motif.

DNA origami DNA nanostructures in which a single long strand of DNA is folded so that it runs through every *double helix. The term is by analogy with the Japanese art of paper folding in which a single sheet of paper is folded, without cuts, to create elaborate three-dimensional shapes. Conceptually, there are two types of DNA origami. In single-stranded DNA origami, the structure is formed entirely by a single long strand and the folding is due to interactions of the long strand with itself. In scaffolded DNA origami, the folds are formed by the interaction of a long single strand and hundreds of short DNA strands called "staples". Single-stranded origami structures have the advantage of being clonable, potentially exponentially reproducible like living things. So far, an almost-single-stranded DNA octahedron with just five staples has been reported. Scaffolded DNA origami structures have been easier to design and synthesize; a half dozen two-dimensional shapes including a star, rectangle, triangle and smiley face, as well as a three-dimensional hexagonal tube have been reported. Scaffolded DNA origami structures are easily connected to other nanoscale objects such as carbon nanotubes and proteins, and might be used to organize them into complex nanocircuits or nanofactories. PAUL ROTHEMUND

DNA polymerase DNA nucleotidyl transferase. Transmission of the *genetic information from the parental DNA strand to the offspring is crucial for the survival of any living species. In nature this process is catalyzed by the replication machinery in which DNA polymerases are essential for the entire DNA synthesis. In addition, DNA polymerases are important enzymes in molecular biological applications and techniques such as the *polymerase chain reaction, *cloning and *DNA sequencing.

DNA polymerases catalyze proceeding DNA synthesis in a template-directed manner. In nature, DNA- and RNA-dependent enzymes are known. All DNA synthesis required for *DNA replication, *recombination and repair depends on the ability of DNA polymerases to recognize the *template and correctly insert the *complementary nucleotide. Proceeding DNA synthesis is promoted by DNA polymerases through *catalysis of nucleophilic attack of the 3′-hydroxyl group of the 3′-terminal *nucleotide at the *primer strand to the α-phosphate of an incoming 2′-deoxynucleoside-5′-O-triphosphate leading to substitution of *pyrophosphate. This phosphoryl transfer step is believed to be promoted by two magnesium ions that stabilize a pentacoordinated *transition state through complexation of the phosphate groups and essential carboxylate moieties in the *active site.

Even relatively simple organisms like *Escherichia coli* or other bacteria possess several DNA polymerases (e.g. for *E. coli* at least five DNA polymerases are known). The replication process in eukaryotes or higher organisms is more complex. More than a dozen DNA polymerases are known in humans. Apart from acting as the cellular and mitochondrial enzymes involved in DNA replication, several human DNA polymerases fulfill tasks in *DNA repair and the immune response.

DNA polymerases are presented with a pool of four structurally similar *dNTPs from which the sole correct (i.e. *Watson–Crick base-paired) substrate must be selected for incorporation into the growing DNA strand. DNA polymerases that are believed to be involved in DNA replication processes show low error rates (as low as only one error within 1 million synthesized nucleotide linkages) while certain enzymes that are competent to bypass *DNA lesions (e.g. caused by sunlight) exhibit high error rates of up to one error within 1–10 synthesized nucleotide linkages. Some recently discovered enzymes exhibit features like high error propensity when copying undamaged DNA or the ability to bypass DNA lesions that block the replicative enzymes. These DNA polymerases are believed to be involved in DNA repair and the immune response. Furthermore, it has been shown unambiguously that *mutations of DNA polymerases that alter their properties can be involved in the development of various cancers. ANDREAS MARX, KARL-HEINZ JUNG

Proposed transition state for DNA polymerase-catalyzed nucleotide insertion.

DNA polymorphism → polymorphism

DNA primase → primase

DNA probe → nucleic acid probe

DNA profiling → DNA fingerprint

DNA–protein interaction Plays a vital role in a variety of cellular processes like *gene expression, DNA packaging, DNA *recombination, *DNA replication and *DNA repair. Proteins can interact with DNA either specifically or non-specifically. In the case of non-specific interactions, the sequence of *nucleotides does not matter, as far as the binding interactions are concerned. *Histone–DNA interactions are an example of such interactions, and they occur between *functional groups on the protein and the *sugar–phosphate backbone of DNA. Specific DNA–protein interactions, however, depend upon the sequence of bases in the DNA and on the orientation of the bases that can vary with twisting and writhing. These DNA–protein interactions are strong, and are mediated by *hydrogen bonding, ionic interactions, and *van der Waals and *hydrophobic interactions. The *major groove of *B-DNA, being wider than the *minor groove, can accommodate larger structural motifs, and the pattern of *base pairs that are exposed in the floor of the grooves is more specific and discriminatory for the major groove. Thus, the majority of sequence-specific protein–DNA interactions occur in the major groove. There is a collection of motifs that provide a scaffold for particular protein secondary structures, usually an α-helix, to recognize and bind to DNA. Common motifs for DNA–protein interactions are: zinc fingers, leucine zippers, *helix–loop–helix motifs and other less well-defined motifs. Important techniques for a biochemical characterization of DNA–protein interactions include filter binding assay, *band shift assay, gel chromatography/sedimentation, nuclease protection assay, *footprinting, chemical *interference mapping, chemical *crosslinking, *fluorescence spectroscopy, electron microscopy, *nuclear magnetic resonance spectroscopy and X-ray crystallography. TOBIAS RESTLE

DNA repair Any enzymatic process that reduces the *mutation rate by reconstituting the original DNA sequence from a structure which contains at least one element which, when serving as a *template in *DNA replication is either miscoding or non-coding. DNA repair can be classified in three categories. (a) Direct reversal of the damage as exemplified by photoreactivation of cis-syn *thymine dimers and the demethylation of O^6-me-G and N^1-me-A residues by Ada and AlkB, respectively. (b) Mechanisms by which the aberrant structure is nucleolytically removed from one DNA strand and a shorter or longer stretch of the affected strand resynthesized by a repair *DNA polymerase with extraction of the original information from the opposite strand. (c) Mechanisms, known under the generic name of recombinational repair, by which the correct information is extracted from a separate or (during replication) the sister DNA molecule. Recombinational repair normally does, but needs not necessarily, involve DNA intermediates with double-strand breaks. Mechanisms of type (b) can be further subdivided on the basis of tract length of the DNA polymerase reaction or the mechanism of nucleolytic removal of the aberrant structure. As to the former, a repair mechanism is considered of the short patch type if a stretch of a single up to several dozens of nucleotides is resynthesized, whereas long patch repair is characterized by tract lengths of up to several thousands of nucleotides. With respect to mechanism of damage removal, there are at least four different types. (a) Incision repair (or strand incision repair; also called "nucleotide incision repair") in which the affected DNA strand is directly incised endonucleolytically at the site of the damaged nucleotide (to its 5'-side). (b) *Base excision repair with hydrolytic removal of the damaged base by a *DNA glycosylase as the first step. (c) Nucleotide excision repair (which is better called "oligonucleotide excision repair") by which two endonucleolytic

cuts are set in the same DNA strand, in a coordinate fashion, *upstream and *downstream of the damaged site, which leads to the removal of an *oligonucleotide encompassing the erroneous structure. A prototype example of nucleotide excision repair is the removal of *cis-syn*-thymine dimers from ultraviolet-irradiated DNA by the *E. coli* UvrABC repair system. (d) *Mismatch repair in the course of which a strand incision can occur hundreds of nucleotides away from the aberrant structure (in this case the opposition of two natural nucleotides).
HANS-JOACHIM FRITZ

DNA replication Generating a duplicate of the DNA *template. Since the *DNA double helix consists of two *complementary strands, duplication of each strand and formation of double helices with their parent strands leads to two new DNA double helices which are identical to their originals.

DNA replication is performed in the *replisome which contains all required protein complexes. To enable replication, supercoiling has to be removed by *DNA topoisomerases. The replication of DNA starts at the *origin, a specific sequence of *nucleotides. The bacterial DNA of the

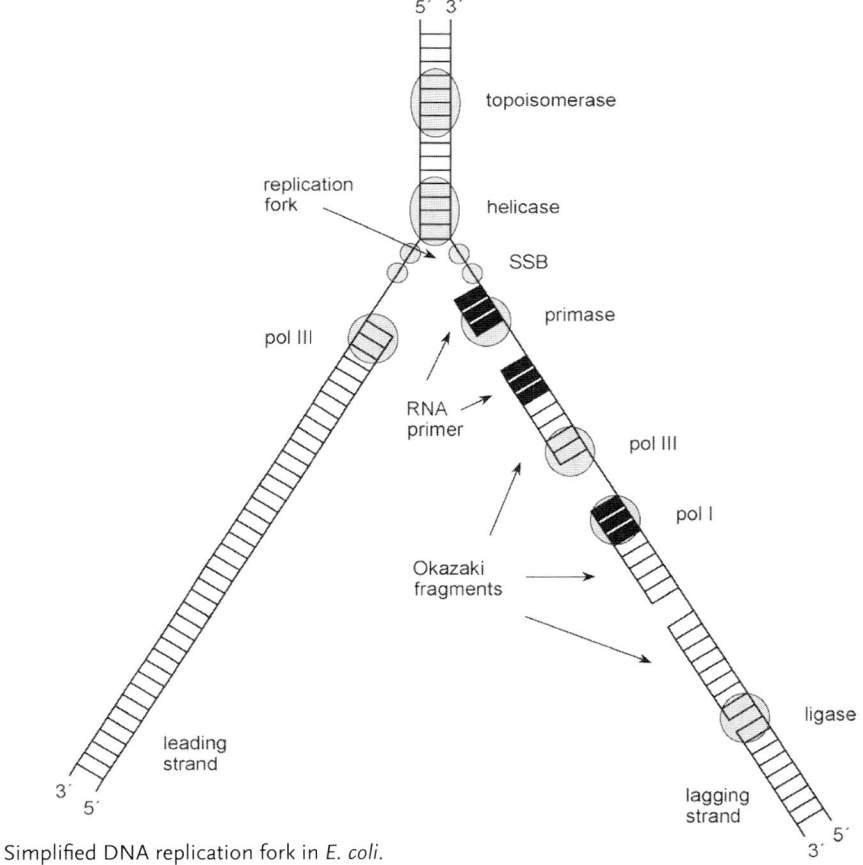

Simplified DNA replication fork in *E. coli*.

Escherichia coli *chromosomes is circular and has one origin (oriC). *Bidirectional replication is performed forming two *replication forks and is terminated when the two forks meet together. The chromosomes of eukaryotes and higher organisms have multiple origins along their DNA strand. The origins are recognized by an origin recognition complex of proteins (ORC).

At first, *helicase unwinds the two strands of the DNA double helix by breaking the *hydrogen bonds and opens the replication fork to form two separated strands which are stabilized by single-stranded binding proteins to prevent rewinding. As *DNA polymerases cannot initiate DNA synthesis, but can only elongate nucleic acid polymers, *primase is required. It initiates and performs the synthesis of short *RNA primers at the leading and lagging strand, which are then elongated by a replicative DNA polymerase (e.g. DNA polymerase III complex in *E. coli*). Since DNA polymerases perform DNA synthesis exclusively in the 5′–3′ direction, the synthesis of DNA is semidiscontinuous, i.e. continuous at one strand, the *leading strand, and discontinuous in shorter pieces (→Okazaki fragments) at the other strand, the *lagging strand. When the synthesis of the Okazaki fragments is finished, the *primers are removed by DNA polymerase I which hydrolyzes the *ribonucleotides with the help of its 5′–3′ exonuclease activity and replaces them by *2′-deoxyribonucleotides. Finally, the DNA fragments are connected by a *DNA ligase. ANDREAS MARX, KARL-HEINZ JUNG

DNA–RNA hybrid Double-stranded molecule consisting of a DNA chain base paired to a *complementary RNA chain. DNA–RNA hybrids appear as intermediate states during *replication and *transcription. They can be produced *in vitro* by *hybridization.

DNase (deoxyribonuclease) Enzyme that cleaves DNA. DNase may be double- or single-strand specific and an *endo- or exonuclease. Important representatives are DNase I and *restriction endonucleases. DNA polymerases also exhibit 5′–3′ or 3′–5′ DNase exonuclease activity. ULI HAHN

DNA self-assembly *Hybridization of DNA molecules to form target structural species. Intramolecular self-assembly leads to the formation of particular motifs, such as branched *junctions or DNA double-crossover (DX) tiles (→DNA nanoarchitectures). Intermolecular self-assembly is performed by putting single-stranded *overhangs (called "*sticky ends") on the ends of the helical domains, particularly in branched species. These sticky ends are complementary to each other and interact specifically; not only is the affinity of sticky ends predictable, following the rules of *Watson–Crick base pairing, but they are known to form the classical *B-DNA structure when they cohere. The programmability of DNA interactions by sequence design is one of the most important reasons for using it. Other examples of DNA-based self-assembly are the assembly of derivatized metallic nanoparticles, the attachment of active species to DNA arrays and the operation of sequence-dependent DNA nanomechanical devices (→DNA nanotechnology). Virtually all operations involving DNA in nanotechnology and most operations involving DNA in computation entail the self-assembly of DNA molecules. NADRIAN C. SEEMAN

DNA sequence →nucleotide sequence

DNA sequence analysis Determination of the sequence of monomers (bases)

forming a DNA molecule. Methods for the sequence analysis of DNA were first developed 1977 with two approaches known as (a) *Sanger sequencing that is based on enzymatic DNA synthesis from a *single-stranded DNA *template with chain terminating *dideoxynucleotides, and (b) *Maxam–Gilbert sequencing which involves the chemical degradation of DNA fragments. Both methods rely on the electrophoretic separation of labeled DNA fragment mixtures on high-resolution polyacrylamide gels and reading of the sequence in a staggered ladder-like fashion. Sanger sequencing is technically easier and faster, and thus became the main DNA sequencing method for the majority of applications and predominantly for *de novo* sequencing. Originally, sequence reading was based on the random incorporation of radiolabeled nucleotides and autoradiography of the sequencing gels. With the development of highly efficient *fluorescent dyes that are attached to either *primer or chain-terminating *nucleotides, time-consuming autoradiography was replaced by fluorescence detection. Fluorescence can be read directly during *electrophoresis using a variety of detection set-ups that can be combined with software-based signal recognition systems. Furthermore, if four spectrally different dyes coding for A, G, C or T are employed, all fragments can be analyzed in one single gel lane instead of four lanes required with the radiolabel approach. Automatic *DNA sequencers that are based on fluorescence reading were first presented in 1986, and since then have greatly improved the ease, speed, accuracy and reliability of DNA sequence analysis, thereby mostly replacing the manual approach. Highly parallelized automatic sequence analysis also enabled large-scale sequencing projects such as the resolution of the complete human genome (approximately 3×10^9 bp, →Human Genome Project) or the *genomes of various viruses, bacteria, archaea, eukaryotes, mammals and plants.

DNA sequence analysis may also be carried out using methods that avoid DNA *amplification and electrophoresis that are essential for Sanger sequencing. One prominent technique is oligonucleotide microarray-based (→DNA chip) hybridization analysis. In this approach, every possible 10–25mer sequence fragment of the DNA region of interest is represented in an array of *oligonucleotides. Fluorescently labeled target DNA is hybridized to the array giving fluorescence signals that are assessed relative to that of a reference DNA. *DNA microarrays have been shown to be useful for screening new as well as known *mutations (polymorphisms) in specific *genes of individuals and of populations.

*Pyrosequencing quantitatively measures the *pyrophosphate released during DNA copying by a *polymerase. This method significantly accelerates sequence analysis (one base per minute, parallel analysis of 96 samples), but is suited only for resolving relatively short DNA fragments (up to 200 bases). Alternatively, short sequence fragments (up to 50 bases) can be analyzed by *mass spectrometry.

A future approach that may further accelerate sequence analysis is under construction for the sequence determination of single DNA molecules. This technique requires DNA that is labeled at each position with a base-specific *fluorescent label and subsequent stepwise degradation of the completely labeled molecule with an *exonuclease. Although not yet realized for practical application, single-molecule sequencing promises that *de novo* sequencing will be limited only by the rate of the enzymatic reaction. SUSANNE BRAKMANN

DNA sequence analysis by mass spectrometry *Oligonucleotides are separated by their m/z values after ionization with either matrix-assisted laser-desorption ionization (MALDI) or electrospray ionization (ESI) (→mass spectrometry: ionization methods). Certain kinds of mass spectrometers, e.g. ion traps, time-of-flight (TOF)/TOFs, Fourier transform ion cyclotron resonance (FTICR) or triple quadrupole mass spectrometers, allow an isolation of the mass separated ions, followed by fragmentation. The resulting fragments and thus the tandem mass spectra/MS^n spectra differ depending on the kinetic energy of the ions (E_{kin}) and the energy which is transferred. Bond cleavages of the *sugar–phosphate backbone result in structure- and sequence-specific fragment ions. For fragment ion nomenclature according to McLuckey et al. [(J. Wu and S. A. McLuckey, *Int J Mass Spectrom* **2004**, *237*, 197), the article also describes other, less common and flexible nomenclatures.], see figure.

Powerful radiative fragmentation methods are blackbody infrared radiative dissociation (BIRD) and infrared multi-photon dissociation (IRMPD). Energy transfer and activation of the isolated ions via radiation leads to a rather slow heating and gentle fragmentation resulting in the cleavage of the weakest bonds. a-(a-B)- and w-ions are formed. These fragment ions, which are important for structure and sequence analysis, are often accompanied by fragment ions formed through additional loss of nucleobases. BIRD is mainly used for determination of physicochemical properties. IRMPD usually leads to good sequence coverage. Both methods can be used in FTICR instruments only.

The most common fragmentation method is collision-induced dissociation (CID; also known as collision-activated dissociation). The activation of the precursor ions is achieved by collisions, either with the residual gas or short gas pulses. Usually nitrogen or a noble gas like He, Ar or Xe is used as collision gas. Depending on the

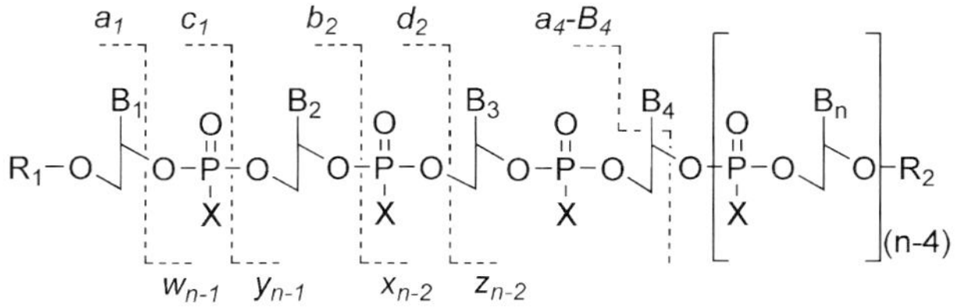

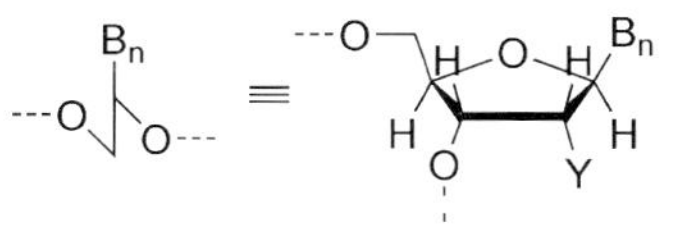

Schematic view of a single-stranded oligonucleotide. The dotted lines indicate possible bond cleavages; the nomenclature for the 5′-fragment ions is shown above, for the 3′-fragment ions below the strand. In RNA Y is OH, while in DNA it is H. For unmodified oligonucleotides R_1, R_2 and R_3 are H (or X = OH). The same nomenclature can be used for derivatized RNA/DNA, then R_1, R_2, R_3, X or Y may differ. Modified nucleobases B can also be implemented.

kinetic energy (E_{kin}) and the energy uptake per collision, the resulting fragment ion spectra disclose different fragmentation pathways. Slow vibrational activation with fragment ion patterns similar to those obtained by radiative methods are obtained in ion traps or employing sustained off-resonance irradiation (SORI)-CID in FTICR instruments. Higher energy uptake with fragment spectra similar to those obtained by electron capture dissociation and lots of internal fragments (neutral loss from both sides, i.e. 3′ and 5′, of the oligonucleotide) are observed in TOF/TOF instruments. An intermediate position is occupied by triple quadrupole mass spectrometers, showing mainly *a*-, *b*-, *w*- and *y*-(*a-B*)-ions, and internal fragment ions.

Although more prominent in peptide/protein analysis, the application of electron capture dissociation (ECD) in FTICR MS or electron transfer dissociation (in linear ion traps) for structural and sequence analysis of DNA-derived analytes may be of interest: the initial charge reduction accompanied by a fast transfer of energy leads to non-ergodic, direct bond cleavages. The fragment spectra, mainly *d*- or *w*- and *z*•- or *a*•-ions, but also (*a-B*)- and (*c-B*)-ions, are orthogonal to those obtained by the mild fragmentation methods, IRMPD or CID. ANDREAS SPRINGER

DNA sequencer Fully automated machine for the determination of DNA base sequences. The sequencing technology employed in these machines relies on fluorescence detection and was possible with the development of efficiently fluorescing dyes that can be attached to DNA either at the *sugar–phosphate backbone (5′-terminal modification of primers) or directly at the dideoxy nucleotides via alkyl linkers to the bases (→oligonucleotide labeling). Sequencers consist of a combination of *electrophoresis and laser-based fluorescence detection units that are computer controlled. Fluorescence signals arising from labeled DNA fragments are recorded in real-time during electrophoretic resolution, providing data that can be directly analyzed or entered into programs for mutation analysis, sequence editing and sequence assembly. The underlying sequencing chemistry used in DNA sequencers is conventional *Sanger sequencing with chain-terminating *nucleotides except that either the *primers or the terminators are dye labeled. The *fluorescent dyes employed so far show diverse chemical structures and emit fluorescence in the visible or near-infrared range. Among the dyes emitting in the visible range, rhodamines play an important role. The introduction of energy transfer dyes based on rhodamines (BigDyes) has significantly improved the signal-to-noise ratio and decreased the detection limit to approximately 10^{-18} mol per band.

If labeled primers are used in combination with normal dideoxynucleotides, four different reactions must be analyzed (four-lane approach), whereas use of dye-labeled terminators employs four different dyes to identify A, G, C and T that can be run on a single lane of the sequencing gel. The dye-primer chemistry provides the user with even signal strength across all four bases, whereas dye-terminator chemistry is less tedious because all terminators can be combined in a single-tube reaction. Automated DNA sequencers can be divided into two groups: those using polyacrylamide slab gels and those using *capillary electrophoresis. Slab-gel-based systems separate fluorescently labeled fragments by electrophoresis through denaturing polyacrylamide gels that vary in length from 14 to 60 cm. The number of bases that can accurately be recognized in a particular run depends on gel length and run time. Typical read lengths vary from 750 to 1200 bases per lane, giving maximal throughputs of approximately 80 000 to 130 000 bases in 24 h. Capillary-based systems use an array of capillaries or a single capillary filled with polyacrylamide or specially developed polymers as the stationary phase for electrophoresis. These systems are advantageous because they allow for greater automation (sample loading, electrophoresis, analysis) and show read lengths of 550–700 bases. Thus, 10 000 to a maximum of 600 000 bases can be called in 24 h using one machine. Capillary sequencers are preferentially used in large genome projects. SUSANNE BRAKMANN

DNA sequencing →DNA sequence analysis

DNA shuffling →shuffling, →gene shuffling, →exon shuffling, →intron shuffling

DNA structures →DNA conformations

DNA superstructure Organization level of *DNA structure that is more complex than the relaxed, linear state. Superstructures like negative and positive *supercoils are introduced into DNA in order to render it more compact and allow for storage in the nucleus, for example. →superhelix, →supercoil, →DNA nanoarchitectures, →DNA nano-objects, →DNA nanotubes, →DNA self-assembly. JÖRN WOLF

DNA synthesis →oligonucleotide synthesis, →DNA replication, →reverse transcription

DNA synthesizer Instrument for automated assembly of *oligonucleotides. DNA fragments up to 150 *nucleotides can be synthesized. The same instrumental procedure is used also for the synthesis of RNA fragments. →oligonucleotide synthesis. RONALD MICURA

DNA topoisomerase Topoisomerases control the interconversion of the topological isomers (topoisomers) of DNA. They can increase or reduce the degree of supercoiling. Thus, they are involved in knotting and unknotting DNA strands. By means of strand breaking they are also able to allow crossing of DNA strands over one another and to catenate or decatenate DNA molecules. Topoisomerase I reduces supercoiling and acts by breaking one strand of the *DNA double helix, winding it around the other strand, and religates the nicked strand without the consumption of *ATP. Topoisomerase II (e.g. gyrase) removes *supercoils and catenanes with involved double-strand breaks; after *replication, it can make negative supercoils so that the DNA can be packed into the cell. ANDREAS MARX, KARL-HEINZ JUNG

DNA topology Geometry and orientation of DNA molecules. DNA topology is quite complex and is controlled by different enzymes. *Topoisomerases, for example, break the DNA strands and reconnect the ends. MATTHÄUS JANCZYK

DNA triplex →triple helix, →H-DNA, →triplex-forming oligonucleotides

DNAzyme →deoxyribozyme

dNDP Abbreviation for deoxynucleoside diphosphate

dNMP Abbreviation for deoxynucleoside monophosphate

dNTP Abbreviation for deoxynucleoside triphosphate

dot blot Direct punctual *hybridization of DNA, RNA or protein onto a carrier (→blotting, →dot blot method).

dot blot method RNA or DNA sample is applied punctually onto a membrane (nitrocellulose or nylon) and fixed by ultraviolet light or heating. Detection is achieved by *hybridization of a labeled oligonucleotide probe with *complementary sequence to the DNA or RNA to be analyzed (→blotting). This method is easy to handle, because a separation of the biomolecules via *gel electrophoresis followed by the blotting technique is not required. However, this method gives no information about the size of detected oligonucleotides. Also, no conclusions can be made about the number of detected molecules if the DNA or RNA probe or the antibody targets more than one oligonucleotide. Irene Drude

double helix →DNA double helix

double-strand →double-stranded DNA, →double-stranded RNA

double-stranded DNA (dsDNA) Any DNA molecule consisting of two *complementary strands that interact via *Watson–Crick base pairs. Usually, the two strands have opposite 3′→5′ polarity. dsDNA occurs in different conformations (→DNA conformation). Sabine Müller

double-stranded nucleic acid →double-stranded DNA, →double-stranded RNA

double-stranded RNA (dsRNA) Either the product of endogenous *transcription of repetitive DNA elements or viral sequences or of deliberate exogenous expression. dsRNA is a precursor of *small interfering RNAs and leads in most eukaryotic cells to *gene silencing via the *RNA interference pathway. Nicolas Piganeau

down mutation *Mutation causing the weakening of a *phenotype to a low but finite level; often a mutation in a regulatory DNA sequence such as a *promotor (→null mutation, →up mutation). Hans-Joachim Fritz

downstream To the 3′-side of a particular position in the directional nucleic acid strand.

Drosha Eukaryotic *double-stranded RNA-specific *endonuclease containing two tandem RNase III domains and generating 2-nucleotide 3′-protruding ends at the cleavage site. Drosha is involved in the maturation of *pri-miRNAs into *pre-miRNAs (→micro-RNA). Nicolas Piganeau

dsDNA →double-stranded DNA

dsRNA →double-stranded RNA

duplex Two nucleic acid strands that interact via *Watson–Crick base pairing forming a *double helix.

dUTP system mutagenesis →Kunkel mutagenesis

E

Eadie–Hofstee plot Similar to the *Lineweaver–Burk plot, the Eadie–Hofstee plot represents a transformation of the *Michaelis–Menten equation into a form which allows analysis by linear regression:

$$v = v_{max} - K_M \cdot v/c_S$$

According to this equation the rate of product formation, v, is plotted versus the ratio of this rate and the substrate concentration, c_S. The intercept yields the maximum rate, v_{max}, and the slope the *K_M value. It is somewhat superior to the *Lineweaver–Burk plot as it does not compress the data points at high concentrations so much. However, the Eadie–Hofstee plot does not separate the variables, which is a disadvantage. CHRISTIAN HERRMANN

early genes *Genes of a virus or bacteriophage that are first expressed after infecting a host cell.

edge-to-edge interaction of nucleobases
RNA molecules appear in a large variety of complex structures with the *nucleobases being involved in long-range RNA–RNA interactions forming canonical *Watson–Crick as well as *non-Watson–Crick *base pairs. Eric Westhof et al. have suggested a *geometric nomenclature and classification of RNA base pairs based on geometry, particularly on the base edges participating in the interaction. Thus, each nucleobase has been assigned a Hoogsteen edge (for purine bases) and a "C–H" edge (for pyrimidine bases), a Watson–Crick edge, and a sugar edge.

Twelve families of edge-to-edge base pairs are formed (see table on page 94) being defined by the relative orientation of the *glycosidic bond of interacting bases and the edges taking part in the interaction. SABINE MÜLLER

editing →RNA editing

E-DNA Specific *conformation of DNA that contains no *guanine bases, so far only observed *in vitro*. It forms a *right-handed double helix containing 7.5 residues per *helical turn. SABINE MÜLLER

EGS →external guide sequence

electron hopping In principle, DNA-mediated charge transport processes can be categorized as either oxidative hole

Nucleic Acids from A to Z: A Concise Encyclopedia. Edited by Sabine Müller
Copyright © 2008 WILEY-VCH Verlag GmbH & Co. KGaA, Weinheim
ISBN: 978-3-527-31211-5

Edge-to-edge interaction of nucleobases.

Glycosidic bond orientation	Interacting edges	Local strand orientation
cis	Watson–Crick/Watson–Crick	antiparallel
trans	Watson–Crick/Watson–Crick	parallel
cis	Watson–Crick/Hoogsteen	parallel
trans	Watson–Crick/Hoogsteen	antiparallel
cis	Watson–Crick/sugar edge	antiparallel
trans	Watson–Crick/sugar edge	parallel
cis	Hoogsteen/Hoogsteen	antiparallel
trans	Hoogsteen/Hoogsteen	parallel
cis	Hoogsteen/sugar edge	parallel
trans	Hoogsteen/sugar edge	antiparallel
cis	sugar edge/sugar edge	antiparallel
trans	sugar edge/sugar edge	parallel

transport or reductive electron transport. In contrast to the broad knowledge about the highest occupied molecular orbital-controlled oxidative hole transport, the mechanistic details of the lowest unoccupied molecular orbital-controlled excess electron transport are not completely clear. This lack of knowledge has been filled at least partially during the last 5 years, but a well-defined and suitable donor–acceptor system for time-resolved spectroscopic measurements is still lacking. Meanwhile, the mechanisms of hole hopping and transport processes have been transferred to the problem of excess electron hopping and supported by experimental evidence. It is likely that such an electron hopping involves all *base pairs (TA and CG) and the *pyrimidine radical anions as intermediate electron carriers. This proposal is based on the trend for the reducibility of DNA bases, which makes clear that the *pyrimidine bases C and T are reduced more easily than the *purine bases A and G. However, electron hopping via the *thymine radical anion seems to be more favorable since protonation of the *cytosine radical anion by the *complementary DNA bases or the surrounding water molecules probably interferes with the electron hopping. As a result, the proton-coupled electron hopping over CG base pairs decreases the electron transport efficiency and rate, but does not stop electron migration in DNA. The results of the mechanism of electron hopping together with the proposed lack of covalent *DNA damage as a chemical result of electron trapping support clearly the idea that this type of DNA-mediated charge transport has a high potential for molecular diagnostics, as the development of electrochemical *DNA chips, and for molecular electronics, like nanodevices which are based on DNA or DNA-inspired architectures. →charge transfer in DNA.
HANS-ACHIM WAGENKNECHT

electron paramagnetic resonance (EPR)
Spectroscopic technique that, complementary to *nuclear magnetic resonance (NMR) spectroscopy, uses the spin of unpaired electrons to gather structural and dynamic information about RNA (also called electron spin resonance). Since EPR is like a magnetic resonance experiment, a magnet is used to split the energy levels of the

a) 3' GCUGACUAUAGUCAGC
    5' CGACUGAUAUCAGUCG

U = [structure of spin-labeled uridine with nitroxide]

b) [PELDOR pulse sequence diagram with ν_1, ν_2, and T]

c) [PELDOR time trace: norm. echo amplitude vs t [ns]]

d) [distance distribution: intensity [arb. u.] vs r [nm]]

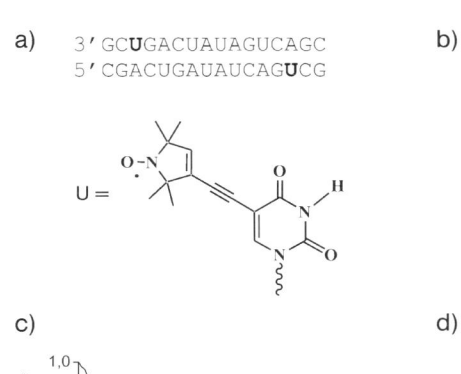

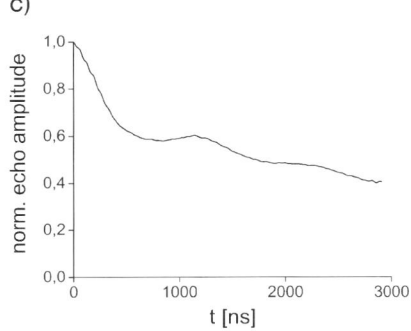

Electron paramagnetic resonance (EPR): (a) Sequence of a 2-fold spin-labeled duplex RNA, (b) the PELDOR pulse sequence, (c) the corresponding PELDOR time trace of the RNA shown in (a) and (d) its distance transformation. The distance obtained from PELDOR is 3.9 nm. Molecular dynamic simulations also yielded a distance of 3.9 nm. The small peak at about 4.3 nm corresponds to an end-to-end stacking of two helices.

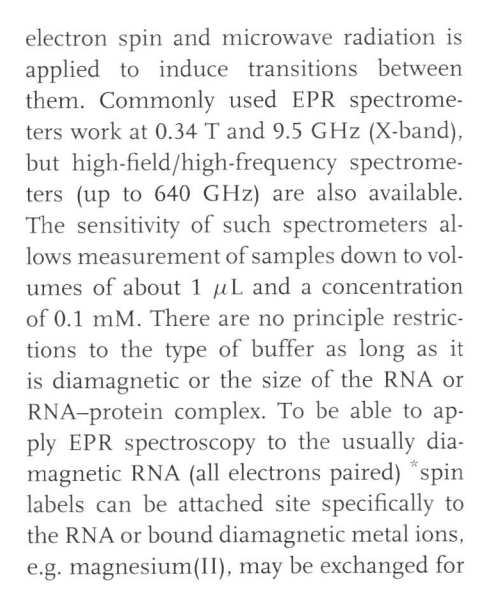

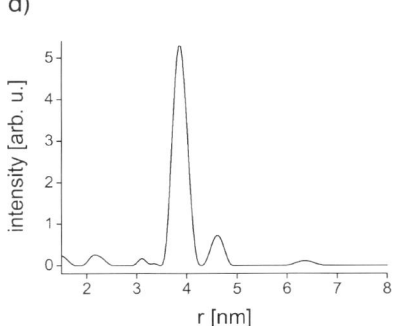

electron spin and microwave radiation is applied to induce transitions between them. Commonly used EPR spectrometers work at 0.34 T and 9.5 GHz (X-band), but high-field/high-frequency spectrometers (up to 640 GHz) are also available. The sensitivity of such spectrometers allows measurement of samples down to volumes of about 1 μL and a concentration of 0.1 mM. There are no principle restrictions to the type of buffer as long as it is diamagnetic or the size of the RNA or RNA–protein complex. To be able to apply EPR spectroscopy to the usually diamagnetic RNA (all electrons paired) *spin labels can be attached site specifically to the RNA or bound diamagnetic metal ions, e.g. magnesium(II), may be exchanged for paramagnetic ones like manganese(II). For example, titrating manganese(II) ions into a RNA solution and following the intensity of the manganese(II) continuous wave EPR signal allows quantification of the metal(II) binding sites and determination of the corresponding binding constants. In addition, the electron spin of manganese(II) is coupled to magnetic nuclei, which are located within a radius of 10 Å of the manganese center. For nuclei with $I > 0$, e.g. $I(^{31}P) = 1/2$, this gives rise to hyperfine coupling parameters and for nuclei with $I > 1/2$, e.g. $I(^{14}N) = 1$, also to quadrupole coupling parameters. The corresponding splitting of the EPR lines is in the case of manganese, however, too small to be resolved in a continuous wave

EPR spectrum. Therefore, electron nuclear double resonance (ENDOR) in its continuous wave or pulsed version or other pulsed EPR experiments like electron spin echo envelope modulation (ESEEM) or the two-dimensional hyperfine sublevel correlation experiment (HYSCORE) are used to resolve and assign these couplings. Selective isotope labeling and density functional theory (DFT) methods might than be applied to locate the binding sites and to translate the gathered EPR parameters into structural information.

*Spin labels covalently attached to RNA in a site-specific manner are mainly used to study the dynamics of RNA and how the dynamics change upon binding of proteins, metal ions or small organic *ligands. This can be done in a qualitative way just by comparing the width of the nitroxide spectra with and without binder or in a quantitative way using simulation programs and maybe high-field/high-frequency relaxation measurements. The changes observed in the first case can be used to locate *binding sites, whereas the second case yields in addition detailed information about the motion of domains or *loops.

If two spin labels are attached to a RNA, the distance between them can be measured via dipolar electron–electron coupling using a pulsed EPR sequence called pulsed electron double resonance (PELDOR). This method, which is complementary to *FRET, yields distances and distance distribution up to 80 Å in a very precise way, e.g. making it possible to study *secondary structure elements or the arrangement of domains.

EPR spectroscopy is also employed to identify and follow the course of phosphate, sugar and base radicals formed upon interaction of RNA–DNA with high-energy radiation (e.g. ultraviolet, X-rays or γ-radiation). OLAV SCHIEMANN

electron spin resonance → electron paramagnetic resonance

electron transport in DNA → charge transfer in DNA, → electron hopping

electrophoresis Method to separate charged particles or *macromolecules in an electrical field, taking advantage of their differences in net electrical charge, shape and size. A wide variety of different electrophoresis techniques have been developed. *Gel electrophoresis and *capillary electrophoresis are most frequently used in biochemistry. VALESKA DOMBOS

electrostatic catalysis Stabilization of charged *transition states by a *catalyst. Here, residues in the *active site form ionic bonds or partial ionic charge interactions with the intermediate. These bonds can either come from side-chains of the catalyst or from metal cofactors. DENISE STROHBACH

electrostatic interaction Type of interaction that is based on mutual attraction of groups or molecules carrying opposite charges.

elongation One of the three steps of *translation during which efficient polypeptide synthesis proceeds after the *ribosome is assembled and moves along the *mRNA. Elongation refers also to the process of *transcription, when the *transcriptome is assembled and travels along the DNA *template for RNA synthesis. STEFAN VÖRTLER, SABINE MÜLLER

elongation factor → translation factor

elongator tRNA All *transfer RNA not involved in initiation (→ initiator tRNA) or suppression (→ suppressor tRNA) and supporting the elongation steps of protein biosynthesis (→ translation). STEFAN VÖRTLER

end labeling (or terminal labeling) Incorporation of labels at either the 3'- or 5'-end of DNA and RNA (→oligonucleotide labeling).

endonuclease Single- or double-strand-specific nuclease that cleaves nucleic acids within the polymer chain. The best-known representatives of this class of enzymes are *RNases A and T1 and DNA *restriction endonucleases. ULI HAHN

engineering of nucleic acids for crystallization The chemical space that is theoretically available for crystallization trials is enormous. The observation that biological *macromolecules crystallize under some of the very limited number of conditions selected for tests suggests that these molecules can crystallize under many more conditions than will ever be discovered. Therefore, rather than screening ever-increasing numbers of reservoir formulations, changing the characteristics of the target molecules has proven a more successful alternative. This strategy is also widespread in protein crystallization, where it includes removal of flexible domains or sections or the screening of the naturally available sequence diversity in orthologs. A general principle observed is that crystal packing of nucleic acids can exploit the same interaction principles that also contribute to intrinsic nucleic acids structures. Short double-helical DNA or RNA *duplexes often form quasi-infinite *helical stacks in crystals. Similarly, *helix stacking is a major structuring principle in complex nucleic acids as initially shown for *transfer RNA. Thus, it turned out to be beneficial to vary the length of the duplexes for crystallization in order to adapt the helical screw symmetry to a crystal symmetry. This principle can be further pursued by introducing (self-complementary) *sticky ends into the nucleic acids of interest that can mediate crystal packing by complementary *base pairing. In the first crystal structure of a *hammerhead ribozyme it was observed that crystal packing is mediated by specific intermolecular interactions between a *tetraloop in one molecule and a *tetraloop receptor in a neighboring molecule. Again, such tetraloop–tetraloop receptor interactions are known to shape the structures of large RNA molecules. Thus, one strategy for engineering RNA molecules for crystallization is the identification of regions that are functionally not relevant and then introducing specific interaction motifs into these elements, such as a tetraloop–tetraloop receptor pair that can only form intermolecularly.

Often, nucleic acid *tertiary structure is maintained by rather weak interactions, giving rise to a conformational ensemble that is difficult or impossible to crystallize. It has been shown that *in vitro* selection methods can be powerful means of generating variants of the nucleic acid in question that are more stable and conformationally homogeneous. Such samples may form crystals more readily or give rise to better ordered crystals than the wild-type parent molecules.

Another general observation is that the surfaces of complex folded nucleic acids provide rather limited chemical diversity for crystal packing. This situation is decisively different in proteins, where various side-chains or portions of the backbone can be exposed on the surface and provide diverse interaction sites for crystal packing. Thus, it has been found extremely useful in crystallization attempts to derivatize nucleic acid surfaces by adding a protein component. Typically, a protein-binding site is engineered into a functionally inert region and the artificial RNA–protein complex is being crystallized rather than the RNA alone. It has been demonstrated that many critical crystal contacts in such artificial complexes are mediated by the protein subunit and that the functional elements of the RNAs are unaffected. In addition, an engineered protein interaction site stabilizes local RNA structure and would provide some restriction to the overall topology. Finally, it can constitute a powerful phasing vehicle, e.g. in the form of the selenomethionine variant. MARKUS WAHL

enzymatic labeling Modifications, or labels, can be introduced enzymatically into DNA and RNA. Examples include uniform incorporation of isotope labels into *nucleic acids by polymerases and incorporation of labels at either the 3'- or 5'-end of DNA and RNA with kinases and ligases (→oligonucleotide labeling). SNORRI TH. SIGURDSSON

episome *Circular DNA molecule usually found in bacteria that can exist independently in the cell and is capable of autonomous *replication or can integrate itself into the host *chromosome. MAURO SANTOS

EPR →electron paramegnetic resonance

equilibrium centrifugation Special type of centrifugation where molecules or subcellular particles are separated in a density gradient. An example is the sucrose density gradient. Linear or exponential gradients can be formed by mixing two sucrose solutions with different concentrations (densities). MATTHÄUS JANCZYK

equilibrium constant In a reaction, it is defined by the product of the concentrations of the product species divided by the concentrations of the reactants. This means the higher the value of the equilibrium constant, the more product is formed at equilibrium. The unit of the equilibrium constant depends on the stoichiometry of the respective reaction according to the mathematic expression mentioned above. All types of reactions can be characterized by an equilibrium constant, like redox reactions, reactions involving covalent bond formation or formation of complexes established by non-covalent interactions (→dissociation constant).

Equilibrium constants are usually represented by a capital K. They may refer to the overall reaction taking the concentrations of the initial reactants and the final products into account or they may be defined for the equilibrium concentrations in individual reaction steps being part of a more complex reaction mechanism. CHRISTIAN HERRMANN

equilibrium dialysis Method to determine the *association constants of low-molecular-weight *ligands to *macromolecules by using *dialysis. It is a very simple, inexpensive method to examine, for example, antigen–antibody interactions or *mRNA–*aptamer interactions, without the need to introduce a radiolabeled or fluorescent tag. The smaller molecule is able to diffuse through a semipermeable membrane, whereas the large molecule cannot penetrate it. After equilibrium has been established the free ligand is equally distributed on both sides of the membrane and the amount of ligand-associated macromolecule, free macromolecule and free ligand can be quantitatively determined. This method offers the opportunity to study even low-affinity interactions that are not detectable by other methods. VALESKA DOMBOS

error-prone polymerase chain reaction (PCR) Method for introducing random *mutations during PCR reactions (→polymerase chain reaction). It may be used when *DNA libraries are required for the generation of mutant proteins, following *transcription and *translation, and may be considered as an *in vitro* selection method. Normal *polymerases have mutation rates of the order 1 in 10^6 and this error rate may be increased by a number of factors, most of which affect the fidelity of *replication of the polymerase. The most common method for altering polymerase fidelity is the use of metal ions; increasing the magnesium ion concentration (up to 10mM) or replacement of magnesium by manganese ions (up to 0.5mM) each decrease enzyme fidelity. Using biased *dNTP pools increases the rate of misincorporation, whilst lowering the annealing temperature during the PCR cycle causes both mis-annealing of *primer to *template and increases the misincorporation rate of dNTPs. Increasing the number of PCR cycles also leads to an increase in the rate of mutation. Other methods of error-prone PCR involve the use of mutagenic nucleoside triphosphates, such as 8-oxo-dGTP, which introduces *transversion mutations, and 5-aza-dCTP, which introduces *transition mutations. DAVID LOAKES

E-site → ribosome

ESR → electron paramagnetic resonance

EST → expressed sequence tag

ethidium bromide (homidium bromide)
Nucleic acid *intercalator mostly used in *gel electrophoresis to visualize small amounts of nucleic acids otherwise undetectable under ultraviolet light (260 nm). Ethidium bromide is an orange to dark-red, non-volatile, crystalline solid (melting point 260–262 °C). Due to its intercalating abilities it is a strong *mutagen and therefore may also be carcinogenic and teratogenic.

Excited by ultraviolet light at 254 and 366 nm, fluorescence emission occurs at a wavelength of 590 nm. The nucleic acid–ethidium bromide complex can enhance the fluorescence signal up to about 50 times compared to free ethidium bromide. VALESKA DOMBOS

euchromatin *Chromosome material that is relatively rich in *gene content and undergoes *transcription.

evolutionary conserved RNA structures
Most functional RNA molecules have characteristic *secondary structures that are highly conserved in evolution. Well-known examples include *ribosomal RNAs (rRNAs), *transfer RNAs (tRNAs), *RNase P and MRP RNAs (a *ribonucleoprotein *endoribonuclease that has been shown to cleave mitochondrial *primer RNA sequences from a variety of sources), the RNA component of *signal recognition particles, transfer-messenger RNA (tmRNA), *group I and *group II introns, *microRNAs, and *small nucleolar RNAs. In all these cases, secondary structure is much better conserved in evolution than the underlying sequence.

Given a sufficiently large database of aligned RNA sequences, one can directly infer a consensus secondary structure from the data. The basic idea is that substitutions in the sequence will respect the common structural constraints. Therefore, substitutions in helical regions have to be correlated, since in general only six (GC, CG, AU, UA, UG and GU) out of the 16 combinations of two bases can be incorporated in the helix. A pair of correlated *mutations that replace one type of *base pair by another one (e.g. GC → AU or GC → CG) is called a *compensatory mutation. Similarly, a single mutation that changes the base pair type (e.g. GC → GU) is called a consistent mutation. Compensatory and consistent mutations have a high probability of leaving the structure unchanged; hence, they are frequently neutral. As a consequence, the sequence can change much faster than the functionally required secondary structures.

If sequences that have diverged significantly on the sequence level are still able to form the same secondary structure, then there must be a selection pressure to retain the structure. Correlated substitutions can be used to infer the functional structure. This approach is usually referred to as phylogenetic structure reconstruction. It works well when a large set of related sequences is available, as in the case of tRNAs and rRNAs.

Several computational methods have been developed to infer such *consensus structures from smaller datasets. The most common approach is to augment thermodynamic structure predictions (→RNA folding) with the covariance information present in a given multiple sequence alignment. This, however, is limited to those cases where the RNA sequences can be still be aligned reliably. See RNA structure comparison and alignment for alternatives if sequence alignments are not available. PETER F. STADLER, IVO HOFACKER

evolution *in vitro* →*in vitro* evolution of nucleic acids, →SELEX

excision repair →base excision repair

exon Sequences coding for proteins in eukaryotes. At the *gene level, exons are split into segments by gene sequences called *introns. Introns are removed after *transcription by *splicing. TINA PERSSON

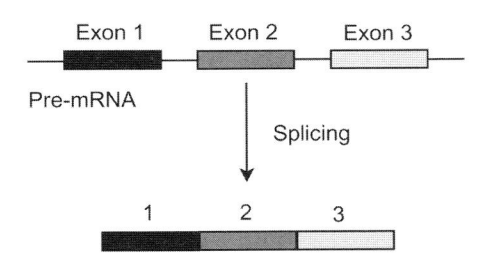

exon shuffling Hypothesis put forward by Walter Gilbert that *exons code for functional units of a protein and that evolution of new *genes has proceeded by *recombination or exclusion of exons, using introns as hotspots for genetic recombination (→shuffling). GEMMA MARFANY, ROSER GONZÀLEZ-DUARTE

exonuclease Single-strand (5' → 3' or 3' → 5') or double-strand specific nuclease that cleaves nucleic acids from their ends (S1 nuclease, λ exonuclease, mung bean nuclease, *DNA polymerase, nuclease Bal 31). ULI HAHN

expressed sequence tag (EST) Sequence tagged site with a direct relationship to an expressed *gene.

expression platform Part of *riboswitches that alters the level of *gene expression depending on *ligand binding to the *aptamer. It can involve an intrinsic *transcription terminator, the *ribosome binding site or could even coincide with the *aptamer itself, if ligand binding leads to structural changes substantial enough to interfere with gene expression. RÜDIGER WELZ

external guide sequence →guide sequence

extinction →absorbance

extinction coefficient (Synonym: absorption coefficient, absorptivity) →absorptivity.

extra arm →variable loop

extra chromosome Additional *chromosome outside the organism's normal chromosomal makeup (karyotype). If one extra complete chromosome is present, this is known as trisomy; the most common in human populations is the trisomy of chromosome 21 that causes Down's syndrome. MAURO SANTOS

extra DNA Can refer to copy number variation of DNA sequences. Genetic variation in the human genome takes many forms and recently it has been discovered that there is an abundance of copy number variation of DNA segments ranging from kilobases to megabases. The functional significance remains to be ascertained. MAURO SANTOS

ex vivo **gene transfer** In the context of *gene therapy, gene transfer to cells removed from the patient, grown and modified *in vitro*, and then returned to the patient. The *ex vivo* approach allows better assessment of the efficiency of the gene transfer and *gene expression than the *in vivo* gene transfer. →gene transfer.
GEMMA MARFANY, ROSER GONZÀLEZ-DUARTE

F

FAD →flavin adenine dinucleotide

FADH Semichinon form of FAD (→flavin adenine dinucleotide).

FADH₂ Fully reduced form of FAD (→flavin adenine dinucleotide).

Fenton reaction Iron-salt dependent decomposition of dihydrogen peroxide generating the highly reactive hydroxyl radical:

$$Fe^{2+} + H_2O_2 \rightarrow Fe^{3+} + OH^{\cdot} + OH^{-}$$

Addition of a reducing agent such as ascorbate or dithiothreitol leads to a cycle that increases the damage to biological molecules. PASCALE ROMBY, PIERRE FECHTER

Fill-in reaction Used to create *blunt ends on a DNA fragment produced by a *restriction endonuclease that leaves 5'-*overhangs. The 5' → 3' DNA synthesis can be catalyzed by T4 *DNA polymerase or DNA polymerase I (→Klenow fragment). ANNEGRET WILDE

FISH Abbreviation for fluorescence *in situ* hybridization (→*in situ* hybridization).

flanking sequence Untranscribed DNA sequence that follows after the 5'- or 3'-end of the transcribed regions of *genes. Flanking sequences have often functions in the regulation of *gene expression. SLAWOMIR GWIAZDA

flavin adenine dinucleotide (FAD) In contrast to its name, FAD represents not a real *dinucleotide, but only a *nucleotide derivative. It consists of *flavin mononucleotide (FMN, riboflavine-5'-phosphate) and adenosine-5'-phosphate (→adenosine phosphates). Due to the flavin chromophore, FAD and FMN exhibit a yellow color. The corresponding vitamin is riboflavin (lactoflavin, vitamin B₂) which represents the biosynthetic precursor of FMN and FAD. In a kinase-dependent reaction the riboflavin is phosphorylated to FMN and subsequently attached to adenosine-5'-triphosphate (ATP) yielding FAD and inorganic diphosphate. The latter reaction is catalyzed by the enzyme FAD pyrophosphatase. FAD and FMN are the redox-active prosthetic groups of several redox-active flavoenzymes such as dehydrogenases, oxidases and reductases. Both flavin-containing cofactors exist in three different oxidation states: (a) the fully oxidized form (FAD), (b) the fully reduced form (FADH₂) and (c) the semichinone form (FADH). The latter state allows enzymes which apply FAD or FMN as redox-active cofactors to operate by both two-electron (polar) or one-electron (radical) pathways (→FMN riboswitch). HANS-ACHIM WAGENKNECHT

flavin mononucleotide (FMN) →flavin adenine dinucleotide

flavin mononucleotide riboswitch →FMN riboswitch

fluorescein Fluorophore with an ultraviolet absorption maximum at 490 nm and an emission maximum of 520 nm. Fluorescein has an isoabsorptive point (equal absorption for all pH values) at 460 nm. Due to the high fluorescence of fluorescein, its

Nucleic Acids from A to Z: A Concise Encyclopedia. Edited by Sabine Müller
Copyright © 2008 WILEY-VCH Verlag GmbH & Co. KGaA, Weinheim
ISBN: 978-3-527-31211-5

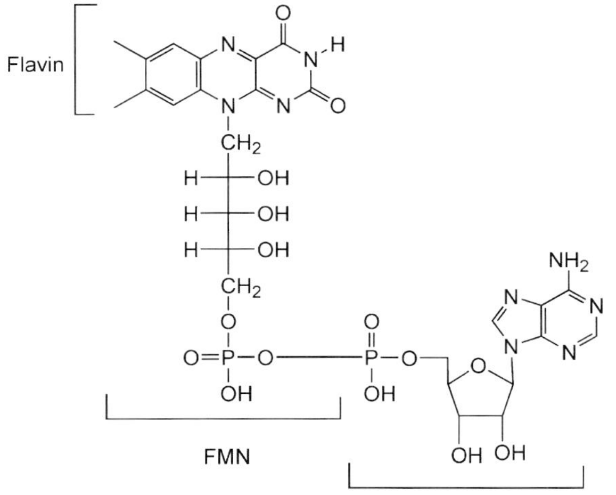

FMN

Flavin adenine dinucleotide

AMP

derivatives are often used for *labeling of DNA and RNA. BETTINA APPEL

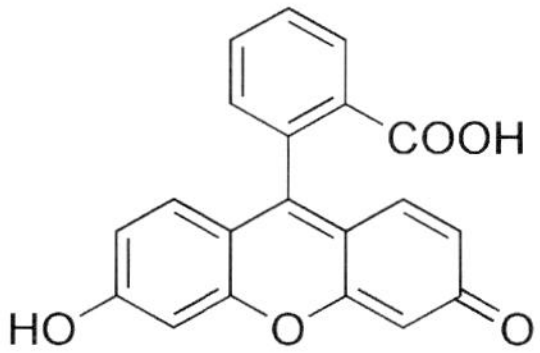

fluorescein isothiocyanate (FITC) Derivative of fluorescein in which the hydrogen in the benzoic acid part in the *para* position to the xanthenyl residue is replaced by an isothiocyanate group. The isothiocyanate reacts with amino groups forming a relatively strong thiourea bridge. This function is used to introduce a fluorescein label into DNA and RNA. The *solid-phase synthesis of DNA and RNA offers the possibility to introduce modifications as long as they are available as *phosphoramidites. Many amino modified phosphoamidite *linkers for the 3'-end (as *solid support), the 5'-end or any other position are available. Trifluoroacetyl is the most common protecting group for the amino function. During the post-synthetic work-up of DNA and RNA, the trifluoroacetyl group is removed by treatment with ammonia. The free amino group can be easily linked to FITC under mild basic conditions (pH 8.5–9) to obtain fluorescein-labeled oligonucleotides. FITC is also used to add a fluorescent group to proteins, e.g. for fluorescent antibody tracing. BETTINA APPEL

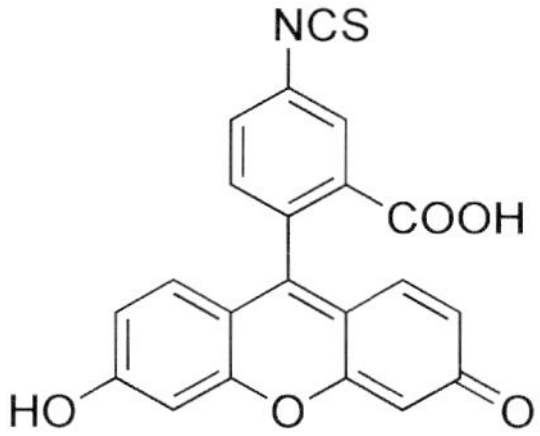

fluorescence correlation spectroscopy (FCS) →fluorescence spectroscopy of nucleic acids

fluorescence cross-correlation spectroscopy (FCCS) →fluorescence spectroscopy of nucleic acids

fluorescence dye → fluorescent dye

fluorescence labeling Incorporation of a *fluorescent dye into DNA and RNA (→ oligonucleotide labeling).

fluorescence spectroscopy of nucleic acids
Fluorescence is a luminescence phenomenon in which the absorption of a photon generates an electronically excited state of the chromophore with a usually short lifetime τ of 10^{-9} to 10^{-8} s (in the simplest case it is defined as 1/e time of single exponential fluorescence decay). Among other deactivation pathways the electron can return to its ground state via spontaneous emission of a characteristic fluorescence photon with a longer wavelength. The energy difference between the absorbed and emitted photons ends up as molecular vibrations and, finally, as heat. The spectral wavelength gap between the maximum of the longest absorption band, λ_A, and the maximum of fluorescence spectrum, λ_F, is called Stokes shift of the particular fluorophore. The existence of the Stokes shift makes a nearly background-free detection with an exquisite sensitivity technically feasible. Together with the fact that a fluorophore can undergo multiple absorption emission cycles within a very short time, laser-induced fluorescence allows even for the detection, identification and characterization of a single fluorescent molecule.

A prerequisite for fluorescence detection is the existence of a suitable fluorophore with a high quantum yield, Φ_F (ratio of the number of photons emitted to the number of photons absorbed). However, the natural *nucleobases in nucleic acids have been selected in evolution for photostability because competing internal conversion processes lead to a fast radiationless deactivation of the chemically reactive electronic excited state. Thus its lifetime is very short (less than 10^{-12} s), which results in a low fluorescence yield (less than 10^{-4}) and a low probability to undergo an irreversible photoreaction leading to non-fluorescent product. Therefore, either a non-natural nucleobase such as *2-aminopurine or ethenoadenosine or a efficient *fluorescent dye (e.g. rhodamine or cyanine dyes) has to be incorporated as marker in nucleic acid research, which can also be used as "reporter" on its local environment. In ensemble as well as in single-molecule multi-parameter measurements, five observables of the chromophore can be deduced from the time-resolved detection of the "chromophore parameters": spectral properties of absorption and fluorescence, $F(\lambda_A, \lambda_F)$, fluorescence brightness and quantum yield, Φ_F, fluorescence lifetime, τ, and fundamental anisotropy, r_0.

As the electron density of the chromophores has certain symmetry, most molecules absorb light along a preferred direction. Therefore, it is possible to probe the orientation and the mobility of the chromophores with linear polarized excitation light (photoselection). The polarization state of fluorescence is characterized by the anisotropy, which is defined as the difference between the fluorescence components parallel and perpendicular to the polarization of the excitation light normalized by the total fluorescence intensity emitted in full space. In chemistry, fluorescence anisotropy assays the rotational diffusion of a molecule from the decorrelation of polarization in fluorescence. By measuring the temporal decay of the fluorescence anisotropy, the obtained decorrelation times contain information on the rotation of the molecule as a whole or the rotation of a part of the molecule relative to the whole. From these one can estimate the rough shape and flexibility of a *macromolecule (e.g. single- to double-strand transition). The measurement of anisotropy belongs to the most important assay techniques for binding of proteins to labeled nucleic acids. Sensitivities down to nanomolar concentrations in a volume of less than 100 μL can be reached in conventional steady-state fluorescence spectrometers. The increase or decrease of the fluorescence quantum yield of dyes in the presence of nucleic acids is also used frequently for their detection in *gel electrophoresis or in *polymerase chain reaction: (a) binding of dyes to the *minor or *major groove (e.g. SYBR Green) and *intercalation between *base pairs (e.g. ethidium bromide, TOTO, YOYO), and (b) identification of nucleobases by "intelligent" fluorescent dyes.

Considering molecular systems, changes in fluorescence parameters of a single coupled fluorophore sometimes do not provide enough information for molecular identification or for more detailed investigations of molecular interactions. Further information can be obtained by the use of more chromophores or study of the fluorescence fluctuations as a function of macroscopic time. (a) By having more than one fluorophore per particle involved, thereby increasing the possibilities to determine stoichiometries and interactions of individual particles from photon densities and coincidences. (b) Optical coupling between two chromophores via *Förster fluorescence resonance energy transfer (FRET). (c) Transport properties (rotational and translational diffusion) and faster movements (conformational fluctuation) of fluorescently labeled molecules can be studied via their resulting signal fluctuations in confocal measurements at nanomolar concentrations or lower. Many different methods have been developed to derive information about the molecular dynamics from these fluctuations. Fluorescence correlation spectroscopy (FCS) reveals molecular concentrations as well as the molecule's mobility parameters, and two-color fluorescence cross-correlation spectroscopy (FCCS) allows one to detect molecular interactions.

The most attractive feature of the optical coupling of two fluorescing reporters via FRET is the possibility to determine structural features of nucleic acids in a qualitative or quantitative manner by performing single-molecule or ensemble measurements. It has been used to study nucleic acid structure and dynamics: structure and folding of *ribozymes, helicity, *bending and *junctions of DNA and RNA, nucleic acid *hairpins, *ribosomes, nucleic acid–protein complexes, single-molecule trajectories of reaction cycles of various *nucleic acid-binding proteins (e.g. *polymerases, *helicases). In FRET, the energy from an excited donor fluorophore, D, is non-radiatively transferred to an acceptor fluorophore, A, by a strongly distance-dependent dipole–dipole coupling. Via the measured FRET efficiency, long-range molecular distance information can be provided, in a range of 20–100 Å, which is not covered by virtually any other solution technique. However, quantitative analyses are difficult to perform since FRET is dependent not only on distance, but on several time-varying parameters such as molecular orientations, spectral overlap, dye quenching or dye labeling grade and bleaching. In order to provide the high resolution necessary to distinguish between structurally similar species, FRET experiments were performed using multi-parameter fluorescence detection (intensities, lifetimes and anisotropies) with two-dimensional analysis of single-molecule data, which accounts for inhomogeneous broadening due to the above problems. When performed at the single-molecule level, FRET studies can yield information about heterogeneities in terms of conformations and conformational dynamics that are unavailable from ensemble measurements. With a donor–acceptor distance approaching or exceeding 100 Å, the FRET efficiency typically is so small that the fluorescence from the acceptor can be determined only by additional direct excitation such as two-photon or two-color excitation in a continuous wave or alternating mode (ALEX). FRET is also applied as a analytic, highly specific tool in *real-time polymerase chain reaction (PCR) (e.g. Light cycler, TaqMan). These detection schemes are based on FRET with the proximity of D to A (can be also non-fluorescent), which is changing during the PCR cycle. CLAUS SEIDEL

fluorescent dye Organic molecule, such as *fluorescein, that is excited by absorption of light at a certain wavelength and emits light, or "glows", at a higher wavelength as the dye returns to its ground state. The excitation and emitting wavelengths are specific to each fluorescent dye and, therefore, a large number of different colors are available to choose from. Fluorescent dyes are useful for locating and tracking *nucleic acids, as well as other *biopolymers, and are used in *molecular beacons. Fluorescent dyes are essential components of the *FRET technique that can be used to measure long-range distances within large molecules and to study conformational changes of single molecules. SNORRI TH. SIGURDSSON

fluorescent marker *Fluorescent dye that has been incorporated into a molecule under study, such as a *nucleic acid, usually for the purpose of tracking its movements or monitoring its function. SNORRI TH. SIGURDSSON

fluorescent primer Molecular biological applications involving the *amplification of DNA or RNA are usually carried out with quantities of *nucleic acid that are not possible to observe visually. Methods have been developed that enable the detection and manipulation of such small quantities of nucleic acid. The most common method of labeling DNA or RNA is with a radioisotope such as ^{32}P or ^{35}S, which allows for nucleic acid detection by exposure to autoradiography film (→radioactive labeling). However, due to the hazards associated with the use and storage of radioactivity non–hazardous methods have been developed including fluorescence. *Primers can be labeled with fluorescently modified nucleotides or by direct attachment of a fluorophore to an *oligonucleotide during solid-phase chemical synthesis of the primer (→oligonucleotide synthesis). Fluorescent primers have characteristic excitation and emission wavelengths, which are often quite narrow bands, which allows for the use of multiple fluorophores. For example, modern *sequencing methods make use of four different fluorophores corresponding to each of the natural DNA nucleobases and detection of the different fluorophores is carried out simultaneously because the emission wavelengths are distinct. A large number of fluorophores have been developed covering a range of excitation/emission frequencies and fluorescence detection is as sensitive as the use of radioisotopes, although they can be bleached following prolonged exposure to light. Fluorescent primers have great utility for *DNA sequencing, *polymerase chain reaction quantification, *FRET and single-molecule detection methods.
DAVID LOAKES

fluorescent probe Probe used for detection of analytes. It contains a *fluorescent dye so that it emits fluorescent light when excited at the dye appropriate wave length.

5-fluorocytidine and 5-fluoro-2'-deoxy-cytidine (5-FC) Analogs of *cytidine and *2'-deoxycytidine in which the hydrogen atom at C5 is replaced by fluorine. As a *deoxynucleotide, it is used in *DNA methylation studies, in particular for elucidating the mechanism of *methyltransferases that catalyze the transfer of a methyl group onto C5 of cytidine. The reaction proceeds via a covalent adduct, formed reversibly by Michael addition of the enzyme to the C5–C6 double bond of the cytosine residue. Due to the fluorine atom at C5, the last step of the enzymatic mechanism, i.e. release of the enzyme by β-elimination, cannot take place, thus leaving the enzyme covalently attached at the cytidine residue. Using 5-FC in DNA substrates of *DNA cytosine-C5 methyltransferases helped to understand the mechanism of these enzymes and allowed crystallization of the protein–DNA adduct. →DNA methylation, →DNA methyltransferase, →DNA cytosine-C5 methyltransferases.
SABINE MÜLLER

5-fluorouridine (5-FU) Analog of *uridine in which the hydrogen atom at C5 is replaced by fluorine. In the same manner as described for *5-fluorocytidine, 5-FU can be used in *ribonucleotide methylation studies as suicide substrate for methylases.
SABINE MÜLLER

f-Met-tRNA →formylmethionyl-tRNA (→ribosome)

FMN →flavin mononucleotide (→flavin adenine dinucleotide, →FMN riboswitch)

FMN riboswitch Highly conserved *RNA domain frequently found in the *5'-untranslated region of prokaryotic *genes involved in biosynthesis and transport of riboflavin and *flavin mononucleotide (FMN) (also known as the RFN element). It was identified as the *aptamer part of a class of *riboswitches that directly and selectively bind FMN. Bioinformatics revealed that the FMN riboswitch is widely distributed among different bacterial species and that it controls *gene expression either on the transcriptional or translational level. *Ligand binding by the RNA leads to down-regulation of gene expression in all analyzed instances; therefore FMN riboswitches belong to the class of OFF switches. The aptamer *consensus sequence, derived from the phylogeny for this riboswitch, consists of 212 *nucleotides and folds into a complex *secondary structure, which is characterized by five *hairpins that surround a central *loop. This complexity might be needed to discriminate against other FMN-related metabolites like riboflavin, which differs from FMN only by the absence of the phosphate group but is bound by the aptamer with more than 100-fold lower affinity.

Binding of FMN to the riboswitch has been studied in detail *in vitro*, which led to the conclusion that *transcription proceeds too fast for the binding to reach equilibrium. This means that for the riboswitch to function as a genetic switch (e.g. to terminate transcription), higher FMN

FMN riboswitch aptamer consensus sequence

concentrations are needed than the (equilibrium) *dissociation constant implies. In other words, the system works as a kinetic switch and the dissociation constant of the aptamer appears to be tuned to effectively trigger the switch rather than to the actual cellular concentration of FMN. This could be a general feature of riboswitches, since the *in vitro* determined apparent K_D for most riboswitch aptamers is much smaller than the estimated metabolite concentrations *in vivo*. RÜDIGER WELZ

Förster fluorescence resonance energy transfer (FRET) →fluorescence spectroscopy of nucleic acids

footprinting Technique for identifying the *binding site on DNA or RNA bound by *ligands (proteins, RNA, metabolites) or to follow the assembly of *ribonucleoprotein complexes by virtue of the protection of nucleotides against attack towards enzymes or chemical probes. The most commonly used probes are small and are not sensitive to the *secondary structure of RNA since they modify the ribose–phosphate backbone, such as Fe-EDTA or I_2. This latter probe necessitates the synthesis of a *modify RNA using 5′-O-Sp(1-thio)nucleotide analogs. The *nucleotide analog triphosphate is randomly incorporated into a RNA *transcript, where the *phosphorothioate linkage can be then selectively cleaved by the addition of I_2. PASCALE ROMBY, PIERRE FECHTER

foreign DNA Exogenous DNA, i.e. DNA that comes from an origin other than the host cell.

formylmethionyl-tRNA (tRNA$_f^{Met}$) *Initiator tRNA (tRNA$_i$) used in prokaryotic cells, mitochondria and plastids only for the initiation of *translation. It is a specialized tRNA^{Met} aminoacylated by conventional MetRS similar to the situation in eukaryotes and archaea, but carries an additional formylated amino group introduced by methionyl-tRNA-*N*-formyltransferase (or transformylase; EC 2.1.2.9). *N*-formylation can be abolished by gene knock-out resulting in impaired growth in *Escherichia coli*. However, some bacteria like *Pseudomonas aeruginosa* show less severe effects indicating a range of susceptibility. Moreover, native proteins

are in any case rapidly deformylated by peptide deformylase (EC 3.5.1.31). The advantages for such a transient formylation seem to lie in the higher hydrolytic stability of a tRNA$_f^{Met}$ allowing a longer lifespan and lower intracellular tRNA concentrations. *N*-formylation ensures more correct recognition during initiation (positive or negative determinant for IF2 or EF-Tu recognition, respectively) and blocks *P- to *A-site reactions of the free amino group during the first peptidyl transferase step. Still, eukaryotes manage without, which makes all enzymes involved in the formylation cycle interesting antibiotic targets.
STEFAN VÖRTLER

four-way junction Structural element of DNA or RNA. It describes a point in the *secondary structure of a single molecule or a multi-molecular arrangement where the structure branches cross-like into four *double helices. The best known four-way junction is the *Holliday junction.
SLAWOMIR GWIAZDA

fragment *Oligonucleotide or *polynucleotide resulting either from endonucleolytic digestion, physical rupture or chemical digestion, e.g. alkaline hydrolysis, of larger nucleic acids. VALESKA DOMBOS

fragmentation In *mass spectrometry, the fragmentation pattern of a compound characterizes its structure. Nucleic acid fragmentation or *footprinting by hydroxy radicals, endonucleolytic digestion, alkaline hydrolysis, etc., is used to probe nucleic acid structure and specific *nucleic acid–protein interactions. VALESKA DOMBOS

frame shift →reading frame shift

frame shift mutation →reading frame shift mutation

FRET Abbreviation for Förster fluorescence resonance energy transfer (→fluorescence spectroscopy of nucleic acids).

functional cloning As opposed to *positional cloning, identification of disease-causing *genes based on their function and presumptive involvement of the encoded protein in the disease.
GEMMA MARFANY, ROSER GONZÀLEZ-DUARTE

functional groups Parts of a compound (or molecule) where reactivity is exerted; in organic compounds, frequently associated with the heteroatoms (N, O, S).

fusion (fused) gene Genetically engineered *gene in which two *open reading frames have been fused, so that a recombinant chimeric protein is expressed. Most used protein moieties for gene fusions include the *Escherichia coli* glutathione-S-transferase (to obtain and purify recombinant proteins) and the enhanced green fluorescent protein (a reporter).
GEMMA MARFANY,
ROSER GONZÀLEZ-DUARTE

G

G →guanine

GAAA tetraloop receptor RNA structural motif consisting of 11 *nucleotides and providing a specific *binding site for GAAA *tetraloops.

```
    | = |
    C   G
   C     G
  U       U
 A         A
  A       U
    G   U
    | = |
```

gag genes *gag* stands for "group-specific antigen" and refers to those *genes that encode for the core structural proteins of retroviruses. Gag proteins are fast-evolving structural components of retroviruses and *long terminal repeat *retrotransposons, and are essential for particle formation in the viral budding process. Recent *genome-wide analyses have identified 85 human *gag*-like genes that encode about 100 domesticated Gag proteins with different functions. MAURO SANTOS

gametic imprinting Epigenetic modifications on the gametes of eutherian mammal *chromosomes, which are distinctly modified in maternal or paternal meiosis. Imprinted maternal or parental genes are silenced during embryonic development and only the non-imprinted *gene on the other parental chromosome is expressed. This non-equivalent expression of the inherited *alleles is transmitted to daughter cells following cell division. Several human diseases, such as the Angelman and Prader–Willi syndromes (involving chromosome 15), are due to partial monosomies of chromosomal regions that undergo genomic imprinting. GEMMA MARFANY, ROSER GONZÀLEZ-DUARTE

ganciclovir Synthetic analog of *2′-deoxyguanosine. It is an antiviral drug used for treating or preventing cytomegalovirus infections. The mechanism of action runs through the phosphorylation to a *deoxyguanosine triphosphate (dGTP) analog. This competitively inhibits the incorporation of dGTP by viral *DNA polymerase, resulting in the *termination of *elongation of viral DNA. BETTINA APPEL

gap genes *Genes that cause gaps in the segmentation pattern of the embryo when mutated. In *Drosophila*, the anterior–posterior polarity of the egg defines the anterior–posterior polarity of all subsequent developmental stages. Maternal genes expressed in the mother's ovary produce *messenger RNAs that are placed in different regions of the egg. These mRNAs encode regulatory proteins (Bicoid,

Hunchback, Nanos and Caudal) that diffuse and generate anterior–posterior gradients that affect the expression of certain genes in the newly formed zygote. The zygotic genes regulated by these maternal factors are called gap genes. *Hunchback, krüppel, giant, tailless* and *knirps* are some of their charming names. MAURO SANTOS

gapmer →antisense oligonucleotides

gas chromatography–mass spectrometry (GC-MS) A method used for structural analysis and quantification of small and volatile organic compounds. After vaporization the analytes are separated by GC, ionized by electron or chemical ionization (→mass spectrometry: ionization methods) and then analyzed by MS. The GC separation limits this methodology to volatile and non-polar compounds with a molecular weight below approximately 600 Da. In nucleic acids research, GC-MS is mainly used for the determination and quantification of *DNA alkylation, oxidative damage and *photoadducts. The general sample preparation includes enzymatic digestion or hydrolysis to the free (modified) *nucleosides or *nucleobases, succeeded by following chemical derivatization (often silylation) to increase volatility. →liquid chromatography–mass spectrometry, →mass spectrometry: terms and definitions, →mass spectrometry in DNA relevant research. ANDREAS SPRINGER

G-box *Cis*-element (CACGTG) that exists in many *gene *promoters and has critical effects on plant development, hormone responses and fungal infections.

GC–AT rule *Guanine always pairs with *cytosine. *Adenine always pairs with *thymine in a *DNA double helix.
→Chargaff's rule.

GC content The *guanosine–*cytosine content is usually expressed as a percentage. It refers to the amount of GC *base pairs in *double-stranded nucleic acids. The AT content in mol% plus the GC content in mol% equals 100 mol% of any given double-stranded nucleic acid. GC base pairs form three *hydrogen bonds. Therefore GC base pairs are stronger and more resistant to *denaturation, i.e. by high temperature, compared to AT base pairs. Consequently it requires more energy to disrupt GC-rich double-stranded nucleic acids than to disrupt AT-rich double-stranded nucleic acids. Apart from the *melting temperature, the GC content also determines density, torsional flexibility and bending flexibility of nucleic acids. The more GC base pairs a nucleic acid contains, the denser and less flexible it will become. *Vice versa*, high *AT content indicates lower melting temperature, higher flexibility and lower density of a nucleic acid. By measuring the melting temperature photometrically it is possible to determine the GC content of double-stranded nucleic acids. Different densities of nucleic acid, on the other hand, can be used to isolate or separate different classes of nucleic acids with the help of a CsCl density gradient, i.e. separation of *satellite DNA from *chromosomal DNA. Geneticists use the GC content to characterize the *genome of any given organism or any DNA or RNA. In taxonomy, it can be used to classify organisms. Looking at a long region of genomic sequences, the different genes can be characterized by their GC content in contrast to the rest of the genome. *Exons of genes are rather GC-rich, whereas *introns tend to be AT-rich. For *polymerase chain reaction experiments the GC content of DNA *primers is used to determine their *annealing temperature with the *template. VALESKA DOMBOS

GC-MS →gas chromatography–mass spectrometry

gDNA →genomic DNA

G-DNA (also G₄-DNA) Specific DNA conformation, occurring in G-rich DNA strands that form tetrahelices (→G-quadruplex) by interaction of *guanines through their *Hoogsteen and *Watson–Crick edges. SABINE MÜLLER

GDP Abbreviation for guanosine diphosphate (→nucleoside phosphates).

gel electrophoresis *Electrophoresis performed in a gel matrix. Two different matrices are most commonly used: polyacrylamide (polymerized acrylamide) and agarose (a highly purified form of agar). Matrix-assisted *electrophoresis stabilizes the separated moieties against diffusion and allows separation not only by electric charge, but also by shape and by size. Nucleic acids are negatively charged due to their phosphate *backbone and in an electric field migrate towards the plus electrode. Smaller nucleic acid molecules move faster through the gel than larger nucleic acid molecules and the concentration of the matrix can be varied to increase this effect. In order to separate smaller RNA (50–200mers) denaturing polyacrylamide gel electrophoresis (PAGE) is usually applied with urea being the denaturing detergent. Smaller *double-stranded DNA is also separated by PAGE, but under native conditions. For larger nucleic acids, especially *messenger RNAs and *plasmid DNA, agarose gel electrophoresis has proved to be more convenient. VALESKA DOMBOS

gel filtration Chromatographic technique that separates molecules depending on their size. It is usually used to purify *macromolecules such as proteins or nucleic acids. The matrix of a gel filtration column consists of porous beads. The pores have a defined size. Small molecules can migrate through the pores into the beads, thus delaying the movement of the molecules along the column. Larger molecules remain excluded outside the beads and pass the column faster than the smaller ones. Unlike in other chromatographic techniques, the mobile phase may remain the same during the whole procedure. It may, however, differ from the buffer of the applied sample, allowing a simple and quick exchange of the sample buffer. SLAWOMIR GWIAZDA

gel mobility shift assay →band shift assay

gene Physical and functional unit of heredity that carries information from one generation to the next. In molecular terms, a gene is the entire DNA sequence necessary for the synthesis of a functional peptide or RNA molecule. It includes regions preceding and following the *coding region (leader and trailer), and non-coding intervening sequences (→introns) between individual coding segments (→exons). BEATRIX SÜß

gene activation Switch-on of the *transcription machinery of a *gene in response to specific cellular signals (→induction of gene expression).

gene amplification Increase in the number of copies of a given *gene. In the context of cancer, this term applies to the allelic dosage increase of a particular chromosomal gene that makes the tumoral cells selectively advantageous or more resistant to drugs. GEMMA MARFANY, ROSER GONZÀLEZ-DUARTE

gene array → DNA chip

gene bank Gene database containing DNA (→ genomic DNA, → copy DNA) sequences.

gene chip → DNA chip

gene cloning Isolation and introduction of a *gene containing DNA *fragment into a *vector where it will be replicated (→ cloning).

gene conversion Nonreciprocal transfer of *genetic information between homologous non-sister *chromatids in individual meiotic tetrads. Gene conversion has been proposed to account for the shift of small contiguous segments of DNA among different sequences or haplotypes. MAURO SANTOS

gene duplication Doubling of a DNA region either by local events that generate tandem duplications, larger-scale events that can duplicate entire *chromosomes or genome-wide events that result in polyploidization (complete genome duplication). Forty years ago Susumo Ohno stated that gene duplication was the single most important factor in evolution, simply because without duplicated genes the transition from unicellular to multi-cellular organisms would have been impossible. MAURO SANTOS

gene exchange *Genes can move between cells or organisms by a number of different methods. A remarkable process is so-called "horizontal gene transfer" where an organism transfers genetic material to a non-offspring cell. The sequence of many *genomes has indicated that considerable horizontal *gene transfer has occurred between prokaryotes, but its occurrence has also been documented among higher plants and animals. MAURO SANTOS

gene expression Overall process by which the information encoded in a *gene is converted into an observable *phenotype (most commonly production of a protein). Gene expression is a multi-step process that begins with *transcription, *messenger RNA processing/transport and *translation, and is followed by protein folding, post-translational modification and protein targeting. The amount of protein that a cell expresses depends on the tissue, the developmental stage of the organism and the metabolic or physiologic state of the cell and can be controlled at each step (→ gene regulation). BEATRIX SÜß

gene family Set of *genes (called paralogous) derived by duplication and variation of an ancestral gene. Genes are generally categorized into families based upon shared sequence motifs and similarities in structure. MAURO SANTOS

gene frequency In population genetics, used to express the proportion of an *allele in a population. For example, sickle-cell disease is frequent in sub-Saharan Africa and is caused by the sickle hemoglobin allele HbS, which can have a frequency of up to 20% in those populations. MAURO SANTOS

gene inactivation Switch-off of the *transcription machinery of a *gene in response to specific cellular signals.

gene library Group of *cDNA *clones from an organism or tissue previously cloned into phage or *plasmid *vectors, and which can be amplified and screened for specific sequences (→ cDNA library). GEMMA MARFANY, ROSER GONZÀLEZ-DUARTE

gene localization → gene mapping

gene locus *Chromosome location of a *gene that may be occupied by any of its *alleles.

gene locus control region *Cis*-acting DNA *fragments that are capable of enhancing the expression of a linked *gene that is integrated in ectopic *chromatin sites. The LCR was first identified in transgenic mice and characterized phenotypically by conferring tissue-specific and physiological levels of expression on linked genes; by activating the *transcription of *transgenes in a position-independent, copy number-dependent manner; and by determining the time and *origin of *DNA replication. MAURO SANTOS

gene manipulation Modification of the genetic material (DNA) of a *gene using recombinant DNA techniques.

gene map Graphical representation of the relative arrangement of *genes on *genomic or *plasmid DNA. Gene maps are generated using specific computer software and can be linear (chromosomal maps) or circular (plasmid maps). SUSANNE BRAKMANN

gene mapping Refers to the specific position of a *gene on a given *chromosome. There are two types of gene mapping: genetic mapping and physical mapping. Genetic mapping traces back to Alfred Sturtevant who developed the first genetic map in *Drosophila*, also called a linkage map because it estimates the tendency of *alleles at two or more loci to be inherited together. In other words, a linkage map shows the relative position of genes or genetic markers of a species on each chromosome. Physical mapping uses more recent techniques to determine the absolute position of a gene on a chromosome. If the complete sequence of a particular species is not known, genes can be mapped by *in situ *hybridization. MAURO SANTOS

gene mutation Any change in the sequence of a *gene. Can be *point mutations when they exchange a single *nucleotide for another, *insertions (usually caused by *transposable elements) when one or more nucleotides are added into the gene or *deletions when two or more nucleotides are removed. Most common point mutations are *transitions that exchange a *purine (A ↔ G) or a *pyrimidine (C ↔ T) by another, and less frequent are *transversions that exchange a purine by a pyrimidine or *vice versa*. *Silent mutations are point mutations that do not alter the amino acid in the encoded protein of the gene. Insertions in the *coding region may cause a shift in the *reading frame or alter the *splicing of the *messenger RNA. Deletions can also alter the reading frame. →mutation. MAURO SANTOS

gene pool In population genetics, used to mean all the *genetic information distributed among an interbreeding group of individuals in a population at a particular time. Since in diploid individuals *alleles are sorted and shuffled in the gametes that combine during fertilization to produce a new set of genotypes, it is useful to consider populations of *genes rather than populations of individuals. MAURO SANTOS

general acid–base catalysis *Functional groups in the *active site of the *catalyst serve as an acid or base to assist proton transfer. This principle is used by small *ribozymes to mediate *transesterification. During *catalysis, the attacking 2′-hydroxyl group is deprotonated by a functional side-chain, thus enhancing its nucleophilicity. In parallel, a second side-chain, acting as acid, stabilizes the 5′-hydroxyl leaving group by donation of a proton. DENISE STROHBACH

gene rearrangement Refers to a particular order of *genes in a *chromosome. For instance, a structural change such as an inversion, where a chromosome segment has rotated 180° relative to other regions on either side, produces a gene order that is reversed relative to a standard reference orientation. MAURO SANTOS

gene regulation Cellular control of the amount and timing of a functional *gene product. It gives the cell the control over structure and function, and is the basis for cellular differentiation, morphogenesis, and the versatility and adaptability of any organism. It includes all mechanisms involved in regulating *gene expression. Gene expression can be modulated at several molecular steps including *chromatin structure, *transcription, *RNA processing, *RNA transport, *transcript stability, *translation and post-translational modification. In eukaryotes, the regulation of gene expression is controlled nearly equivalently from many different points. In prokaryotes, gene regulation mainly includes *transcriptional control [positive or negative control by *induction (activation), *repression, *antitermination or *attenuation]. BEATRIX SÜß

gene silencing Term generally used to describe the switching off of a *gene by epigenetic processes of *gene regulation. Gene silencing can occur at the transcriptional level by relocating a gene near *heterochromatin (a phenomenon called position-effect variegation) or at the post-transcriptional level as the result of *messenger RNA of a particular gene being destroyed. *RNA interference is a common mechanism of post-transcriptional gene silencing. MAURO SANTOS

gene splicing Removal of *introns and joining of *exons to produce mature *messenger RNA from the original RNA that results after *transcription. *Alternative splicing is an important mechanism to increase the *coding capacity of a *gene, allowing the synthesis of several structurally and functionally distinct proteins. *Pre-mRNA splicing takes place within the *spliceosome, a large molecular complex composed of four *small nuclear ribonucleoproteins (U1-, U2-, U4/U6- and U5-snRNPs) and about 50–100 non-snRNP splicing factors. In genetic engineering, gene splicing refers to any external process for manipulating genes. MAURO SANTOS

gene synthesis *In vitro* chemical synthesis of the DNA sequence (→ genetic information) of a particular *gene (→ DNA synthesis).

gene tagging Technique that allows us to tag a *gene in order to identify it. This technique usually involves the use of *mutagenesis caused by known *transposable elements, which randomly insert into the *genome. The *phenotype of interest is selected and the altered gene is identified by means of tracking the transposable element. Another name for the same technique is transposon tagging. Protein tagging, instead, is the recombinant addition of an epitope or peptide (gene fusion) *in-frame, in order to easily detect the gene encoded protein. GEMMA MARFANY, ROSER GONZÀLEZ-DUARTE

gene technology General term referring to recombinant DNA techniques.

gene therapy In the strict sense, defined as the transfer of new genetic material to the cells of a patient with the intention of a therapeutic benefit to that individual. In the broader sense, these approaches do not only include the delivery of *genes, but comprise the application of any type of nucleic acids (e.g. *antisense strategies employing *antisense oligonucleotides, *ribozymes, *triplex-forming oligonucleotides or *small interfering RNAs to induce *RNA interference). The first clinical trial exploring the potential of gene therapy was initiated in 1990 to treat children with severe combined immunodeficiency disease (SCID). In the following 15 years, more than 1000 clinical trials have been carried out according to the *Journal of Gene Medicine* Clinical Trial Database (www.wiley.co.uk/genmed/clinical).

For most of the gene therapeutic approaches viral *vectors are employed to deliver the genetic material into the target cells. Approximately half of the clinical trials have been performed with either retroviruses or adenoviruses. Additional viruses used as vectors for gene therapy include the herpes simplex virus and adeno-associated viruses. Moreover, methods for non-viral transfer of genes are being developed. Gene therapy can be carried out *ex vivo* (out of the body) into cells explanted from the patient or *in vivo* (in the body) by direct injection of the vector into the bloodstream.

Approximately two-thirds of the gene therapeutic trials attempt to treat various types of cancer. In addition, monogenetic, vascular and infectious diseases are indications intensively dealt with. Among the most successful approaches was treatment of children with X-linked SCID by *ex vivo* transduction of their bone marrow cells with retroviruses. Later on, however, several of the treated children developed a leukemia-like condition most likely due to insertional mutagenesis. In a different trial, a patient died after delivery of the gene encoding ornithine transcarbamylase with an adenoviral vector which triggered a severe immune response. Progress in gene therapy can be expected from newly developed vectors with improved properties. Early-phase clinical trials showed promising results, including treatment of hemophilia B or the delivery of angiogenic factors to treat vascular and coronary diseases. Furthermore, conditionally replicating oncolytic viruses have been developed for the treatment of cancer.

For ethical reasons only somatic gene therapy employing adult cells of persons known to have a disease has been carried out to date, whereas modifications of the germ line which will be passed on to further generations have not been performed. Additional ethical considerations deal with the question how to ensure that gene therapy will only be employed to treat severe diseases rather than improving physical abilities or intelligence of healthy individuals (→gene transfer). JENS KURRECK

genetic code In contemporary organisms most reactions are catalyzed by enzymes made of proteins. The genetic code refers to a table whereby *genetic information stored by DNA or RNA can be translated into proteins (enzymes and structural proteins). There is the linear *nucleotide sequence of *genes and a corresponding amino acid sequence in proteins. In the simplest form one gene corresponds to one protein. Read in a specified (5′–3′) direction, the *messenger RNA (mRNA) molecule instructs a collinear synthesis of the encoded protein. One *triplet of nucleotides (→codon) specifies one amino acid and contiguous, non-overlapping triplets specify contiguous, non-overlapping amino acids.

There are $4^3 = 64$ possible triplets and there are 20 amino acids. The genetic code is a table showing which codon corresponds to which amino acid.

The genetic code is unambiguous (one triplet sequence encodes only one type of amino acid), but degenerate (several different triplets, up to six, can encode the same amino acid). In contemporary organisms the genetic code is implemented in the

genetic code

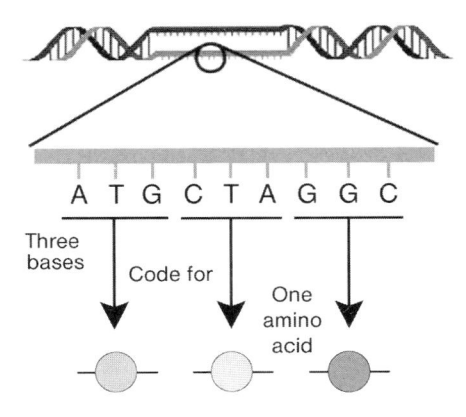

process of *translation on the *ribosome. Each amino acid is charged to a cognate *transfer RNA (tRNA) molecule by a corresponding *aminoacyl-tRNA synthetase enzyme. Normally there are only 20 different such synthetases, but more tRNAs, partly matching the degenerate nature of the genetic code. *Codons in mRNA pair with *anticodons of tRNAs on the ribosome. Normally there are three *stop codons without an amino acid assignment: they instruct the ribosome to quit translation. The genetic code was thought to be universal, but this is not the case. In the figure below we show what is now called the "canonical" version from which there are minor deviations in various organisms (including mitochondria, *Mycoplasma* and ciliates). Occasionally, additional amino acids can be incorporated by stop codon capture, such as selenocysteine (codon UGA) and pyrrolysine (codon UGA). The origin of the genetic code is a notoriously difficult problem; many believe that it goes back to the *RNA world. Indeed, selected RNA sequences can implement every type of step of contemporary translation and the formation of the peptide bond

First position (5' end)		U	C	A	G	Third position (3' end)
U		UUU ⎤ Phe UUC ⎦ UUA ⎤ Leu UUG ⎦	UCU ⎤ UCC ⎥ Ser UCA ⎥ UCG ⎦	UAU ⎤ Tyr UAC ⎦ UAA Stop UAG Stop	UGU ⎤ Cys UGC ⎦ UGA Stop UGG Trp	U C A G
C		CUU ⎤ CUC ⎥ Leu CUA ⎥ CUG ⎦	CCU ⎤ CCC ⎥ Pro CCA ⎥ CCG ⎦	CAU ⎤ His CAC ⎦ CAA ⎤ Gln CAG ⎦	CGU ⎤ CGC ⎥ Arg CGA ⎥ CGG ⎦	U C A G
A		AUU ⎤ AUC ⎥ Ile AUA ⎦ AUG Met	ACU ⎤ ACC ⎥ Thr ACA ⎥ ACG ⎦	AAU ⎤ Asn AAC ⎦ AAA ⎤ Lys AAG ⎦	AGU ⎤ Ser AGC ⎦ AGA ⎤ Arg AGG ⎦	U C A G
G		GUU ⎤ GUC ⎥ Val GUA ⎥ GUG ⎦	GCU ⎤ GCC ⎥ Ala GCA ⎥ GCG ⎦	GAU ⎤ Asp GAC ⎦ GAA ⎤ Glu GAG ⎦	GGU ⎤ GGC ⎥ Gly GGA ⎥ GGG ⎦	U C A G

Amino acid names:

Ala = alanine
Arg = arginine
Asn = asparagine
Asp = aspartate
Cys = cysteine

Gln = glutamine
Glu = glutamate
Gly = glycine
His = histidine
Ile = isolevcine

Leu = leucine
Lys = lysine
Met = methionine
Phe = phenylalanine
Pro = proline

Ser = serine
Thr = theonine
Trp = tryptophan
Tyr = tyrosine
Val = valine

in ribosomes is catalyzed by *ribosomal RNA rather than ribosomal proteins even today. EÖRS SZATHMÁRY

genetic fingerprint → DNA fingerprint

genetic imprinting Phenomenon of parent-of-origin *gene expression – the expression of a gene depends upon the parent who passed on the gene. A process known as reprogramming occurs during passage through the sperm or egg with the result that the parental and maternal *alleles have different properties in the very early embryo. Imprinting is achieved through *DNA methylation or *chromatin structure which prevents *transcription of the gene. Relatively few genes in humans are known to be imprinted. These genes tend to be clustered in the *genome. The *gene silencing acts in much the same manner as *mutation or deletion of one copy of a gene, except that it is not a permanent heritable change. Mutation or deletion of the active copy of an imprinted gene may result in disease.

Several of the genes in region on *chromosome 15q11–13 are subject to *genomic imprinting. Two different disorders – Prader–Willi and Angelman syndrome – are due to deletion within this region. The Prader–Willi syndrome results when the paternally contributed region is absent, whereas the Angelman syndrome results when the maternally contributed region is missing. BEATRIX SÜß

genetic information Hereditary information stored by organisms in nucleic acids. There are other types of hereditary information carriers (prions, epigenetic systems and memes) that are not regarded as genetic. Inheritance is one of the prerequisites of evolution. Genetic information is carried by *chromosomes in bacteria and eukaryotes. In eukaryotes the bulk of this information is in the nucleus, but mitochondria and plastids also have a number of *genes. EÖRS SZATHMÁRY

genetic polymorphism Presence in a population of two or more different forms or types of individuals produced when different *alleles are segregating.

genetic print → DNA fingerprint. Also, genetic modifications that allow us to identify or track one particular sequence or state of a *gene.

gene transfer Transport of *genes into cells. For research purposes genes are frequently transferred into bacteria or cultured eukaryotic cells. Electroporation and the use of lipids as carriers (lipofection) are widely used methods for these applications. For gene transfer into humans with therapeutic intention, a basic distinction can be made between delivery of genes into germline tissue (eggs or sperm) and genetic modifications of somatic cells. Alterations obtained by germline gene transfer will be passed on to future generations and have not yet been performed due to ethical considerations. In contrast, corrections obtained by somatic gene transfer are restricted to the treated patient.

In order to obtain efficient gene transfer into the target cells, carriers need to be employed. For gene therapeutic applications (→gene therapy) viral *vectors are frequently used. In these viruses essential genes are replaced by therapeutic genes so that the viruses lack the information needed for *replication. Five main classes are important in clinical settings: oncoretroviruses, lentiviruses, adenoviruses, adeno-associated viruses (AAVs) and herpes simplex-1 viruses. Oncoretroviruses and lentiviruses are two subclasses of retroviruses with RNA *genomes that generate DNA by *reverse transcription that is integrated as a provirus into a host *chromosome and therefore ensures long-term expression of the therapeutic gene. A major problem associated with the use of retroviruses is the risk of insertional *mutagenesis upon integration of the genetic material into the host chromosome. In contrast, the genetic material of the other viruses remains mainly episomal and thus results only in transient synthesis of the desired protein. More recently developed variants of AAVs and adenoviruses, however, allow prolonged *gene expression. A major disadvantage of adenoviruses is their immunogenic potential that prohibits high-dose or repeated applications. To circumvent this problem, helper-dependent adenovirus vectors devoid of all viral *coding sequences ("gutless") have been designed that cause only low inflammation and show sustained *transgene expression.

Due to the adverse reactions caused by viral vectors, techniques for non-viral gene transfer are being developed as well. These methods comprise delivery of *naked DNA or the use of lipid carriers. A major limitation of these approaches to date is their low transfer efficiency. Furthermore, ballistic injection of DNA with a gene gun is developed mostly for vaccination.

Gene therapy can either be carried out *ex vivo* (out of the body) or *in vivo* (in the body). For the former approach, which is also designated *indirect gene transfer, cells (e.g. hematopoietic stem cells) are isolated from the patient, transduced (usually with a retroviral vector) and re-infused into the donor. For *in vivo* gene transfer, the vectors are injected into the bloodstream of the patient. Direct application of the vector into the affected tissue (e.g. tumors) can also be referred to as *in situ* gene transfer. JENS KURRECK

genome Total *genetic information carried by a cell or organism.

genome analysis Investigation of *genomes and genomic data using a variety of analytical techniques. Genome analysis covers the detection of *genetic *polymorphisms at the *nucleotide level (mutational analysis, *single-nucleotide polymorphisms) and at the level of *gene and genome *fragments (→DNA fingerprinting, *restriction fragment length polymorphisms, *amplification fragment length polymorphisms), the *sequence analysis of genomes as well as their structural, organizational and functional analysis (genomics). In addition to molecular techniques, genome analysis needs the application of bioinformatic tools for the assembly of sequence data, for *sequence *alignment and comparison, for detection of *open reading frames and genetic control elements, and for *genome mapping. Genome analysis is important for medicine (diagnostic and therapy), e.g. for screening specific *gene mutations or genetic factors determining the susceptibility to certain diseases, for identifying bacterial and viral infections, for the development of vaccines, and for *gene therapy of mutant or deficient genes. Beyond medical applications, genome analysis is used for the study and comparison of organisms at the genome level, and for detecting the evolutionary relationships of genes, *gene families and species. SUSANNE BRAKMANN

genome array →DNA chip

genome sequencing Determination of the *base sequence of complete *genomes. During the past decade, several hundred viral genomes and the genomes of diverse bacteria, archaea and eukaryotes have been completed. The resolution of the human genome was one of the largest genome sequencing projects that was published in 2001 (→Human Genome Project). The large-scale genome sequencing projects have a great impact on scientific progress in many disciplines, such as the development of sequencing technology, the initiation of functional genome analysis (genomics) and a future personalized medicine. SUSANNE BRAKMANN

genomic DNA Complete set of *chromosomal DNA.

genomic imprinting →gametic imprinting.

genomic library Usually refers to the pool of sequences produced from fragmentation of *genomic DNA for *sequencing purposes. →shotgun sequencing, →Human Genome Project, →DNA sequence analysis. SABINE MÜLLER

genotype Genetic constitution of an individual in the form of DNA, usually referring to specific characters under consideration.

geometric nomenclature and classification of RNA base pairs The most fundamental interaction in *nucleic acids is the *edge-to-edge hydrogen bonding of *nucleobases, with the canonical *Watson–Crick base pair being the prototype. While in Watson–Crick base pairs nucleobases interact through their *Watson–Crick edges, with the *glycosidic bond oriented in *cis* relative to the axis of interaction, other modes of interaction have been defined in the large number of *RNA structures that have been solved over the past 10 years. Eric Westhof et al. have proposed a nomenclature that classifies RNA base pairs based on their geometry and have suggested *isostericity matrices summarizing the geometric relationships in 12 families of pairwise combinations of the four natural bases, considering the type of *edge-to-edge interaction as well as the relative orientation of the glycosidic bond (N. B. Leontis, et al. *Nucleic Acids Res* 2002, *30*, 3497–3353). SABINE MÜLLER

germline gene transfer →gene transfer

GlcN6P riboswitch The *glmS* ribozyme is a natural *self-cleaving RNA that is found in several Gram-positive bacteria. It is located in the *5′-untranslated region of *glmS* *messenger RNA and selectively activated by glucosamine-6-phophate (GlcN6P), the metabolic product of the GlmS enzyme. The *ribozyme functions as a *riboswitch by repressing the glmS *gene in response to increasing GlcN6P concentrations.

The proposed *secondary structure derived from an 18 member phylogeny consists of four *stem–loop regions (P1–P4) forming at least one *pseudoknot (P3a). Domains P1/P2 contain most of the *conserved nucleotides and form the *catalytic core of the ribozyme, whereas domains P3/P4 enhance activity but are dispensable. While there is little conservation 5′ to the cleavage site, nucleotides immediately 3′ to it are highly conserved and probably interact with the catalytic core of the ribozyme (e.g. via alternative base pairing). The minimal form of the ribozyme that still shows cleavage activity encompasses *nucleotides from the −1 position through to the P2 domain. Similar to other natural *small nucleolytic ribozymes, the *glmS* ribozyme catalyzes an internal phosphoester transfer reaction forming products with 2′,3′-cyclic phosphate and 5′-hydroxyl termini, and, by dissection in the P1 loop, can be engineered to cleave substrates *in trans* with *multiple turnover.

In the absence of GlcN6P or other related amine-containing compounds the P1–P4 *Bacillus cereus glmS* ribozyme exhibits a cleavage rate of around 10^{-5} min^{-1}, which is similar to the rate of an uncatalyzed RNA *transesterification reaction. Saturating concentrations of GlcN6P (10 mM) accelerate the reaction by approximately 5 orders of magnitude to a rate of around 1 min^{-1}, with an half-maximal rate at 30 μM GlcN6P. Interestingly, the cleavage rate in Tris-buffered solutions is elevated to around 10^{-3} min^{-1}, suggesting that Tris can substitute for GlcN6P in activating the ribozyme. However, the ribozyme discriminates against closely related metabolic compounds like glucosamine (GlcN), which induces RNA cleavage with a reduced rate (3×10^{-2} min^{-1}) and affinity (half-maximal rate at 5mM GlcN or higher).

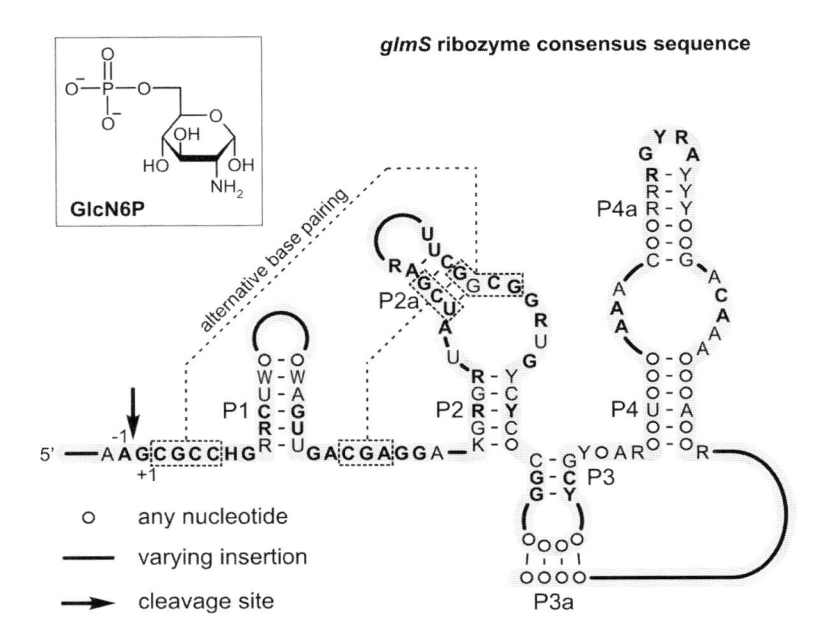

Although the mechanism for *glmS* ribozyme activation by its *ligand has not yet been elucidated, experimental data support the hypothesis of GlcN6P being a coenzyme rather than an allosteric effector, meaning that it is directly or indirectly involved in reaction chemistry rather than influencing ribozyme activity by changing its structure. RÜDIGER WELZ

***glmS* ribozyme** → GlcN6P riboswitch

glucosamine-6-phosphate riboswitch
→ GlcN6P riboswitch

glycine riboswitch *Riboswitch found in many bacterial species *upstream of *genes that encode proteins involved in glycine utilization as energy source. It controls *gene expression on either transcriptional or translational level and since the presence of high *ligand concentrations triggers increased expression of these proteins, the glycine riboswitch is one of the rare "ON" switches.

In most instances two similar glycine *aptamers are joined by a short conserved linker that binds two glycine molecules in a cooperative fashion. For instance, in the case of the double aptamer sequence found in *Vibrio cholera*, ligand binding at one aptamer increases affinity of glycine to the other aptamer by a factor of 1000 and *vice versa*. The *Hill coefficient (the degree of cooperativity) for the *V. cholera*

Glycine riboswitch aptamer consensus sequence

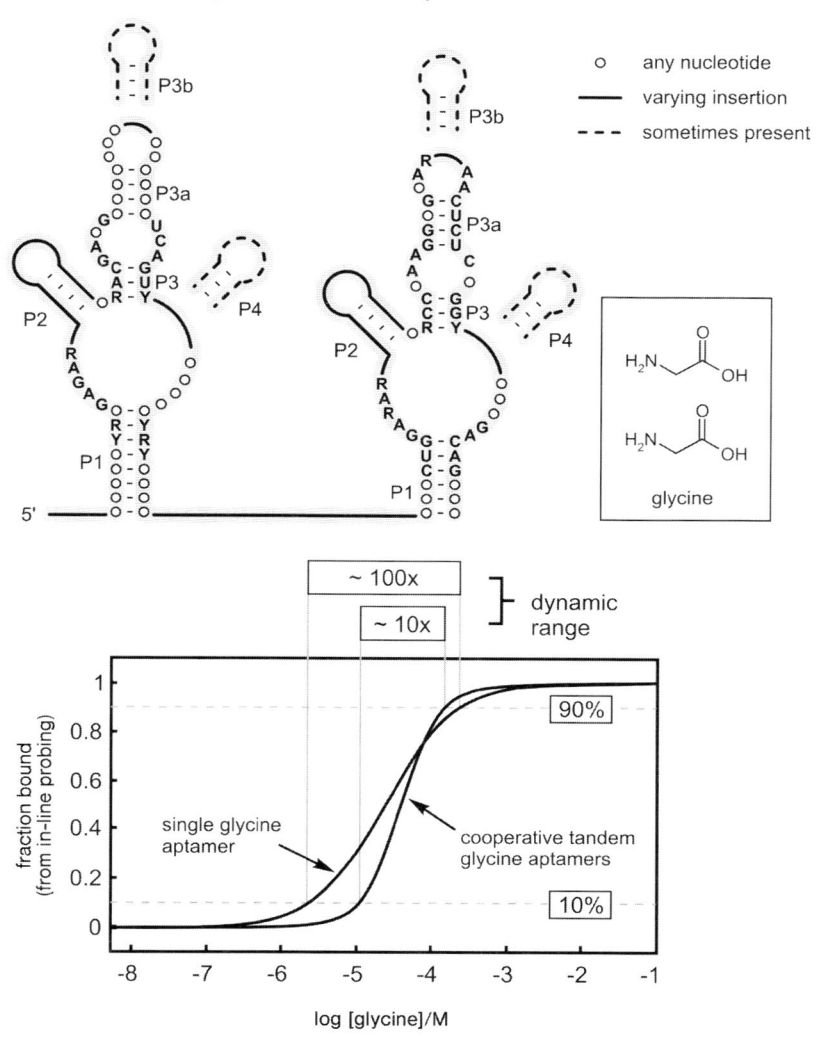

double aptamer system was determined to be 1.64 – a level of cooperativity similar to that known from oxygen binding to hemoglobin. Due to the cooperative behavior the double aptamer shows higher sensitivity to changes in glycine concentration compared to single aptamers (see graph). The concentration change to alter the amount of RNA in bound state from 10 to 90% is reduced from around 100-fold for a single glycine aptamer to around 10-fold for the cooperative tandem aptamer. This probably enables cells to rapidly adjust to changing glycine concentrations, e.g. to increase expression of glycine cleavage genes at higher glycine concentration to efficiently use it as energy source, or reduce gene expression at low glycine concentration to ensure sufficient levels for protein synthesis, respectively. In addition, the glycine riboswitch shows high discrimination against related natural compounds like L-alanine or L-serine, despite the fact that it binds a ligand of only 10 atoms. →riboswitches. RÜDIGER WELZ

glycol nucleic acids (GNAs) Minimalistic analogs of naturally occurring DNA or RNA, since they contain only a three-carbon backbone which consists of propylene glycol *phosphodiester units. They are capable of forming antiparallel *duplexes according to the *Watson–Crick base pair rule and are possible predecessors of RNA in a pre-*RNA world. JÖRN WOLF

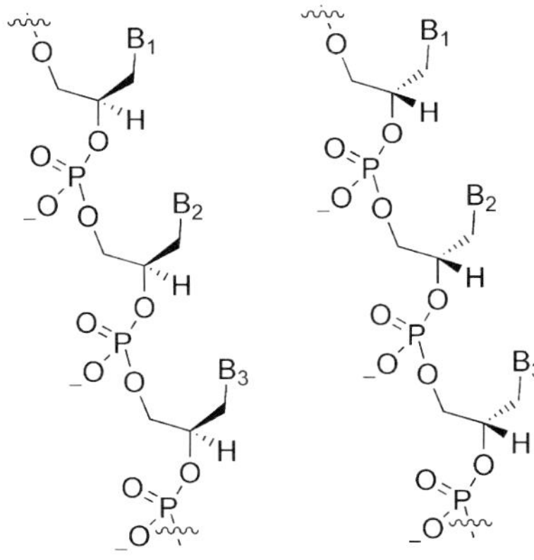

(R)-GNA   (S)-GNA

glycoside Acetal formed from the cyclic semiacetal of a saccharide through reaction with an alcohol or phenol at its anomeric OH group (→glycosidic bond).

glycosidic bond Bond formed between the hemiacetal group of a saccharide (or a molecule derived from a saccharide) through its anomeric carbon and another moiety. Specifically, an O-glycosidic bond is formed to the hydroxyl group of some alcohol or other saccharide. Likewise, in S-glycosidic bonds the anomeric carbon of a sugar is bound to some other group via a sulfur atom. In *nucleosides, each base is attached to the anomeric carbon of the sugar by an N-*glycosidic bond to N1 of *pyrimidines and N9 of *purines. Substances containing N-glycosidic bonds are also known as glycosylamines. KLAUS WEISZ

glycosidic torsion angle (κ), Describes the rotation around the N-*glycosidic bond in *nucleosides between the anomeric carbon of the sugar and N1 of *pyrimidines or N9 of *purines. It is defined by the torsion angle O4′–C1′–N9–C4 for purines and O4′–C1′–N1–C2 for pyrimidines. The two ranges found for this angle in nucleosides are designated *syn* and *anti* (→*syn* conformation, →*anti* conformation). KLAUS WEISZ

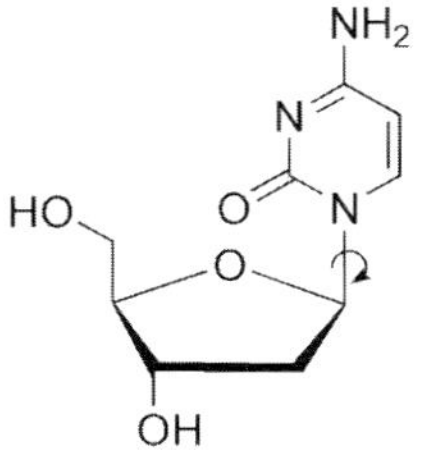

syn deoxycytidine

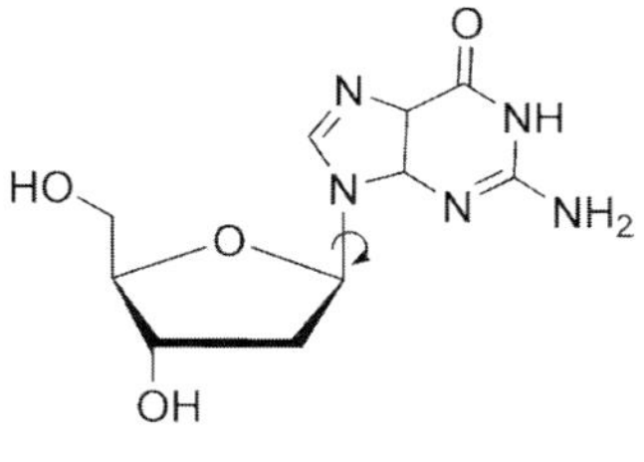

anti deoxyguanosine

glycosylase →DNA glycosylase

GMP Abbreviation for guanosine monophosphate (→nucleoside phosphates).

GNA →glycol nucleic acid

G-quadruplex Guanine-rich nucleic acid sequence that forms a specific four-stranded structure. Four guanines interact through their *Watson–Crick and *Hoogsteen edge, thus forming a square structure. Monovalent cations further stabilize G-quadruplexes by complexation. G-quadruplexes can be formed intra- or intermolecularly, being composed of one, two or even four different strands (→telomeres). MATTHÄUS JANCZYK, SABINE MÜLLER

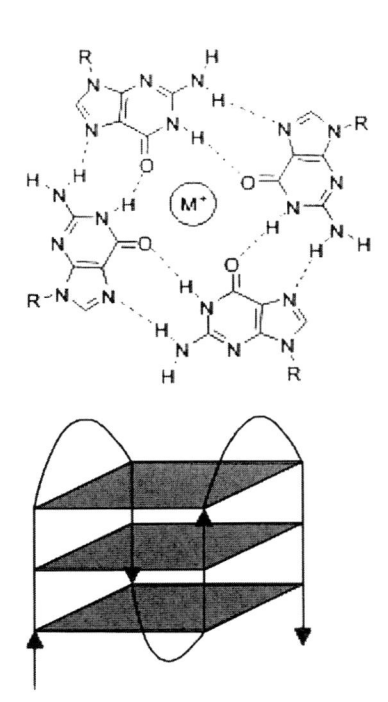

gRNA →guide RNA

group I intron Group I introns were the first class of *catalytic RNAs to be discovered, by Cech et al. in the early 1980s in the *Tetrahymena* *ribosomal RNA (rRNA) precursor (→ *Tetrahymena* ribozyme). Due to their size, ranging from a few hundred to around 3000 *nucleotides, and their reaction mechanism, they belong to the group of large *ribozymes. More than 2000 group I introns have been sequenced to date, interrupting genes for rRNAs, *transfer RNAs and proteins in bacteria, viruses, lower eukaryotes and organelles. Overall, group I introns are now regarded as selfish *mobile genetic elements that are widespread due to self-integration into various *genes (homing), but because of *self-splicing at the RNA level are non-deleterious to their hosts.

Group I introns consist of nine to 10 paired helices with a conserved *catalytic core of about 100 nucleotides (a, Fig. 1). The core structure forms by assembly of three main domains (P4/5/6, P3/7/8/9 and P10/1/2), with the *helical stack of P10/1/2 oriented in parallel to P8/3/7 (c, Fig. 1). P1 contains the 5'-substrate strand and P10 is formed during alignment of the 3'-splice site (a and b, Fig. 1). The *active site is created by juxtaposition of elements P1, P3, P4, P6, P7, J4/5 and J8/7. Peripheral elements branch from this core and by long-range tertiary interactions stabilize its architecture. Some group I introns further depend on proteins for folding into an *active conformation.

Group I introns accomplish *splicing by a two-step *transesterification reaction initiated by an exogenous *guanosine (or its mono- or triphosphate) cofactor (a, Fig. 2). For the first step of the reaction, the 5'-splice site is positioned by *base pairing to an *internal guide sequence within the intron, resulting in helix P1, and the exogenous guanosine binds to the G site at helix P7. The 3'-hydroxyl group of the guanosine then performs a nucleophilic attack on the 5'-splice site phosphate and attaches to the 5'-end of the intron. In the second step, the free 3'-hydroxyl of the 5' *exon attacks

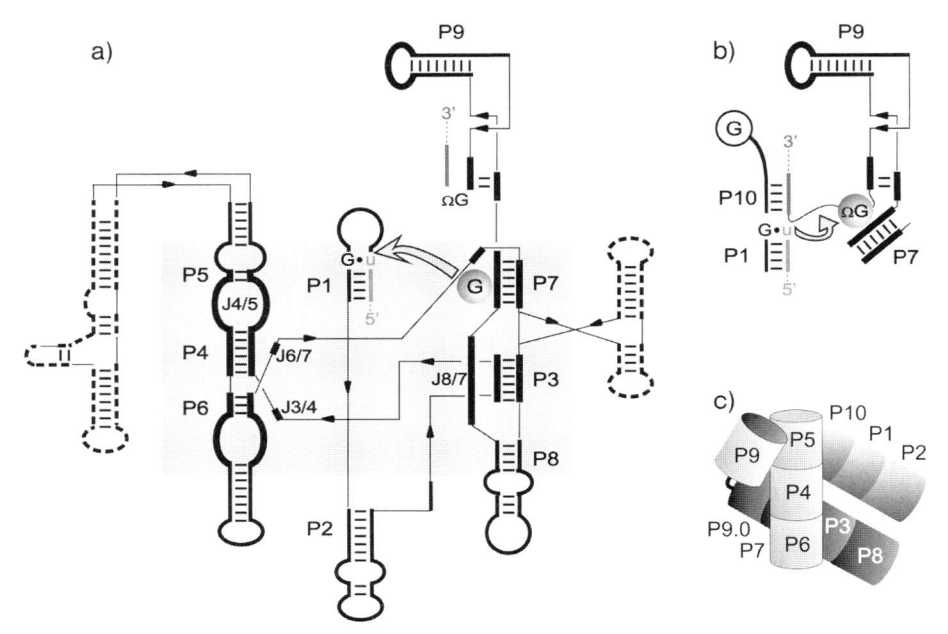

Figure 1: Group I intron structure. (a) Two-dimensional structure before the first transesterification; grey box, catalytic core; open arrow, nucleophilic attack of the G cofactor on the 5'-splice site; thick lines, regions of marked sequence conservation; broken lines, examples of peripheral elements variable among group I introns; grey lines, exon sequences. (b) Structural features before the second cleavage step, with the G cofactor covalently attached to the 5'-end of the intron and the additional P10 helix formed by base pairing to the 3' exon. (c) Conserved arrangement of major helical stacks in the core.

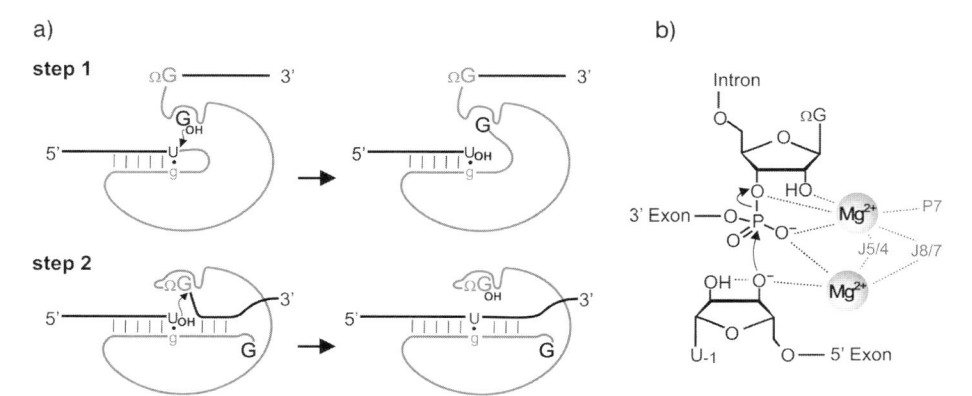

Figure 2: (a) Universal two-step splicing pathway of group I introns. (b) Proposed two-metal ion mechanism of transition state stabilization during group I intron-catalyzed transesterification. The coordination of the metal ions to several phosphate oxygens in structural elements of the catalytic core (J5/4, J8/7 and P7) illustrates the complex architecture of the group I intron *active site.

the phosphate at the 3'-splice site, ligating the exons and excising the intron. According to current models, two magnesium ions appear to directly contribute to *catalysis (b, Fig. 2).

For therapeutic applications, group I introns have been engineered for repair of *messenger RNAs by *trans-splicing, i.e. cutting out and replacing the defective target sequence. ROLAND K. HARTMANN, MARIO MÖRL, DAGMAR K. WILLKOMM

group II intron Less abundant (than the *group I-type) self-splicing *intron in mitochondrial genes. Like the nuclear splicing reaction, splicing of group II introns involves two *trans-esterification steps and starts with a nucleophilic attack by an intronic *branch point A residue on the 5'-splice site. Then, the new free 3'-hydroxyl group of the preceding *exon attacks the 3'-splice site. Consequently, here too the intron is released in the form of a *lariat. As compared to *pre-mRNA splicing, however, this self-splicing reaction differs in that no *spliceosome or other factors are required. Rather, a sophisticated *secondary structure with six *stem–loop domains brings the two splice sites into close spatial vicinity. Base pairing within the stem of domain 6 results in a bulged branch point A residue that is responsible for the first nucleophilic attack on the 5'-splice site. Therefore, at the level of RNA structure, the *catalytic center of group II introns basically resembles those of nuclear introns. Since the number of *phosphodiester bonds remains constant, no further energy supply is necessary. Generally, it appears that the mechanism of nuclear pre-mRNA splicing has evolved from such an autocatalytic reaction observed with group II introns. One has to keep in mind, however, that most group II introns are not autocatalytic, i.e. catalysis is not detectable in the absence of protein factors. BERND-JOACHIM BENECKE

Grunstein–Hogness method →colony blot

G-tetrad →G-quadruplex

GTP Abbreviation for guanosine triphosphate (→nucleoside phosphates).

guanine →nucleobases

guanine arabinonucleoside →arabinonucleosides

guanine ribonucleoside →nucleosides

guanine riboswitch Shares main characteristics with the *adenine riboswitch and hence both are described as *purine riboswitches.

guanine xylonucleoside →xylonucleosides

guanosine Natural building block of RNA (→nucleosides).

guanosine diphosphate (GDP) →nucleoside phosphates

guanosine monophosphate (GMP) →nucleoside phosphates

guanosine triphosphate (GTP) →nucleoside phosphates

guide RNA Small 3'-oligo-uridylated RNA molecules involved in U insertion/deletion type *RNA editing of some *messenger RNAs, *transfer RNAs and *ribosomal RNAs. Guide RNAs are complementary to short sections of the edited *transcripts. The editing information is contained in A and G guiding *nucleotides which can form *base pairs with the inserted U residues. Guide RNAs are bound to one or more proteins, forming *ribonucleoprotein complexes. Other guide RNAs (→small nucleolar RNA) direct the chemical modification

of ribosomal precursor RNA molecules.
BERND-JOACHIM BENECKE

guide sequence Special part of a RNA which hybridizes to a *messenger RNA and facilitates *splicing of a RNA *intron. There are *external guide sequences (e.g. *small nuclear RNAs) and *internal guide sequences. External sequences hybridize to *exons, internal sequences bind *introns to form a splicing complex (→ guide RNA).
MATTHÄUS JANCZYK

gyrase → DNA gyrase

H

hairpin loops Composed of an *antiparallel helical stem and "single-stranded" *nucleotides that form a closing loop. Their inherent attribute is to reverse the direction of a RNA chain. Among all *secondary structural elements – except for canonical helical stems – they represent the dominating structural motif. In principle, the loop part can be made up of any number of unpaired nucleotides. The smallest loop is therefore composed of two nucleotides and referred to as the biloop.

Biloops studied by *nuclear magnetic resonance (NMR) spectroscopy of the cUNg type are highly abundant in *ribosomal RNA. In biloops of the type cUUg the two *uracil nucleotides are single stranded (or unpaired) and a base pair between the C and the G nucleotide is closing and connecting the loop to the stem.

Triloops of the sequence cUUUg have been studied by NMR spectroscopy. The three U nucleotides are highly dynamic and the uracil nucleobases point directly into the solvent. This is probably due to the fact that no stabilizing interactions by *base stacking are present – this is in line with former studies of single-stranded poly(U) and UU where the lack of base stacking has also been described.

Pentaloops have also been described. However, the loop structure of uUUCUGa could also be described as a triloop that is extended by a GU base pair. Similar to the pentaloop the structural description of hexaloops as tetraloops extended by *non-canonical base pairs is valid. The conserved sequence cGUAAUAg is primarily found in rRNA and it behaves like an UAAU *tetraloop closed by a GA base pair.

Among all hairpin loops the most prominent one is surely the *anticodon loop of the *transfer RNA that is composed of seven nucleotides. In all characterized tRNA molecules, it is mainly composed of conserved and semiconserved nucleotides – exchanges of those mostly occur for isosteric nucleobases. These conservations are necessary for the establishment of functional canonical anticodon hairpin structures that include the first motif that has been characterized in RNA molecules, which is called the *U-turn motif.

Loops with eight or more unpaired nucleotides are at least identified on the basis of biochemical experiments; it is probable that those are stabilized by non-canonical base pairs or by intracatenaric *hydrogen bonds.

The most abundant hairpin loops are those composed of four unpaired nucleotides (→tetraloops). BORIS FÜRTIG, HARALD SCHWALBE

hairpin ribozyme Naturally occurring RNA that catalyzes sequence-specific cleavage and ligation of RNA following a mechanism that is common to all *small nucleolytic ribozymes. It is derived from the minus strand of the *satellite RNA from tobacco ringspot virus. In the virus, the hairpin ribozyme is responsible for cleavage and likewise ligation of intermediates occurring during *rolling-circle amplification of the viral *satellite RNA. The cleavage reaction proceeds via *in-line attack of the 2′-hydroxyl group on the adjacent

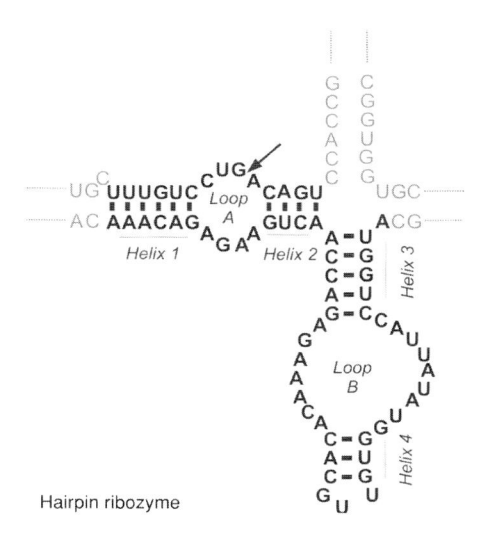

Hairpin ribozyme (diagram showing Helix 1, Loop A, Helix 2, Helix 3, Loop B, Helix 4)

phosphorous atom, leading to the departure of the 5′-oxygen atom on the adjacent *ribose and to generation of a *2′,3′-cyclic phosphate (→small nucleolytic ribozymes). The same chemical reaction is catalyzed by the *hammerhead ribozyme, the *hepatitis delta virus ribozyme and the *Varkud satellite ribozyme. Typical rates of *multiple turnover reactions are between 0.01 and 1 molecule per minute at physiological salt concentrations, consistent with a substantial 10^7- to 10^9-fold rate enhancement over uncatalyzed non-specific RNA cleavage. Also the reverse reaction can be catalyzed. In this case a 5′-oxygen atom attacks a *2′,3′-cyclic phosphate group leading to ring opening of the cyclic phosphate and consequently to *ligation. The hairpin ribozyme is particularly efficient in catalyzing the ligation reaction. Strikingly, *ligation is favored if the hairpin ribozyme is folded into a stable three-dimensional structure, while cleavage occurs from ribozyme-substrate complexes that are less stable, yet stable enough to fold into a catalytically competent structure. The *secondary structure of the minimal catalytic motif consists of four Watson–Crick base-paired helices separated by two *internal loops A and B. In the natural RNA context the hairpin ribozyme is embedded into a RNA *four-way junction, which assists tertiary folding.

Intimate association of loops A and B, called docking, is stabilized by divalent metal ions and generates the local environment in which catalysis occurs. The docked ribozyme is further stabilized by a *Watson–Crick base pair between a loop A guanosine and a loop B cytidine as well as a *ribose zipper involving nucleotides of both loops. A minimal version of the ribozyme in which the four-way junction is replaced by a simple *phosphodiester bond is active, but requires a 1000-fold higher Mg^{2+} concentration to induce folding (→RNA folding). Upon binding to its substrate the hairpin ribozyme undergoes extensive structural rearrangement, which results in distortion of the substrate RNA that primes it for cleavage. To achieve catalysis the ribozyme is likely employing exclusively RNA *functional groups. The crystal structure of a precursor form of the RNA as well as biochemical experiments have suggested that an *active site adenosine and an active site guanosine might adopt the role of a general acid or base, respectively. Metal ions or other catalytic cofactors are not used for active site chemistry. SABINE MÜLLER

hairpin structure →stem–loop structure, →hairpin loops

hammerhead ribozyme Smallest and the most intensely studied of the naturally occurring *ribozymes. Hammerhead RNA motifs were originally found in virus *satellite RNAs and viroids. During *rolling-circle replication of the circular RNA pathogens, they process the long linear *concatamers to monomeric units. After self-cleavage, the hammerhead ribozymes can also catalyze the reverse reaction, leading to circularization of the monomers. Recently, hammerhead motifs have been identified in *genomes of all kingdoms of life. Due to their widespread occurrence and simplicity, multiple emergence of hammerhead RNAs throughout evolution has been proposed.

The name derives from the hammerhead-shaped *secondary structure (a, below): a minimal hammerhead, consisting of around 40 *nucleotides, contains three helices that radiate from a core of *conserved nucleotides containing the *active site. According to crystal structures, helices II and III stack upon one another, and the core of invariant nucleotides includes a uridine turn (→U-turn) motif CUGA [C3 to A6] adjacent to the scissile phosphate. In natural hammerhead RNAs, helix I and II are capped by internal or apical *loops (L1 and L2) not conserved in sequence (b, below). These loops are drawn together by *tertiary interactions, resulting in an *active conformation at Mg^{2+} concentrations as low as 0.1 mM. Minimal hammerhead ribozymes which lack these loops operate at millimolar magnesium concentrations *in vitro* only, with cleavage rates 1–2 orders of magnitude lower than those of natural hammerheads and rarely exceeding 1 min^{-1}. The crystal structure of a full-length hammerhead ribozyme from *Schistosoma mansoni*, reconciling numerous biochemical data, finally provided the basis to mechanistically explain hammerhead catalysis. Hallmarks inferred from the structure are: (1) a bent and partially unwound stem I, (2) an overwound stem II to position G12, A13 and A14 against × 17 such that the 2′-OH nucleophile of C17 is positioned for in-line attack, (3) an extensive hydrogen bonding network arranging the invariant nucleotides of the hammerhead core, (4) the proximity of G8 and G12 to the scissile

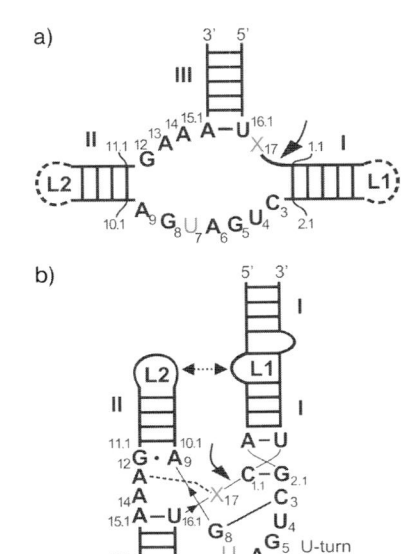

Secondary structure representations of (a) minimal hammerhead ribozymes and (b) of a catalytically competent hammerhead conformer as inferred from the crystal structure of S. mansoni full length hammerhead ribozyme. Nucleotides of the catalytic center are indicated (with conventional numbering; conserved nucleotides in bold; X = C, [U, A]); single-headed arrows depict the cleavage site, the two-headed arrow indicates the stabilizing tertiary loop–loop interaction of natural ribozymes. In bimolecular constructs of minimal hammerheads, either of the terminal loops may be missing, as indicated by the broken lines, and also variants with stem III closed by a loop have been constructed.

phosphate, consistent with their suggested role in acid-base catalysis, (5) lack of evidence for a specific role of metal ions in hammerhead catalysis, and (6) the closest nonbridging oxygen distance between the phosphates 5′ of A9 and 3′ of C17 reduced to 4 Å rather than 20 Å as in the minimal hammerhead structure.

Like the other small ribozymes known, the hammerhead ribozyme catalyzes

Proposed reaction mechanism of hammerhead ribozyme cleavage.

sitespecific self-cleavage that results in products with 2′,3′-*cyclic phosphates and 5′-hydroxyl termini. Cleavage chemistry obeys an S_N2 (in-line) mechanism, based on a nucleophilic attack at the scissile phosphate by the vicinal *ribose 2′-oxygen. At very high ionic strength, hammerhead cleavage chemistry can occur in the absence of divalent cations. However, under low salt conditions, at least one divalent metal ion has been proposed to form an *inner-sphere contact with the pro-Rp non-bridging oxygen at the scissile phosphodiester. Linear dependence of the log of reaction rate on pH has been interpreted to indicate that a hydroxide coordinated to the same metal ion deprotonates the 2′-hydroxyl for nucleophilic attack. Also a conformational rearrangement postulated for effective formation of the pentacoordinated *transition state has been proposed, and the number and precise role of metal ions is still under debate.

Technical and therapeutic applications of the hammerhead ribozyme exploit the fact that it can be specifically tailored to cleavage of a RNA substrate *in cis* or *in trans*. In constructs used for *trans*-cleavage of target RNAs, loop 1 is disrupted (see a, above) and the "substrate" strand contributes the 5′-parts of helices III and I to the hammerhead structure. In such a bimolecular set-up, the remaining portion of the hammerhead may become a genuine *trans*-acting ribozyme capable of *multiple turnover. Also, by fusion to a *riboswitch, minimal hammerhead ribozymes have been converted into *allosteric ribozymes functionally dependent on small *ligands (→Varkud satellite ribozyme, →hepatitis delta virus ribozyme, →hairpin ribozyme).
ROLAND K. HARTMANN, MARIO MÖRL, DAGMAR K. WILLKOMM

H-DNA Intramolecular triple-stranded structure (triple helix) formed by homopurine/homopyrimidine sequences with mirror repeat symmetry in *supercoiled DNA. *In vitro* two main structural motifs have been defined in which the third strand runs parallel or antiparallel to the duplex *purine strand. A *Watson–Crick duplex accommodates in its major, deep groove an extra homopyrimidine or homopurine strand which is hydrogen bonded to the Hoogsteen base-pairing side of the homopurine. Parallel triplex formed by pyrimidine-rich third strands contains isosteric triplets $C^+ \times G$–C and $T \times A$–T (– and × denote Watson–Crick and Hoogsteen pairing, respectively; →edge-to-edge interaction of nucleobases). The protonation of the third-strand cytosines makes the stability of these systems pH dependent. Purine-rich third strands, on the other hand, bind antiparallel to the duplex purine strand through reverse Hoogsteen bonds, forming the pH-insensitive and non-isosteric triplets $G \times G-C$, $A \times A-T$ (− and × denotes Watson–Crick and reversed Hoogsteen pairing, respectively). Also, $T \times A-T$ triplets may be incorporated in antiparallel triplexes through reverse Hoogsteen bonds. The affinity and specificity of third-strand binding is affected by a number of factors such as sequence length and composition, the presence of pyrimidine–purine inversions in DNA duplex target, and solution conditions, including pH and mono- and divalent counterion concentration. Although *triple helices form with considerable sequence specificity, they are generally much less stable than their duplex counterparts. Kinetic studies on DNA triplexes have shown that these structures form very slowly, with association rate constants of 10^3 $M^{-1}s^{-1}$, about three orders of magnitude slower than DNA duplex. The formation of intermolecular triplexes offers a means for targeting unique DNA sequences and has potential for use in antigene therapy (→triplex-forming oligonucleotides).

In H-DNA, the pyrimidine-rich strand from one half of the mirror repeat folds back and forms *Hoogsteen base pairs with the homopurine strand of the other part of the repeat (intramolecular triplex). An extra half of the homopurine strand remains single-stranded. Long homopurine–homopyrimidine sequences cloned in the *plasmid were sensitive to single-strand-specific S1 nuclease, suggesting the occurrence of alternative H-DNA structure under negative *supercoiling. Magnesium ions were shown to promote a switch between the conformation containing $C^+ \times G-C$ triplets (pyrimidine-rich third strand) to the one containing $G \times G-C$ triplets (purine-rich third strand). Homopurine–homopyrimidine regions of several hundred base pairs long are more abundant in eukaryotic *genomes than expected from their *base composition. These regions with the potential for forming various types of intramolecular triplexes (H-DNA) may function in some processes of genome regulations such as *replication rate and timing, *recombination and *chromosome folding by absorbing the negative supercoils, and modulating the local structure of DNA and the global topology of *chromatin domains. NINA DOLINNAYA

HDV ribozyme →hepatitis delta virus ribozyme

heat shock The heat shock response is a universal stress response in which a cell induces the expression of *genes encoding heat shock proteins (HSPs) after a temperature upshift. Other conditions that lead to the accumulation of misfolded proteins, e.g. recombinant protein production and exposure to antibiotics or toxins, often also induce a heat shock response. HSPs are highly conserved proteins in prokaryotes and eukaryotes. Most HSPs are either chaperones assisting protein folding or proteases responsible for degradation of misfolded proteins. FRANZ NARBERHAUS

heat shock element (HSE) Regulatory sequence *upstream of eukaryotic *heat shock genes that is bound by the *heat shock factor.

heat shock factor (HSF) *Transcription factor that binds as a trimer to the *heat shock element *upstream of eukaryotic *heat shock genes. The HSF is phosphorylated as a consequence of *heat shock and

activates *transcription of heat shock genes.
FRANZ NARBERHAUS

heat shock gene Heat shock genes code for heat shock proteins (HSPs) that are induced under *heat shock conditions and counteract protein misfolding and aggregation. The expression of heat shock genes is tightly regulated by a number of transcriptional and post-transcriptional mechanisms, which vary between different organisms. Some heat shock genes are expressed at low levels under all conditions because of their essential role in protein folding.

Eukaryotic heat shock genes are preceded by a *heat shock element that is bound by a *heat shock factor. The *transcription of heat shock genes in *Escherichia coli* and related bacteria is controlled by alternative *sigma factors. Other microorganisms use *repressor proteins to block transcription of heat shock genes at low temperatures or in the absence of misfolded proteins. *Translation of heat shock gene transcripts can be controlled by *RNA thermometers.
FRANZ NARBERHAUS

heat shock promoter Many bacterial *heat shock genes are transcribed from specialized *promoters that are recognized by alternative *sigma factors, e.g. the σ^{32} (RpoH) factor in *Escherichia coli*.
FRANZ NARBERHAUS

helical junction Structural motif in DNA and RNA. Junctions are formed as the meeting point of double-helical arms. The most simple helical junction is composed of just two helical arms that meet in a *junction loop. Particularly in RNA, junctions are also observed where three (→three-way junction), four (→four-way junction, →Holliday junction) or even more helical arms meet. SABINE MÜLLER

helical stack Structural motif in *nucleic acids that is formed by two or more helices interacting by *base stacking along their helical axes (→coaxial stack).

helical turn →helix pitch

helical twist Describes the number of *helical turns in *double-stranded nucleic acids.

helicase →DNA helicase

helix →DNA double helix

helix–coil transition Phase transition of DNA that takes place upon rising temperature. Ideally, the DNA structure is transformed from a double-helical state with lower entropy to a totally unordered coil upon reaching the *melting temperature. The helix-coil transition also depends on charge, size and hydration shell of the cations that are present. JÖRN WOLF

helix–loop–helix (HLH) motif Protein structural motif found in proteins involved in *gene regulation, consisting of a short α-helix connected by a loop to a second, longer α-helix. This motif should not be confused with the *helix–turn–helix motif. The flexibility of the loop allows one helix to fold back and pack against the other. The two-helix structure binds both to DNA and to the HLH motif of a second HLH protein. As with leucine zipper proteins, the second HLH protein can be the same (creating a homodimer) or different (creating a *heterodimer). In either case, two α-helices extending from the dimerization interface make specific contacts with the DNA. Some HLH proteins lack the α-helical extension responsible for binding to DNA. These truncated versions function as negative regulators since they can form heterodimers with full-length HLH proteins, but the heterodimers are unable to bind DNA tightly because they form only half of the necessary contacts. This provides a way to mediate selective pairing among members of a large *transcription factor family involved in cell fate determination. TOBIAS RESTLE

helix pitch Distance traveled along the helical axis for complete 360° rotation. The pitch height relates the number of base pairs in one turn and the nucleotide height, defining the translation per residue (*base pair) along the helical axis.
NINA DOLINNAYA

helix stacking →helical stack

helix–turn–helix motif Protein structural motif consisting of a recognition helix and a stabilizing helix separated by a short loop. Originally identified in bacterial proteins, this motif has since been found in hundreds of *DNA-binding proteins from both eukaryotes and prokaryotes. The two α-helices are held at a fixed angle, primarily through interactions between the two helices. The more C-terminal helix is called the recognition helix because it fits into the *major groove of DNA. Its amino acids, which differ from protein to protein, play an important part in recognizing the specific DNA sequence to which the protein binds. TOBIAS RESTLE

hemimethylated DNA →hemimethylation, →DNA methylation, →DNA methyltransferase

hemimethylation *Methylation state of *DNA sequences in which bases in one strand are methylated and bases in the *complementary strand are not. Hemimethylated sequences perform several important tasks. Most *DNA methyltransferases methylate *palindromic sequences (e.g. GATC from *Escherichia coli*).

$$\begin{array}{c} \overset{CH_3}{} \\ 5'-G\overset{|}{A}TC-3' \\ 3'-CTAG-5' \end{array}$$

After *replication, these sequences are hemimethylated and bound by a special protein (SeqA) to block them for remethylation and obstruct additional rounds of *replication within the same cell cycle. Furthermore, hemimethylation helps the repair protein MutH during *mismatch repair on newly synthesized DNA strands to discern which strand is the newly synthesized one and therefore is to be cleaved and repaired when *mismatches occur.
JÖRN WOLF

hepatitis delta virus (HDV) ribozyme

HDV is a human pathogenic satellite virus that needs the hepatitis B virus to infect a cell. It carries a *single-stranded RNA *genome that replicates by a *rolling-circle mechanism, leading to a long concatemeric antigenomic RNA that is cleaved into monomers subsequently circularized for the rolling-circle amplification of the genomic RNA. Concatemeric genomic *transcripts are also cleaved into monomeric RNA molecules. Transcript cleavage is catalyzed by a small catalytic RNA structure of roughly 85 nucleotides, the HDV *ribozyme, that exists in an antigenomic and a genomic form with remarkably similar *primary and *secondary structures.

Crystal and nuclear magnetic resonance structures of the HDV ribozyme are available. Interestingly, the *active site of the ribozyme lacks a high-affinity divalent metal ion-*binding site as found in other ribozymes. Consistent with this observation, the HDV ribozyme shows high activity even at very low Mg^{2+} concentrations (below 0.1 mM). The *active center shows a pocket consisting of helix I and the loop region LIII, which is completed by the joining region JIV/II and a crossover of helices I and III. At the bottom of this pocket, a C residue (αC76 in the antigenomic form,

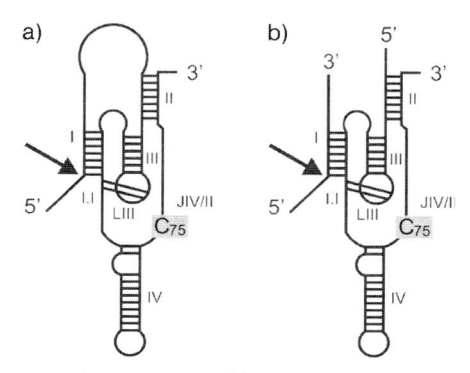

Secondary structure of the genomic version of the HDV ribozyme (a). The 5'-part of helix I can be provided by a RNA molecule added *in trans* which is then efficiently cleaved at the indicated position (b). The cleavage position is indicated by the arrow. The catalytic residue γC75 (grey box) is located in the junction between helices IV and II (JIV/II).

γC75 in the genomic form) in JIV/II plays a key role in catalysis: its N3 function is thought to act as a general base for activating the nucleophilic 2'-hydroxyl at the cleavage site. A hydrated metal ion bound nearby may then act as a general acid by protonating the 5'-oxyanion leaving group. Other catalytic strategies, such as destabilization of substrate binding in the ground state, are still under debate.

The cleavage reaction produces an *upstream RNA molecule ending with a 2',3'-*cyclic phosphate and a *downstream RNA with a 5'-hydroxyl end. Similar to the *Varkud satellite ribozyme, the HDV ribozyme shows almost no upstream sequence requirements (there is a slight preference of A, C and U relative to G at the position 5' of the cleavage site). Therefore, a *cis*-acting (→*cis*-active) HDV ribozyme is often fused downstream of a RNA of interest to generate *in vitro* *transcripts with precisely defined 3'-ends. Transplanting the 5'-part of helix I from the ribozyme to the 3'-end of a second RNA molecule even permits to use the HDV ribozyme as a *trans*-acting *catalyst (→*trans*-active) for the same purpose. Furthermore, HDV variants have been developed for which activity can be efficiently switched on by a *riboswitch or by the addition of imidazole (→Varkud satellite ribozyme, →hepatitis delta virus ribozyme, →hammerhead ribozyme). ROLAND K. HARTMANN, MARIO MÖRL, DAGMAR K. WILLKOMM

heritability Statistical quantity measuring the proportion of phenotypic variation that is explained by the underlying genetic variation of a *quantitative trait. If the genetic variation is the total genetic variance the ratio is referred to as "broad sense" heritability. If the genetic variance is the so-called additive genetic variance (i.e. that due to the *alleles that are transmitted from parents to offspring) the ratio is defined as the "narrow sense" heritability and is of the greatest importance in breeding programmes. MAURO SANTOS

heterochromatin Highly condensed and genetically inactive region of *chromosomes. Traditionally, it has been considered that heterochromatin is a desert in terms of *genes or that contains genes that are not genetically expressed. However, recent studies have shown that heterochromatin performs important cellular functions and carries essential genes: about 450 predicted genes have been identified in *Drosophila melanogaster*. MAURO SANTOS

heterochromatin silencing Transcriptional *gene silencing as the result of relocating a *gene near *heterochromatin (a phenomenon called position-effect variegation). MAURO SANTOS

heterochromosome *Chromosome composed mainly of *heterochromatin.

Heterogametic sex chromosomes are heterochromosomes.

heterodimer Aggregate of two different molecules (e.g. proteins or DNA strands). A *transcription factor, for example, is only active if a dimer of two different molecules with different sequences is formed. MATTHÄUS JANCZYK

heteroduplex *Double strand in which two single strands of different origins are hybridized.

heterogeneous nuclear ribonucleoprotein (hnRNP) Complex consisting of *heterogeneous nuclear RNA and specific *RNA-binding proteins, sometimes designated as a "hnRNA particle". The multimeric hnRNP structure looks like a string of pearls. Mild *RNase A treatment leads to release of 40S monomeric particles. These 40S particles contain about 500 nucleotides of hnRNA and six abundant core proteins, A1, A2, B1, B2, C1 and C2, each with a molecular mass of 30–40 kDa. Furthermore, usually a variety of 10–15 additional proteins are associated with the 40S particle. BERND-JOACHIM BENECKE

heterogeneous nuclear RNA (hnRNA) Eukaryotic nuclear precursor RNA molecules for the mature cytoplasmic *messenger RNA (mRNA), which in turn provides the *template for protein synthesis. *Transcription of hnRNA is by *RNA polymerase II. *Transcripts reveal considerable size variation, ranging from a few hundred to more than 100 000 nucleotides in length. Conversion of hnRNA to messenger occurs within the nucleus, prior to the export into the cytoplasm and mainly involves three steps: addition of the 5'-*cap structure, *splicing of *introns and 3'-*polyadenylation. Capping by the RNA polymerase II holoenzyme complex is a very early event occurring at transcript lengths of less than 10 nucleotides. Removal of non-coding introns is accomplished by the *spliceosome. Polyadenylation is achieved by the poly(A) polymerase complex that recognizes a specific RNA sequence, i.e. the *polyadenylation signal "AAUAAA" of the mRNA precursor. BERND-JOACHIM BENECKE

hexitol nucleic acid (HNA) *Nucleic acid in which the sugar residue is replaced by 1,5-anhydrohexitol. HNA is used for studies of basic biological pathways. Due to high affinity to *complementary RNA, HNAs might be potent *antisense oligonucleotides for therapeutic application. BETTINA APPEL

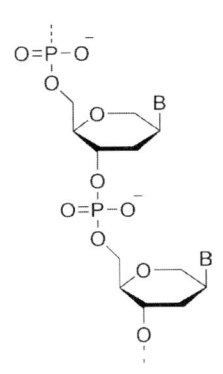

highly repetitive DNA *Nucleotide sequence that is repeated very often in a *genome. It contains short sequences (5–100 bp) that can be repeated up to a million times. In contrast to this, a middle repetitive DNA sequences is repeated only up to 10 000 times. These sequences are often clustered in a specific chromosomal region. MATTHÄUS JANCZYK

Hill coefficient Measure for the cooperativity of multiple *ligand binding to a *biomolecule obtained experimentally from the *Hill plot. The higher the value, the higher the cooperativity. The most famous example is oxygen binding to hemoglobin, which is a tetramer of four identical protein subunits. The cooperativity coefficient is 2.8 for this system. A value of the cooperativity coefficient equal to 1 means that there is no cooperative binding (all binding sites are independent of each other), whereas a value smaller than 1 indicates negative cooperativity. CHRISTIAN HERRMANN

Hill plot Method for determining the cooperativity of a *ligand binding to a *biomolecule with multiple, identical *binding sites. Most often the biomolecule represents a homo-dimer, homo-tetramer or other oligomer (e.g. hemoglobin). Cooperativity means that binding of a ligand changes the ligand affinity at the other binding sites of the complex. Positive cooperativity results in increasing the affinity, whereas negative cooperativity means lowering the affinity for the next ligand. Positive cooperativity leads in extreme cases to the situation that the biomolecule either binds no ligand or is occupied with a ligand at each of its binding sites. Species with less than the maximum number of ligands bound occur at negligible concentrations. Similar to the *dissociation constant for a binary complex, a constant can be defined for a complex consisting of the biomolecule, A, and n ligand molecules:

$$K = c_A \cdot c_L^n / c_{complex}$$

This can be rearranged similar to the dissociation constant which leads to the Hill equation:

$$\Theta = c_L^n / (K + c_L^n)$$

Here, the fractional binding, Θ, corresponds to the concentration of bound ligand divided by the total concentration of biomolecule and the number of binding sites, n. In contrast to the hyperbolic form of the plot of Θ versus c_L for a binary complex ($\rightarrow$ dissociation constant), the shape of the curve is sigmoidal (S-shaped) in the case of multiple, cooperative binding, reflecting lower affinity of the ligand at low concentration that increases after the first binding site has been occupied.

The logarithm of the rearranged Hill equation:

$$\log\{\Theta/(1-\Theta)\} = n \cdot \log(c_L) - \log K$$

allows for a linear plot of $\log\{\Theta/(1-\Theta)\}$ versus $\log c_L$, i.e. the Hill plot. The slope yields the *Hill coefficient or cooperativity coefficient which may equal the number of binding sites at maximum. As this plot is not linear over the full range of ligand concentrations, the slope is determined in the range between 10 and 90% saturation. Above and below, respectively, the slope is almost linear and corresponds to n. CHRISTIAN HERRMANN

hinged DNA → H-DNA

His operon *Operon containing all the *genes necessary for histidine biosynthesis. It is transcribed from a unique *promoter, localized 300 bp *upstream of the first gene. The region corresponding to the untranslated 5'-end of the *transcript, named the *his* leader region, displays the typical features of the T-box transcriptional *attenuation mechanism. This mechanism allows a 10- to 30-fold regulation in *Bacillus subtilis*; 6000-fold regulation of the *his* operon is reached in *Escherichia coli* resulting from the combination of the control of *initiation and *elongation of *transcription. *Transcription initiation is under the control of ppGpp, the effector of the stringent response, and its elongation is regulated by attenuation. In addition, post-transcriptional regulation occurs by *messenger RNA processing and the activity of the first enzyme of the pathway (HisG) could be stimulated or inhibited by several elements. BEATRIX SÜß

histone Major class of proteins associated with *chromatin acting as nucleating structures around which *double-stranded DNA is wound. So far, six classes of histones are known comprising the H1, H2A, H2B, H3, H4, H5 and archaeal histones. Two histones of H2A, H2B, H3 or H4 form the octameric *nucleosomes that form the core to wrap 146 bp of DNA termed the nucleosomes which, spaced by approximately 50 bp of DNA, form the major assembly unit of *chromosomes. Histones can be chemically modified by cellular histone-modifying enzymes, which is related to the control of *gene expression. Histones are found in the nuclei of eukaryotic cells, whereas bacteria lack histones though histones can be found in Euryarchaea. Histones are involved in epigenetic *gene regulation during which they are modified post-translationally thereby altering the mode of DNA condensation. To a major extent this is facilitated via histone methylation which is associated with down-regulation of gene expression, whereas histone acetylation enhances *transcription and, thus, gene expression. The interactions between histones and DNA include the following types: (a) a net positive charge of helix dipoles from α-helices in H2B, H3 and H4 interacts with the negatively charged *phosphate backbone of DNA, (b) *hydrogen bonds between the *DNA backbone and the amine groups of the backbone of histone proteins, (c) non-polar interactions including the *ribose of DNA, (d) ionic interactions and hydrogen bonding involving basic amino acids and phosphate oxygens of DNA, and (e) non-specific *minor groove insertions of the N-terminal domains of H3 and H2B. GEORG SCZAKIEL

HNA →hexitol nucleic acid

hnRNA →heterogeneous nuclear RNA

hnRNP →heterogeneous nuclear ribonucleoprotein

hole hopping In principle, DNA-mediated charge transport processes can be categorized as either the highest occupied molecular orbital-controlled oxidative hole transport or the lowest unoccupied molecular orbital-controlled reductive electron transport. Hole transport in DNA is biologically relevant since it represents one of the primary processes on the way to oxidative *DNA damage which may cause apoptosis, *mutations or cancer. Hole transport on the long range can be described by the hopping mechanism. Among the four different DNA bases, G can be most easily oxidized. Hence, the G radical cation plays the role of the intermediate charge. After photochemical oxidation of the DNA, the positive charge hops from G to G and can finally be trapped at a suitable charge acceptor. If each single hopping step occurs over the same distance then the dynamics of hopping displays a shallow distance dependence with respect to the number of hopping steps. Each hopping step itself is a superexchange process through the intervening AT base pairs, but only if the AT stretch is not too long. The rate for a single hopping step was determined to be 10^6–10^8 s^{-1}. Recently, it was proposed by experimental evidence that *adenines can also play the role of intermediate hole carriers. Such A-hopping can occur if G is not present within the sequential context, mainly in longer AT stretches (at least four AT base pairs). The oxidation of A by the guanine radical cation is endothermic. Once the adenine radical cation has been generated, the A-hopping proceeds fast. The rate of A-hopping has been determined to be 10^{10} s^{-1}. Moreover, it could be shown that hole transport over eight AT base pairs is nearly as efficient as the hole transport over two AT base pairs.

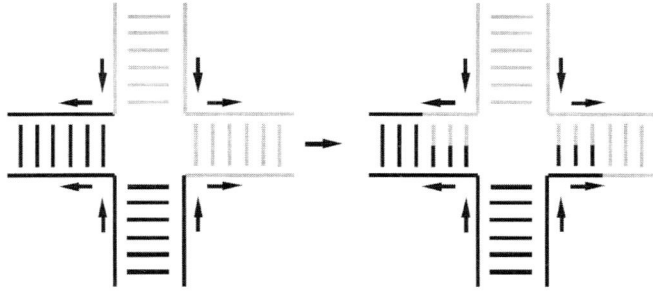

Holliday-junction

In comparison to G-hopping, A-hopping proceeds faster, more efficiently and almost distance independently. →charge transfer in DNA, →electron hopping.
HANS-ACHIM WAGENKNECHT

Holliday junction Dynamic *junction formed of four DNA strands during the process of *homologous recombination. It is named after Robin Holliday, who in 1964 proposed this type of DNA dynamics. Holliday junctions are composed of homologous DNA sequences and thus can slide along the DNA helices (→branch migration). This process is facilitated by the assistance of proteins and consumes energy in form of ATP. Resolution of the junction results in two *duplexes either restoring the parental configuration or leading to genetic *crossing over.
SLAWOMIR GWIAZDA, SABINE MÜLLER

homeobox DNA sequence of 180 bp first found in *genes involved in the regulation of development in the fruit fly Drosophila melanogaster and subsequently identified in many other groups of animals, including mammals. MAURO SANTOS

homeodomain Protein domain encoded by the *homeobox. Homeodomain proteins are a family of *transcription factors characterized by the 60-amino-acid homeodomain that binds to certain regions of DNA and are critical in specifying cell fates.
MAURO SANTOS

homeotic genes *Genes defined by *mutations (homeotic mutants) in which one structure of the body is replaced by another. A well-know example is the *Antennapedia* (*Antt*) mutant in *Drosophila*, which has legs rather than antennae in the head. The homeotic genes generally encode *transcription factors involved in the specification of the body parts. In the fruit fly *Drosophila melanogaster* two regions on *chromosome 3 contain most of the homeotic genes. One is the "Antennapedia complex" where the homeotic genes responsible for the segmental identities of the head and thorax are located: *labial* (*lab*), *Antennapedia* (*Antt*), *sex combs reduced* (*scr*), *Deformed* (*dfd*) and *proboscipedia* (*pb*). The second region of homeotic genes is the "bithorax complex", which contains three genes responsible for the identities of the abdominal segments: *Ultrabithorax* (*Ubx*), *abdominal A* (*abdA*) and *Abdominal B* (*AbdB*). MAURO SANTOS

homidium bromide →ethidium bromide

homo DNA DNA sequence that consists of either *purines or *pyrimidines or even of only one of the four *nucleotides. It is also called *homopolymer.

homoduplex Single-stranded nucleic acid molecule that forms double-stranded regions because of *inverted repeats which are complementary to each other. These double-stranded regions are usually linked by a single-stranded *loop (panhandle).
MATTHÄUS JANCZYK

homologous chromosome In diploid organisms the *chromosomes occur in pairs that are called homologous chromosomes because they contain the same *genes although possibly different allelic forms. MAURO SANTOS

homologous gene Gene with a shared ancestry to another *gene not necessarily on the same *chromosome.

homologous recombination Occurs during meiosis when homologous *chromatids go through a breakage-and-reunion process and exchange some distal portion of their DNA content. MAURO SANTOS

homopolymer Polymer consisting of identical monomer units (→homo DNA, →homoduplex).

Hoogsteen base pairs Double-stranded nucleic acids usually form base pairs in Watson–Crick geometry (→Watson–Crick base pairs). Under certain conditions, however, additional types of base pairs can emerge. An AU pair with *hydrogen bonds in Hoogsteen geometry is shown as an example. This type of base pairing also occurs in DNA triple helices (→triplex-forming oligonucleotides, →triple helix, →edge-to-edge interaction of nucleobases). JENS KURRECK

Hoogsteen edge →edge-to-edge interaction of nucleobases

hotspot Site or a segment in a DNA/RNA sequence with increased activity, e.g. a *mutational hotspot is a site in a DNA sequence where the nucleic acid is significantly more frequently exchanged than at other sites. SLAWOMIR GWIAZDA

housekeeping genes Constitutive *genes transcribed at relatively constant levels whose gene products provide basic functions for the sustenance of the cell. MAURO SANTOS

***hox* genes** *Homeotic genes with an *homeobox are now called *hox* genes for short.

H-phosphonate Short term for a *nucleoside building block used during *oligonucleotide synthesis following the *H-phosphonate method.

H-phosphonate method Method for the synthesis of oligonucleotides (→oligonucleotide synthesis) based on properly protected nucleoside 3'-O-(H-phosphonate) building blocks. Activation is achieved with a sterically hindered acyl chloride, which couples the H-phosphonate to the primary hydroxyl group of the support-bound *nucleoside. The resulting H-phosphonate diester is inert to further phosphitylation, such that the nucleoside chain may be extended without prior oxidation. Oxidation of all phosphorous centers is carried out simultaneously at the end of the synthesis. This opens up useful alternatives for the synthesis of chemically modified oligonucleotides. A serious

H-phosphonate method

side-reaction difficult to control is rapid dimerization of the H-phosphonates. Somewhat lower coupling yields may account for less acceptance of the method compared to the *phosphoramidite method.

HPLC Abbreviation for high-pressure liquid chromatography (→ ion-exchange chromatography, → reversed phase chromatography).

HUGO Abbreviation for Human Genome Mapping Organization (→ Human Genome Project).

human artificial chromosome *Artificial chromosome containing human genomic sequences and designed to be maintained in mammalian (human or murine) cells.

Human Genome Project In 1990, an international enterprise called HUGO (Human Genome Mapping Organization) was started in the USA with the aim to sequence the complete human *genome, and analyze its base sequence and structure. Since then, more than 1000 scientists in 40 countries have contributed to the publicly funded Human Genome Project. The *sequence analysis of the human genome comprising 3.08×10^9 bp was intended to be finished in 2005; however, the first complete analysis was published already in 2001. Meanwhile, more than 2.88×10^9 bp that have been determined with a fidelity of 10^{-5} are accessible via public databases. The Human Genome Project revealed the presence of 20000–25000 genes only; originally, the number of genes within the human genome was estimated to amount 100 000. All of these genes have been located on one of the 46 *chromosomes. Among the genes identified so far, 1500 can be associated to certain diseases. Furthermore, it was detected that even the genomes of two human individuals can differ by 0.1%.

The Human Genome Project significantly contributed to the progress of *sequencing technology. Most of the sequence analysis required for this project was performed based on *Sanger sequencing. In the beginning of the Human Genome Project, automated slab-gel sequencing systems were employed. Already these automata have greatly improved the ease, speed, accuracy and reproducibility of DNA sequencing as compared to the traditional approaches of radio-labeling and autoradiography. In the course of the genome sequencing project, technology moved from slab-gel *electrophoresis to *capillary electrophoresis. These machines enable greater automation of the sequencing process, e.g. for sample loading, *electrophoresis and analysis. Highly automated systems with arrays of 96 capillaries produce thousands of *base pairs in a few hours and, thus, promoted the completion of the human genome much earlier than envisaged. SUSANNE BRAKMANN

Hut operon The *operon *hutPHUIGM* of *Bacillus subtilis* (*hut* = *histidine utilization*) encodes four enzymes responsible for degradation of L-histidine to ammonium, L-glutamate and formamide (histidinase *hutH*, urocase *hutU*, imidazolone propionate aminohydrolase *hutI* and formimino- L-glutamate formiminohydrolase *hutG*). The last *gene, *hutM*, encodes a putative histidine permease. Expression of the operon is regulated by histidine induction and carbon catabolite repression. Histidine-dependent induction is controlled by transcriptional *antitermination at a *stem–loop region located between the *hutP* and *hutH* genes, and a positive regulatory protein, *hutP*, which is encoded in the first *open reading frame of the operon. BEATRIX SÜß

hybrid → DNA–RNA hybrid

hybrid DNA DNA *double strand made of two complementary single strands which have different origins. After successive *denaturation followed by *renaturation new hybrid DNA double strands are formed. MATTHÄUS JANCZYK

hybrid gene Chimeric gene, i.e. a *gene that contains sequences from different genes.

hybridization Process whereby RNA or DNA single strands form a *double strand (e.g. DNA–DNA, DNA–RNA, RNA–RNA) because of *complementary base pairing. With hybridization experiments, *sequence homologies of different sequences can be detected. In this case, a suitably labeled *oligonucleotide probe is used for detection of strand *annealing. The formation of a double-strand depends on diverse reaction conditions. MATTHÄUS JANCZYK

hydrogen bond Special type of attractive interaction that exists between an electron-rich acceptor atom with a lone pair of electrons and a hydrogen atom bonded to an electronegative donor atom. Donor and acceptor atoms are usually fluorine, oxygen or nitrogen. However, a CH group can also participate as donor in hydrogen bonding, especially when the carbon atom is bound to several electronegative atoms, as is the case in chloroform $CHCl_3$. The typical hydrogen bond is stronger than *van der Waals forces, but weaker than covalent or ionic bonds. Although often described as an electrostatic dipole–dipole interaction, the hydrogen bond has also some covalent character as shown by its directionality and interatomic distances shorter than the sum of van der Waals radii as well as the observation of nuclear magnetic resonance scalar couplings across hydrogen bonds. The covalent features are generally more significant when acceptors bind hydrogens from more electronegative donors. Hydrogen bond interactions between *nucleobases are major determinants for the formation of *secondary and *tertiary structures in nucleic acids, and for the binding of many *ligands. KLAUS WEISZ

hydrogen bond acceptor
→ hydrogen bond

hydrogen bond donor → hydrogen bond

hydrophobic interaction In principle, hydrophobic interactions between two *macromolecules bringing them into close vicinity, or leading to complex formation express that they colocalize in an aqueous environment where they are not able to provide hydrophilic interactions with water molecules. Since this type of coaggregation avoids water, it is termed "hydrophobic". Thus, if dispersive interactions would dominate over polar interactions when interacting with other molecules, then the substance would be considered to be hydrophobic in character and, hence, hydrophobic forces can be interpreted as being synonymous with dispersive forces. GEORG SCZAKIEL

5-hydroxymethylcytosine →rare base, →rare nucleotide

6-hydroxypurine →hypoxanthine

hyperchromic effect (hyperchromic shift)
Increase in the ultraviolet light *absorption of a nucleic acid solution when it is denatured by temperature increase or degraded by alkali treatment. This effect is due to the alteration of the electronic interactions between the stacked and hydrogen-bound bases during transition from double-stranded to single-stranded nucleic acids. →hypochromic effect. STÉPHANIE VAULÉON

hyperchromic shift →hyperchromic effect

hyperchromycity Measure of the increase of ultraviolet light *absorption upon *denaturation of nucleic acids in solution at a specific wavelength, usually the absorption maximum. →hyperchromic effect. STÉPHANIE VAULÉON

hypermutation (somatic hypermutation)
Introduction of somatic *mutations in a rearranged immunoglobulin *gene, particularly affecting the antibody binding segments. MAURO SANTOS

hypochromic effect (hypochromic shift)
Decrease in the ultraviolet light *absorption of a nucleic acid solution during formation of *duplexes from single-stranded molecules. The opposite, an increase of absorption, is called *hyperchromic effect. STÉPHANIE VAULÉON

hypochromic shift →hypochromic effect

hypochromicity Measure of the decrease of ultraviolet light *absorption upon *hybridization of nucleic acids in solution at a specific wave length, usually the absorption maximum (→hypochromic effect). The opposite, an increase of absorption, is called *hyperchromicity. STÉPHANIE VAULÉON

hypoxanthine (6-hydoxypurine, 6-oxypurine) Occurs as *rare nucleobase in some transfer RNAs. Attached to ribose via its N9 atom, it constitutes *inosine.

hypoxanthinosine →inosine

I

IDP Abbreviation for inosine diphosphate (→inosine phosphates, →nucleoside phosphates).

IEP →intron-encoded protein

I-genes Complex I-genes are mitochondrial *genes that form a large proportion of the mitochondrial *genome and defects of these genes are increasingly being recognized as important causes of respiratory chain disease. MAURO SANTOS

IGS →internal guide sequence

immobilization of nucleic acids Attachment of nucleic acid to an insoluble organic or inorganic solid phase. Immobilized nucleic acids are of great importance in many fields such as nucleic acid synthesis (→phosphoamidite chemistry), analytics (→biochip) or *in vitro* evolution (→SELEX).

Basically, there are two possibilities: nucleic acids can be either synthesized on the solid phase directly or manipulated and immobilized post-synthetically. Today's synthetic methods (→TBDMS, →TOM, →ACE) all have in common that the desired DNA or RNA is synthesized automatically on a *solid support. Eventually, one obtains an oligonucleotide which is bound covalently to the solid phase, from which it is cleaved off in subsequent purification steps.

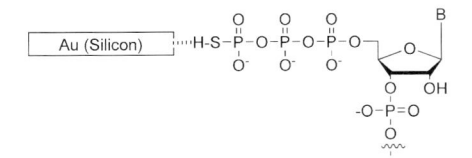

Nucleic acids that carry modifications can be attached to a solid support using strong non-covalent interactions. In the field of biochip arrays, the thiol–gold interaction has proven to be useful. In this method, a thiol group at the 5′-end of a RNA or DNA binds the molecule tightly to a gold-coated silicon support.

Another non-covalent interaction that is used quite often in *in vitro* selection is the *biotin–*streptavidin interaction. Biotin tags can be introduced at the 5′-end by transcription priming. In many selection processes, oligonucleotides (in most cases RNA) that show the desired catalytic activity become labeled with a biotin tag and can be isolated by adsorption on a solid phase that is coated with avidin or streptavidin. *Oligonucleotides that are not biotin labeled can be immobilized by hybridization to an *antisense oligonucleotide carrying a biotin tag. 2′-O-methyl-RNA antisense oligonucleotides have proven to be especially useful for this purpose. JÖRN WOLF

I-motif Tetrameric rectangular-shaped DNA structure formed by antiparallel intercalation of two *cytosine-rich parallel duplexes containing hemiprotonated cytosine-cytosine base pairs.

The four *phosphodiester backbones interact to form two narrow and two wide grooves with close sugar–sugar contacts in the narrow groove. The *deoxyriboses adopt C3′-*endo* conformation. Due to the hemiprotonated *base pairs, the stability of the i-motif is pH-dependent being

Thiol–gold interaction.

R = ribose

at a maximum at a pH value near the pK_a of cytosine. Nevertheless, DNA i-motif stability and its structural bases are yet to be fully understood. In nature this structural motif is found in human telomeric and centromeric DNA structures (→telomere, →centromere). Several proteins bind specifically to C-rich telomeric DNA *fragments probably by recognizing the i-motif structure. The biological significance of this still remains to be investigated. C-rich RNA structures can also form i-motif structures but these proved to be significantly weaker than their DNA equivalents. VALESKA DOMBOS

IMP Abbreviation for inosine monophosphate (→inosine phosphates, →nucleoside phosphates).

imprinting →genetic imprinting

in cell PCR →*in situ* polymerase chain reaction

incision repair →DNA repair

incubation Subjecting of any sample or reaction mixture to defined reaction conditions over a defined period of time.

indicator gene Informative *gene whose activity is easy to detect (→reporter gene).

indirect gene transfer →gene transfer

indirect repeat Short DNA- or RNA-related sequences inversely oriented and generally not tandemly, but closely located in the eukaryotic *genome.

induced-fit model Favored model for enzyme–substrate interaction first proposed by D. E. Koshland in 1958. This model hypothesizes that the reaction between an enzyme and its substrate can occur only following a change in the enzyme's structure, induced by substrate binding. The initial interactions between enzyme and substrate are relatively weak, but they rapidly induce conformational changes in the enzyme that strengthen binding. The theory proposes that (a) a precise orientation of catalytic group is required for enzyme action, (b) the *active site is reshaped by interactions with the substrate and (c) the changes in enzyme structure caused by a substrate will bring the catalytic groups into the proper orientation for reaction, whereas a non-substrate will not. STÉPHANIE VAULÉON

inducer Activator of gene expression (→gene activation, →activator, →repressor).

inducible expression vector *Vector designed to induce the expression of a particular *gene under controllable conditions. *De facto*, it refers to a vector in which the gene of interest is cloned under the control of an *inducible promoter. GEMMA MARFANY, ROSER GONZÀLEZ-DUARTE

inducible gene As opposed to a *constitutive gene, which is usually expressed at steady levels, an inducible gene is expressed under particular conditions that mainly involve positive and/or negative *transcription factors, i.e. *inducers or *repressors, respectively. GEMMA MARFANY, ROSER GONZÀLEZ-DUARTE

inducible promoter *Promoter that responds to particular conditions, thus allowing the expression of the controlled *gene. Once the promoter is recognized by *inducers or *repressors, a shift on the *transcription levels (positive or negative, respectively) is achieved. GEMMA MARFANY, ROSER GONZÀLEZ-DUARTE

induction (activation) of gene expression
Induction is the increase in *gene expression mediated by an *activator, which is in the most cases a *DNA-binding protein that increases the rate of *transcription. *Genes involved in bacterial catabolism (degradation of substrates) are mostly inducible; this means that genes for degrading enzymes are only expressed if the substrate (or a metabolite) is present. Thereby, the bacterium makes sure that the respective enzyme is build only upon demand. BEATRIX SÜß

in frame Alignment according to the actually biologically used *reading frame of a *gene. This state can be disturbed by a deletion or insertion of one or multiple *nucleotides into a sequence, causing a *frame shift. SLAWOMIR GWIAZDA

initial rate The rate of a chemical reaction depends on the mechanism. It is expressed by differential equations, so-called rate laws (→rate constant). They describe the change of concentration with time which is proportional to the respective *rate constant(s) and concentration(s). As the concentration of the reactants decreases with time, the rate becomes slower. Nevertheless, for the beginning of the reaction one can approximate the rate to be constant, i.e. initial rate, which equals in a simple case the product of the rate constant and the initial concentrations of the reactants. Therefore, measuring initial rates allows easy determination of the rate constant and integration of the rate law can be avoided. When initial rates are measured for reversible reactions the backward reaction can usually be neglected since the concentration of the product is low. Measuring enzymatic reactions also implies to measure only the rate of the first 10–20% turnover of the substrate as the change in substrate concentration has an influence on the rate according to the *Michaelis–Menten equation. CHRISTIAN HERRMANN

initiation One of the three steps of *translation in which the translational complex is assembled at the *start codon with the aid of *initiation factors (three in prokaryotes, 10 or more in eukaryotes) (→ribosome). It also refers to initiation of *transcription (→transcription, →transcription initiation). STEFAN VÖRTLER, SABINE MÜLLER

initiation codon →start codon

initiation factor →translation factor, →ribosome

initiator tRNA Protein biosynthesis (→translation) starts always with an N-terminal methionine or in prokaryotes, mitochondria and plastids with a formylated methionine (→formylmethionyl-tRNA), which becomes deformylated once the nascent polypeptide is made. All are incorporated by specialized initiator *transfer RNA (tRNA$_i$) just used for *decoding the AUG *start codon during *intitiation of translation. tRNA$_i$ are aminoacylated by the canonical MetRS (→aminoacyl-tRNA synthetases) like the tRNA^{Met} used during *elongation, but are distinct from it as they interact with initiation rather then elongation *translation factors and enter the *ribosome at the P- instead of the A-site. Formylmethionine is a positive determinant for IF binding, while eukaryotes rely on structural features like the unique U50–A64 and G51–C63 *base pairs in vertebrate tRNA$_i$. Mutation to A50–U64 and U51–A63, the situation found in conventional *elongator tRNA, reduces eIF and promotes eEF1A binding.

Moreover tRNAi contains three consecutive GC base pairs in the stem just above the *anticodon loop, restricting its flexibility and supporting *P-site interaction as any change abolishes binding.
STEFAN VÖRTLER

in-line attack Description of the conformation of reacting groups during *catalysis mediated by small *ribozymes. Following the S_N2 mechanism, the formation of the attacking 2'-hydroxyl group, the scissile phosphate and 5'-hydroxyl leaving group are oriented in-line. DENISE STROHBACH

in-line probing *In vitro* method used to infer structural information about RNA sequences. It is based on the relative instability of RNA *internucleotide linkages due to the fact that the 2'-hydroxyl group can act as an internal nucleophile, attacking the neighboring phosphate in an S_N2 reaction. This requires the attacking 2'-oxygen, the phosphorous and the leaving 5'-oxygen to be in-line. In a typical *A-form RNA double-helix the internucleotide linkages are fixed in a conformation that impedes in-line attack, while single-stranded regions are usually flexible and more likely to pass through an in-line conformation. Therefore, the pattern of spontaneous cleavage for a given RNA in solution (with slightly elevated pH) allows conclusions about its structure. In-line probing has proven to be particularly useful in characterizing *riboswitches, where it is used as a simple method to monitor *ligand-induced changes of *RNA structure in protein-free solutions. RÜDIGER WELZ

inner sphere Direct coordination of a metal ion to a ligand.

inosine β-Glycosidic nucleoside of *hypoxanthine. Inosine is a constituent of the *anticodon of certain *transfer RNAs (→rare nucleotides).

inosine phosphates Phosphoric acid esters of *inosine. Inosine phosphates occur as monophosphate (IMP), diphosphate (IDP), triphosphate (ITP) as well as 3',5'-cyclic monophosphate (cIMP) (→nucleoside phosphates). SABINE MÜLLER

insert DNA segment which is inserted into a *vector by *ligation (→insertion, →insertion element).

insertion Rare, non-reciprocal type of *translocation of a *DNA fragment that is removed from a certain position of a *chromosome and inserted into a broken region of a non-homologous *chromosome. MATTHÄUS JANCZYK

insertion element Discrete DNA segment that can be inserted into *chromosomes, phages or *plasmids. Insertion can occur at defined positions of DNA or at random. Insertion of these elements always produces a *mutation, while their excitation results in a loss of host *genetic information. IS (insertion sequence) elements have a length of 700–1400 bases and contain *genes that are related to insertion function. TN elements (→transposon elements) are larger than 1400 bases and contain genes which are unrelated to insertion function. MATTHÄUS JANCZYK

insertion mutation *Mutation caused by the addition of one or more DNA or RNA residues. Short insertions typically arise during *DNA replication or repair by local *primer/*template slippage, long ones by recombinational events. Insertion mutations can also be caused by *mobile genetic elements such as IS (insertion sequence) sequences and *transposons. →deletion mutation, →substitution mutation. HANS-JOACHIM FRITZ

in situ hybridization Technique for detecting *messenger RNA molecules by *hybridization of *complementary oligonucleotides. This method enables precise localization and identification of individual cells that contain a specific *nucleotide sequence. It involves the hybridization of a specific *nucleic acid probe with a specific nucleic acid target within a tissue section. The technique allows identifying the location of cells that contain the nucleic acid sequence of interest. The requirements on the hybridization conditions are extensive. The sequence should not be too long to avoid hybridization of non-target sites. On the other hand, a long sequence decreases hybridization time. The required *denaturation temperature is 94°C; the hybridization temperature depends on T_m and stringency. A low temperature favors *mismatches, whereas at a too high temperature no hybridization takes place. Other important factors are pH and ion concentration. MATTHÄUS JANCZYK

in situ polymerase chain reaction (PCR) Developed as an aid to *in situ hybridization protocols which have a lower sensitivity limit of about 20 copies of *messenger RNA (mRNA) per cell. In situ PCR (in cell PCR or in situ RT-PCR) is a method for *amplification of mRNA prior to in situ hybridization experiments, the whole process taking place inside cells. Like *reverse transcription polymerase chain reaction (RT-PCR) it is a two-stage process involving first *transcription of the mRNA into *cDNA by a *reverse transcriptase followed by PCR of the resultant cDNA. Prior to in situ PCR, cells need to be fixed and permeabilized to preserve cell morphology and to permit access of the PCR reagents to the intracellular mRNA sequences to be amplified. This process can be carried out either in tubes or on glass plates. The cells are then overlaid with the RT-PCR mix, and the mRNA is transcribed and amplified. The resulting DNA can then be used for various in situ hybridization experiments; it has been used for the detection of single-copy nucleic acid sequences (→single-copy sequence) in cell preparations, and of low copy DNA and RNA sequences in tissue sections. These include the detection of viral or proviral DNA, but it has also been used to identify human *single-copy genes. DAVID LOAKES

in situ RT-PCR →in situ polymerase chain reaction

interaction of nucleic acids with metal ions Nucleic acids carry a high negative charge originating from the bridging *phosphodiester groups of the *backbone and are thus always associated with cations, mainly metal ions. Aside from the *phosphate oxygens, also the *nucleoside residues offer a multitude of possible coordination sites for metal ions, the *purine N7 and carbonyl oxygens being most important. However, M^{n+} ions can coordinate to virtually all N and O atoms in nucleic acids either directly (→inner sphere) or by hydrogen bonding via a coordinated water molecule (→outer sphere). Which type of interaction prevails depends on the kind of metal ion, its size and preference for N or O donors as well as the geometric arrangement of the donor atoms and the size of the *binding pocket (*Coord Chem Rev* **2007**, *251*, 1834). Most metal ions associated with nucleic acids, e.g. Na^+, K^+ and Mg^{2+}, but also Mn^{2+} and Cd^{2+}, bind kinetically labile

Mg^{2+} ion showing inner-sphere coordination to N7 of guanine, as well as outer-sphere binding to O6 and a phosphate oxygen.

(i.e. fast ligand exchange). Only about 10% of the total number of M^{n+} ions needed for charge compensation are located at certain sites, but the vast majority are only loosely bound to the nucleic acid. The specifically bound ions usually show affinities in the range of $K = 10^2–10^4$ M^{-1}, i.e. much lower than in proteins.

Na$^+$ and K$^+$ are the most abundant metal ions freely available in the cell, and thus mainly responsible for charge compensation of nucleic acids. However, both ions can also be distinctly coordinated. Tightly bound Na$^+$ is part of the *spine of hydration in *B-DNA, where it replaces water molecules. K$^+$ stabilizes G and T/U *quadruplex structures in DNA, and some RNA local structural motifs (e.g. *AA-platform) coordinate specifically M$^+$ ions.

In *ribozymes, like in any other complex RNA (e.g. *transfer RNA), specifically bound divalent metal ions are needed for stabilization of local structures, tertiary contact formation, as well as the catalytic process itself. In a typical *phosphodiester cleavage reaction of *ribozymes, Mg^{2+} can activate the nucleophile (a water molecule or a ribose 2'-hydroxy group), electrostatically influence the ground state, stabilize the *transition state or coordinate to the *leaving group, accelerating in all cases the reaction (*Chem Rev* **2007**, *107*, 97). Often the concerted action of several metal ions is required for optimal performance.

The exact positioning of metal ions is crucial for correct functioning of nucleic acids. For example, in *mononucleoside 5'-triphosphates, like *ATP, coordination of two M^{2+} in an α/β and γ fashion leads to cleavage of the terminal γ-phosphate group (kinase activity), whereas an α and β/γ coordination catalyzes the transfer of a nucleotidyl unit (polymerase activity). Generally, *binding sites in more complicated RNA or DNA structures show a high selectivity and specificity for certain M^{n+} ions despite the small number of building blocks (4 nucleotides versus 20 amino acids in proteins). Both properties are achieved by choice of sequence and formation of perfect *binding pockets. Examples are a *riboswitch specific to Mg^{2+} sensing the concentration of this ion in the cytoplasm, the strongly accelerated self-cleavage reaction of the *hammerhead ribozyme in presence of Mn^{2+} and *DNAzymes that have been selected specifically for one kind of metal ion, e.g. Pb^{2+}, opening up their possible use as sensors.

Kinetically inert coordination (i.e. very slow ligand exchange) of metal ions to nucleic acids is not known to occur in nature. However, some of the most widely used anticancer drugs for chemotherapy, e.g. *cisplatin and its derivatives, exhibit their function by coordinating to the N7 positions of two consecutive purine *nucleobases in DNA, thereby inhibiting DNA *replication. ROLAND K. O. SIGEL

intercalation Insertion of atoms or molecules in pre-existing structures, e.g. nucleic acids or membranes. Intercalating molecules are called intercalators or intercalating agents. Intercalators inserting nucleic acids are mostly aromatic, polycyclic molecules like *ethidium bromide, bisbenzimide, proflavin, thalidomide or acridine orange. They interact with *double-stranded DNA or RNA molecules via insertion in between two adjacent *base pairs, thereby distorting the stacked bases and subsequently the double-helical structure of the double-stranded DNA or RNA. As intercalation causes structural changes, it can interfere with *DNA replication, DNA *transcription and lead to *mutations. Intercalators are therefore *carcinogens and *mutagens; proflavin and acridine orange, for example, are known to cause *frame shift mutations. Some DNA intercalators are used as chemotherapeutics to inhibit DNA replication in cancer cells. Others like *ethidium bromide are used as staining agents. Nucleic acids themselves can serve as intercalators, either competitively or by forming a *triple helix. VALESKA DOMBOS

intercalator →intercalation

intercistronic areas *Cistron (a term coined by Seymour Benzer) is also used equivalently to *gene in comprising a unit of inheritance associated with regulatory regions, transcribed regions and other functional sequence regions. Intercistronic areas are regions of DNA with no cistrons. →intergenic region. MAURO SANTOS

interference mapping →nucleotide analog interference mapping

intergenic region Complete DNA sequence between protein-coding regions of two successive *genes.

internal guide sequence (IGS) RNA sequence that binds *introns to form a *splicing complex and to initiate splicing (→guide sequence).

internal loop By definition, internal loops are formed if internal *nucleotides on either side of the *double helix cannot pair to form a *Watson–Crick or GU base pair. By introducing distortions in the helical geometry, the RNA can adopt novel shapes. Such novel shapes are key features in molecular recognition processes. Furthermore, internal *bulges increase the inherent flexibility of domains in the RNA molecule by decreasing the stability of long helical stretches.

As a general common feature in internal loops with an equal number of nucleotides, each pair exhibits non-Watson–Crick *hydrogen bonds and leads to a distortion of the A-form helix (→A-form RNA) that can be expressed as a difference in the P–P distance, a distortion of the *helical twist and as a bending in the helix axis.

The decrease in stability is intimately linked to the structural features of the *loops. There is not only a dependence of the thermodynamics and conformation on the loop nucleotides itself, but also on the sequence context. Unequal numbers of bases on both sides of internal loops lead to unstacking of single nucleotides. Often, unstacked U nucleotides are observed that point towards the solvent. Examples include the P4–P6 domain of the *group I ribozyme and in the RRE element of HIV (→RRE RNA).

Internal loops may act as sites for *tertiary interaction or for RNA–protein recognition (→RNA–protein interaction). In some cases, they may just form if such a tertiary interaction is present – as demonstrated for the internal loop J6a/6b in the group I intron ribozyme that adopts its native shape just within the presence of tertiary interaction with consecutive A nucleotides several positions *downstream the sequence in the same RNA strand (formation of an *A-platform) and that is structured differently in the absence of these tertiary interaction partners. BORIS FÜRTIG, HARALD SCHWALBE

internal ribosome entry site (IRES)
Sequence that can be located on many different positions on an *messenger RNA transcript and allows for *translation initiation far away from the 5'-end. IRES are usually several hundred nucleotides long and gather some of the proteins that are involved in the 5'-cap-dependent translation and are thus recognized by the ribosomal translation machinery. IRESs can be observed in eukaryotic bicistronic transcripts, in which the *gene located at the 5'-end is translated by a 5'-cap-dependent mechanism and the gene that is controlled by the internal ribosome entry site can be translated independently from the 5'-*cap structure. Homologies concerning primary or *secondary structure of the different IRES sequences have not been found yet. JÖRN WOLF

internucleotide linkage →phosphodiester bond

interstrand annealing →annealing of oligonucleotides

interstrand crosslinking →crosslinking of DNA

interval mapping Set of procedures using two adjacent genetic markers to estimate genetic effects and *genome locations of *genes controlling *quantitative traits. MAURO SANTOS

intervening sequence (IVSs; introns and intein-coding sequences) Parts spliced from precursor mRNA (→pre-mRNA) or polypeptide, respectively.

intramer *Aptamer that is used to act intracellularly to target viral proteins, for example. Intramers can be modulated to bind to protein targets with high affinity and selectivity. Moreover, they can be expressed in the cell, conserving their function in the cell's environment. This strategy has been used in treatment of HIV where intramers bind to certain HIV target proteins (HIV-1 *reverse transcriptase, integrase and the Rev protein), obstructing the viral life cycle. JÖRN WOLF

intrastrand annealing Occurs when *double-stranded DNA or RNA molecules contain *inverted repeat sequences which after denaturing and re-annealing form *Watson–Crick base pairs with themselves instead of with the second strand of the original molecule. The resulting molecule usually consists of a base-paired stem and a *loop with a varying number of unpaired bases. These structure elements can be found in *transfer RNA, for example. VALESKA DOMBOS

intrastrand crosslinking →crosslinking of DNA

intron *Gene sequence separating *exon segments. Introns do not contain any useful *genetic information, e.g. are protein non-coding sequences. The number of introns in a gene can vary greatly, but some genes contain more than 50 introns. The introns are usually longer than the exons. Following *transcription the introns are spliced out in a process called *splicing, leaving only the protein-encoding regions in the *pre-mRNA.

Recently, metabolic-binding sites, so-called *riboswitches, have been identified in intron regions in the fungal *genome *Neurospora crassa*. These riboswitches are thought to interfere with the splicing process. TINA PERSSON

intron drift →intron sliding

intron-encoded gene *Gene located within the sequence of *group I or *group II introns and encoding a protein (→intron-encoded protein) that participates in *intron mobility. STÉPHANIE VAULÉON

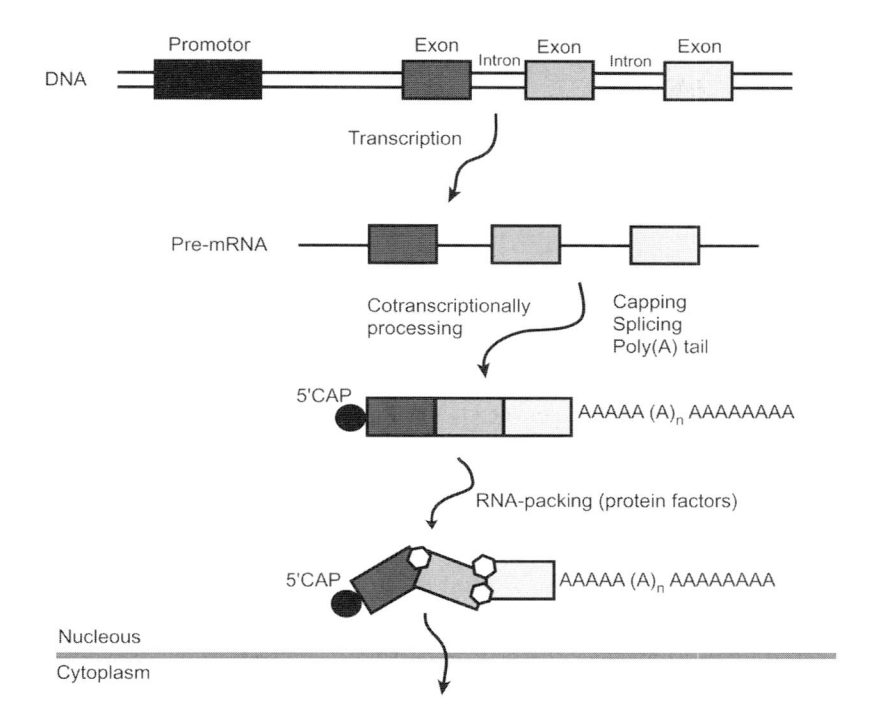

intron-encoded protein Protein encoded within the sequence of *group I or *group II introns and participating in *intron mobility.

intron homing →intron mobility

intron migration →intron sliding

intron mobility (intron homing) Process in bacteria in which *introns (of either *group I or *group II) spread, in replicative fashion, to intron-less *genes. The mechanism of intron homing differs for the two types of introns. Intron homing in group I introns is initiated when an intron-encoded DNA endonuclease (→intron-encoded protein) makes a double-stranded cut at the potential insertion site in the intron-less gene. The lesion is then repaired by enzymes which use the *intron-encoded gene as a *template, so that a copy of the intron is synthesized in the insertion site. For group II introns, intron homing is called *retrohoming because of the involvement of a reverse-transcribed copy of the intron in the process. STÉPHANIE VAULÉON

intron shuffling Altering or rearranging the given array of *introns of a gene such that new combinations of introns are generated (→exon shuffling, →shuffling).

intron sliding Relocation of *intron–*exon boundaries over short distances; also referred to as intron slippage, migration or drift. Intron sliding can occur using several mechanisms. STÉPHANIE VAULÉON

intron slippage →intron sliding

inverse repetitive sequence →inverted repeat, →palindrome

inverted repeat A sequence of DNA repeated in reversed orientation on the same strand. For example, 5′–ACACAT–ATGTGT–3′. It is called a *palindrome when there is no intervening sequence.
MAURO SANTOS

in vitro (Latin: "in glass") Describes reactions carried out in test tubes, outside the cell.

***in vitro* evolution of nucleic acids** *In vitro* evolution mimics Darwinian evolution, with the limitation that "survival-of-the-fittest" is carried out at the level of individual molecules.

A starting pool may contain up to 10^{15} different DNA or RNA sequences that have been prepared by chemical synthesis (→oligonucleotide synthesis) or *in vitro* transcription. All molecules contain regions of defined sequence that are necessary for molecular replication and regions that may either be of a particular length of random sequence or may be based on a particular sequence with mutations artificially introduced at a desired frequency. The pool then is subjected to the desired selective pressure. Surviving sequences are amplified and retransferred into subsequent pools for repeated rounds of selection and *amplification (→SELEX). SABINE MÜLLER

***in vitro* selection of nucleic acids** →SELEX

***in vitro* transcription** Cell-free *transcription used to transcribe *genes into their corresponding *transcript, using specially prepared cell extracts and specific transcription vectors. Such vectors contain promotors for *RNA polymerases flanking polycloning sites. Any *foreign DNA inserted into one of the polycloning sites will be transcribed under control of the *promoter. The transcript accumulates to high concentrations. Simple assays where a DNA template is transcribed into RNA by RNA polymerase in the presence of an appropriate reaction buffer are also called *run-off transcription assays. STÉPHANIE VAULÉON

***in vitro* translation** Cell-free *translation used to translate isolated and purified *messenger RNAs (mRNAs) into their corresponding proteins. The mRNA preparations are mixed with cell extracts from, for example, *Escherichia coli* for prokaryotic mRNA translation or rabbit reticulocyte lysates for eukaryotic mRNA translation. These extracts are devoid of endogenous mRNA, but contain all the components required for translation, like the ribosomal subunits, all *transfer RNA molecules, aminoacyl tRNA synthetases, as well as *initiation, *translocation and *elongation proteins or energy sources.
STÉPHANIE VAULÉON

in vivo (Latin: "in living") Describes reactions carried out inside the cell and therefore also refers to all the biological processes taking place in a cell.

***in vivo* gene transfer** In the context of *gene therapy, refers to the transfer of a therapeutic *gene directly into the patient (i.e. systemically or topically) without the need for removing target cells from the body (→gene transfer).

***in vivo* structure mapping of RNA**
Structure-specific chemical probes (→nucleic acid probe) are unique tools to map *RNA structure *in vivo* under different cell growth conditions. The use of probes is limited by their inability to penetrate the cell wall and membrane due to size structure and/or charge. The reagents that have gained widespread use for *in vivo* RNA probing are dimethylsulfate and lead(II)-induced cleavage.
PASCALE ROMBY, PIERRE FECHTER

ion-exchange chromatography Chromatographic technique that separates molecules depending on their charge and, together with *reversed-phase chromatography, the chromatographic method of choice for purification of chemically synthesized nucleic acids. Since they carry a net negative charge because of the *phosphodiester backbone, anion-exchange chromatography is preferable over cation-exchange chromatography. Columns used for nucleic acid purification have resins carrying positively charged quaternary amine residues. When loaded onto the column, nucleic acids compete with negatively charged ions in the buffer for *binding sites on the positively charged resin. The longer a nucleic acid, the more negative is its overall charge and the tighter it binds to the solid phase. When the concentration of negatively charged ions in the eluent is slowly raised during purification (e.g. ClO_4^-), *oligonucleotides are eluted depending on their charge and therefore on their length. Since denaturing agents like formamide or urea are always used in the eluents, purified *oligonucleotides have to be separated from these compounds by *dialysis or *gel filtration. JÖRN WOLF

IRES →internal ribosome entry site

***Ir* (immune response) genes** Involved in the genetic control of antibody responses.

isoacceptor All *transfer RNAs (tRNAs) which become aminoacylated with the same amino acid and belonging to the same *codon family (→wobble hypothesis). Different members of an isoacceptor family are usually present in different cellular concentrations reflecting the *codon usage of the organism which splits the codon family in *major (mostly used for decoding) and *minor tRNAs (very infrequently used). STEFAN VÖRTLER

isoallele *Allele of a *gene that is considered normal, but can be distinguished from another allele by its differing phenotypic expression.

isocyanate Organic molecule containing the isocyanate functional group (NCO) that can be reacted with *oligonucleotides containing aliphatic amino groups. This reaction connects the organic molecule to the oligonucleotide via a urea linkage. SNORRI TH. SIGURDSSON

$$R-N=C=O \xrightarrow{Oligo-NH_2} Oligo-\underset{H}{N}-\overset{O}{\underset{}{C}}-\underset{H}{N}-R$$

isostericity matrices Twelve basic geometric families of *nucleobase interaction have been observed in RNA crystal structures. Isosteric matrices as suggested by Eric Westhof et al. summarize the geometric relationships between the 16 pairwise combinations of the four natural bases for each of the 12 base-pairing families. →geometric nomenclature and classification of RNA base pairs, →edge-to-edge interaction of nucleobases. SABINE MÜLLER

isothiocyanate Organic molecule containing the isothiocyanate functional group (NCS) that can be reacted with *oligonucleotides containing aliphatic amino groups. This reaction connects the organic molecule to the oligonucleotide via a thio urea linkage. SNORRI TH. SIGURDSSON

$$R-N=C=S \xrightarrow{Oligo-NH_2} Oligo-\underset{H}{N}-\overset{S}{\underset{}{C}}-\underset{H}{N}-R$$

isotope labeling Incorporation of isotopes into DNA and RNA (→oligonucleotide labeling). Examples include ^{32}P (used for *radioactive labeling), ^{13}C and ^{15}N (used extensively for studies of nucleic acids with *nuclear magnetic resonance spectroscopy). SNORRI TH. SIGURDSSON

ITP Abbreviation for inosine triphosphate (→inosine phosphates, →nucleoside phosphates).

J

junk DNA Collective label for those DNA sequences in the *genome with unknown function. "Adaptive" versus "junk DNA" theories are two competing theories of genome-size evolution that intend to solve the puzzle of the large fraction of *non-coding DNA (up to 98% in the human genome) found in the *genome of most species, particularly eukaryotes (organisms in which the genetic material is organized into a nucleus). The "adaptive" theories postulate a function for the *extra DNA, while the "junk DNA" theories simply state that the extra DNA is indeed superfluous, useless and maladaptive. Expansions in genome size of eukaryotes mainly reflect changes in non-coding regions associated to *introns, *mobile genetic elements and other forms of intergenic DNA (→intergenic region). Recent ("junk DNA") theory suggests that this is due to the low efficacy of natural selection for removing those useless sequences in species where population size is small. Since population size scales with life complexity (population size is substantially larger in prokaryotes than in eukaryotes), the lower non-coding content of prokaryotes could be readily explained by the higher efficiency of natural selection in eliminating maladaptive DNA when population size is large. MAURO SANTOS

junction →helical junction, →three-way junction, →four-way junction, →Holliday junction

junction loop Structural motif occurring in *branched nucleic acids. Branched nucleic acids consist of double-helical arms, which are connected at the *junction point. The junction point often contains a number of unpaired bases connecting the individual double helical arms, thus forming a *loop structure (→three-way junction, →RNA structures). SABINE MÜLLER

K

kilobase pairs Unit for the number of base pairs in a DNA double strand (1000 bp).

kinase →polynucleotide kinase

kinetoplast DNA DNA network of the kinetoplast organelle (located at the base of the flagellum) of the kinetoplastid unicellular parasites, such as *Trypanosoma*. This catenated network contains up to 50 DNA maxicircles of 20–35 kb in length (containing *genes for energy production) and 10 000 DNA minicircles of 0.5–1.5 kb (involved in *RNA editing). MAURO SANTOS

kink In a DNA/RNA oligomer a kink distorts the *stacking of the adjacent bases. It is therefore most often not favorable, because the energy created by stacking interactions is lost (→base stacking). In organisms, kinks in *double-stranded DNA are often induced by *intercalation of protein factors. Therefore, kinks in DNA play often an important role in the control of *gene expression. SLAWOMIR GWIAZDA

kink DNA →kink

kink-turn motif (K-turn motif) RNA *secondary structure motif. The motif adopts a stem–bulge–stem structure that serves as *recognition element for proteins. The canonical stem C ends in two GC *base pairs. The non-canonical stem NC consists of two GA base pairs followed by a random base pair and a GC base pair. Both stems are connected through a three nucleotide long asymmetric internal bulge.

Interactions of stem C, stem NC and the two first *nucleotides of the *bulge lead to a bend of the *phosphodiester backbone

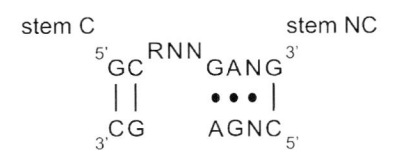

```
      stem C              stem NC
    5'   RNN           3'
      GC      GANG
      ||      •••|
      CG      AGNC
    3'                   5'
```

N = any nucleotide; R = purine

and protrusion out of the strand of the third unpaired nucleotide. The RNA structure adopts a V-shape with an asymmetric *internal loop. VALESKA DOMBOS

kissing hairpins →kissing loops

kissing loops *RNA secondary structure sequence motif in which the two *terminal loops of *hairpin structures interact with each other in form of *Watson–Crick or *wobble base pairs. SLAWOMIR GWIAZDA

Klenow fragment Large fragment (67 kDa) of *Escherichia coli* *DNA polymerase I that has two enzymatic activities. The DNA polymerase activity catalyzes the condensation of *nucleoside triphosphates with the hydroxyl group at the 3'-end of the *nascent DNA strand under release of *pyrophosphate. The 3'–5' exonuclease activity catalyzes the hydrolysis of *nucleotides with wrong *base pairs at the 3'-end of the nascent DNA strand. →Kornberg polymerase. ANDREAS MARX, KARL-HEINZ JUNG

K_M determination K_M is determined from the dependence of the rate of product formation, v, on the concentration of the substrate, c_S (→Michaelis–Menten model). A plot of v versus c_S results in a hyperbolic shape and a non-linear fit to the data

Nucleic Acids from A to Z: A Concise Encyclopedia. Edited by Sabine Müller
Copyright © 2008 WILEY-VCH Verlag GmbH & Co. KGaA, Weinheim
ISBN: 978-3-527-31211-5

according to the *Michaelis–Menten equation yields the K_M value. Alternatively, linearized forms of this equation can be used like the *Lineweaver–Burk plot or *Eadie–Hofstee plot.

Experimentally, the rate of product formation is measured for a set of different substrate concentrations which should cover at least one (better two) order(s) of magnitude and which, more importantly, should be selected to lie below and above the K_M value. As the rate decreases with decreasing substrate concentration measurement of the rate should only be done within the first 10–20% of substrate turnover (→initial rate). The concentration of the enzyme is usually selected to be the same in all experiments; it can be adapted to the time resolution of the experiment. Exact knowledge of the concentration of the enzyme is not important for the determination of its K_M value.

For various technical reasons it is often not possible to measure the rate close to or above the K_M value (e.g. the rate becomes too fast or required substrate concentrations are too high). Nevertheless, at low substrate concentrations (much smaller than the K_M value) the Michaelis–Menten equation can be approximated by $v = c_S \cdot c_E \cdot k_{cat}/K_M$, which means the ratio k_{cat}/K_M can be obtained from the linear slope of the plot of the rate versus substrate concentration. The term k_{cat}/K_M can be regarded as a second-order *rate constant. Even though neither K_M nor k_{cat} can be obtained separately under these conditions the ratio k_{cat}/K_M represents a useful number, i.e. the catalytic efficiency of the enzyme in respect to the respective substrate (the higher k_{cat} and the lower K_M, the higher the efficiency). As this number also allows us to compare competing substrates the k_{cat}/K_M ratio is also called the specificity constant. CHRISTIAN HERRMANN

K_M value The *Michaelis constant K_M characterizes the apparent affinity of an enzyme/substrate couple. The lower the value the higher the apparent affinity of the enzyme for the substrate. According to the *Michaelis–Menten equation the catalytic activity (rate of product formation) increases with substrate concentration. Formally, half maximum activity is reached at a substrate concentration equal to the K_M value (→K_M determination, →Michaelis–Menten model). CHRISTIAN HERRMANN

Kornberg polymerase *Escherichia coli* *DNA polymerase I (Kornberg polymerase) was the first DNA polymerase discovered by Arthur Kornberg in 1957. It harbors three enzymatic activities in one enzyme. Through partial proteolytic digest the enzyme can be cleaved into two fragments. The small fragment (36 kDa) has 5′–3′ *exonuclease activity; the large fragment (67 kDa, →Klenow fragment) has 3′–5′ exonuclease and DNA polymerase activity. *E. coli* DNA polymerase I is involved in *DNA repair, the hydrolysis of the *RNA primers and their replacement by DNA. ANDREAS MARX, KARL-HEINZ JUNG

K-turn →kink-turn motif

Kunkel mutagenesis Method used for the induction of *mutations in a *single-stranded DNA (also called deoxyuridine mutagenesis, dUTP system mutagenesis). It is a site-specific mutagenesis that is used for selection of mutated *genes from wild-type. BETTINA APPEL

L

labeling → oligonucleotide labeling

Lac operator → Lac operon

Lac operon *Operon containing all of the *genes required for the transport and metabolism of lactose in *Escherichia coli* and some other enterobacteria. The three structural genes are *lacZ*, *lacY* and *lacA*. *lacZ* encodes β-galactosidase (LacZ), an intracellular enzyme that cleaves the disaccharide lactose into glucose and galactose. *lacY* encodes β-galactoside permease (LacY), a membrane-bound transport protein that pumps lactose into the cell. *lacA* encodes β-galactoside transacetylase (LacA), an enzyme that transfers an acetyl group from acetyl-CoA to β-galactosides. Only *lacZ* and *lacY* appear to be necessary for lactose catabolism.

The regulatory response to lactose requires an intracellular regulatory protein called the lactose *repressor LacI. The *lacI* gene encoding the repressor lies nearby the *lac* operon and it is always expressed from its own *promoter (referred to as constitutive expression). If lactose is missing from the growth medium, the repressor binds very tightly as a tetramer to a short DNA sequence just *downstream of the promoter near the beginning of the *lacZ* gene called the *lac* operator. Repressor bound to the operator interferes with binding of *RNA polymerase to the promoter, and therefore *messenger RNA encoding LacZ and LacY is only made at very low levels. When cells are grown in the presence of lactose, a lactose metabolite called allolactose binds to the repressor, causing a change in its conformation. Thus altered, the repressor is unable to bind to the operator, allowing RNA polymerase to transcribe the *lac* genes and thereby leading to high levels of the encoded proteins. Expression of the *lac* operon shows also catabolic sensitivity. When glucose level drops the second messenger cAMP binds then the cAMP repressor protein Crp, which is then activated to bind to the *crp* *binding site and acts as *activator of the operon. BEATRIX SÜß

Lac promotor → Lac operon

Lac repressor → Lac operon

***lacZ* gene** First *gene of the *Lac operon and encodes β-galactosidase. This enzyme cleaves the disaccharide lactose in glucose and galactose. The enzyme is not lactose specific and can also act on simple galactosides including 2-nitrophenyl-β-D-galactopyranoside (ONPG). The *lacZ* gene is widely used as a *reporter gene to study promoter activity in different tissues or developmental stages by attaching it to a *promoter region of particular interest as transcriptional or translational fusion. The phenotypic expression is easy to monitor by hydrolysis of colorless ONPG in galactose and the yellow chromagenic compound o-nitrophenol. BEATRIX SÜß

lagging strand DNA strand that is synthesized discontinuously in short fragments (→ Okazaki fragments). The synthesis in short fragments is caused through the exclusiveness of *DNA polymerases to catalyzed DNA synthesis in the 5'–3' direction. After hydrolysis of the *RNA primers the

fragments are coupled by a *DNA ligase.
ANDREAS MARX, KARL-HEINZ JUNG

Lambert–Beer law → Beer–Lambert law

large ribosome subunit → ribosome

lariat Circular intronic RNA element generated during the *splicing reaction of *pre-mRNA. Following the nucleophilic attack of the 2'-hydroxyl group of the *branch point A-nucleotide to the 5'-splice site, a 2',5'-*phosphodiester bond is formed resulting in a circular RNA structure. After cleavage at the 3'-splice site about 30 nucleotides *downstream of the branch point, the *intron is released in form of a lariat. Subsequent degradation of that lariat requires a specific *ribonuclease.
BERND-JOACHIM BENECKE

laser scanning microscopy → scanning force microscopy

late genes As opposed to *early genes, the *genes of a virus or bacteriophage that are expressed later, after infection of a host cell.

latent genes *Genes that are not expressed, but have all the elements to be expressed under appropriate conditions.

LC-MS → liquid chromatography–mass spectrometry

LDA → ligation during amplification

leader sequence (a) Transcribed part of a eukaryotic *gene which follows the *cap site and precedes the *start codon. (b) A special sequence at the 5'-end of a prokaryotic *messenger RNA which is not translated into a protein. It contains the *ribosome binding site, also known as the *Shine–Dalgarno sequence. (c) N-terminal amino acid sequence of secretory proteins. This sequence is cleaved off after or during the secretion process. MATTHÄUS JANCZYK

leading strand DNA strand which is synthesized continuously in the 5'–3' direction in *DNA replication is called the leading strand.

left-handed DNA In contrast to right-handed traditional polymorphs of DNA, i.e. the A, B, C and D forms, left-handed DNA (→Z-DNA) represents a left-handed double helix. For the left-handed helix, the main chain spirals in a counter-clockwise direction (→ right-handed DNA, →DNA structures). NINA DOLINNAYA

ligand Molecule that specifically binds to another molecule. In biology the term ligand is used for small molecules that bind to a protein, a receptor or a nucleic acid at a certain *binding site and thereby trigger a response, e.g. inactivation or activation of an enzyme. Ligand binding induces conformational changes in the target molecule, which alters its reactivity. This process is reversible and so should be the induced effect. →riboswitches. VALESKA DOMBOS

ligand-induced riboswitching →riboswitches

ligase Catalyzes the formation of a *phosphodiester bond between neighboring 5'-phosphate and 3'-hydroxyl termini in DNA or RNA. The enzyme joins *blunt and *cohesive ends and repairs single-stranded *nicks in nucleic acids. ANNEGRET WILDE

ligase chain reaction Method to detect *single-base mutations and is of particular use in the detection of *single-nucleotide polymorphisms. A *primer is designed as two short *oligonucleotides which are annealed to a *template with the possible *mutation at the boundary of the two primer *fragments. A *DNA ligase enzyme will ligate the two fragments only if there is an exact match to the template sequence. Subsequent *polymerase chain reaction reactions will amplify only if the primer is ligated. DAVID LOAKES

ligation Enzymatic formation of a *phosphodiester bond to join the ends of two *nucleic acid molecules (DNA or RNA) using *ligases.

ligation during amplification (LDA) Method for specified *mutagenesis of defined *nucleobases on a *plasmid (→site-directed mutagenesis).

Lineweaver–Burk plot Transformation of the *Michaelis–Menten equation into a linear form:

$$1/v = 1/v_{max} + K_M/(v_{max} \cdot c_S)$$

As the reciprocal values of the rate of product formation, v, are plotted versus the reciprocal values of the substrate concentrations, c_S, it is also known as the double-reciprocal plot. v_{max}, the maximum rate (→Michaelis–Menten model, →turnover number) is obtained from the intercept and the $^*K_M$ value from the slope of the straight line through the data. The Lineweaver–Burk plot not only provides a means to evaluate the characteristic constants of an enzyme without the help of a computer, but it also allows us to recognize clearly deviations from the ideal behavior of an enzyme. →Eadie–Hofstee plot. CHRISTIAN HERRMANN

2′,5′-linkage Connection between monomers appearing in z: certain in RNA. A *phosphodiester is formed between the 5′-hydroxyl group of one nucleotide and the 2′-hydroxyl group of another to bridge two neighboring *nucleotides. SABINE MÜLLER

linkage phase Term used to denote *chromatid locations of *alleles of two linked loci. If the two alleles are located on the same *sister chromatid they are linked in coupling phase; however, if they are located on different chromatids they are linked in repulsion phase. MAURO SANTOS

linker Connecting molecule, e.g. between a *fluorescent dye and a *nucleic acid. This is usually a flexible chain of atoms between 3 and 12 atoms in length.
SNORRI TH. SIGURDSSON

linking number Property of a *biopolymer (e.g. DNA double strand) which describes the number of *twists (turns around the central axis of the double helix) plus the *writhing number. The *twisting number is always constant as long as the *double helix remains intact, so it is a characteristic feature of the helix. The following equation can be used to define the linking number: $Lk = Tr + Wr$, where Lk is the linking number, Tw is the twisting number and Wr is the writhing number. →supercoiled DNA. MATTHÄUS JANCZYK

liquid chromatography–mass spectrometry (LC-MS) Used for structural analysis and quantification of small- to medium-sized organic compounds with moderate to high polarity. The high mass cut-off usually is determined by the packing material of the columns used for chromatography. Not only can reversed-phase high-performance chromatography can be used for the LC-MS separations, but also ion chromatography, thin-layer chromatography (TLC) and others. Following the chromatographic separation, the ionization of the analytes is performed by electrospray ionization. Less common ionization techniques in this coupling method are atmospheric pressure chemical ionization or atmospheric pressure photoionization. The ionization method is mainly depending on the polarity of the analytes (→mass spectrometry: ionization methods).

In nucleic acids research, LC-MS is used for the determination and quantification of *DNA alkylation, oxidative damage and photoadducts. In cancer research, the unique power of LC-MS regarding the high sensitivity of the mass analyzer and good separation in chromatography can be very important. The additional possibility of separation in the mass spectrometer is another powerful tool. The structural evaluation of minor and unknown DNA adducts via MS is valuable for obtaining a deeper insight into the mode of action of anticancer drugs when other methods like ^{31}P-post-labeling or *nuclear magnetic resonance spectroscopy deliver unsatisfactory results.

Sample preparation normally includes gentle enzymatic *digestion of the DNA. The resulting (modified) free *nucleobases, *nucleotides or *oligonucleotides are usually analyzed without further modification. →gas chromatography–mass spectrometry, →mass spectrometry terms and definitions, →mass spectrometry in DNA related research. ANDREAS SPRINGER

LNA →locked nucleic acid

L-nucleosides and L-nucleotides
→Spiegelmers

locked nucleic acid (LNA) Numerous chemically modified *oligonucleotides have been synthesized and studied during the past decades. Among these, LNA stands out because of its capability to bind to *complementary nucleic acids with unprecedented binding affinities. A LNA monomer contains a O2′–C4′-methylene-linked bicyclic *ribose unit that is locked in a RNA-like conformation. LNA is thus a structural mimic of RNA.

LNAs can be synthesized by conventional automated synthesis (→oligonucleotide synthesis) furnishing fully modified LNA, LNA-modified DNA, LNA-modified RNA or, for example, LNA-modified 2′-O-Me-RNA oligonucleotides. The term LNA refers to any oligonucleotide containing one or more LNA monomer(s). The fact that LNA monomers can be readily mixed with a variety of natural or synthetic nucleic acid monomers results in great design flexibility. The basic physicochemical properties of LNA are summarized as:

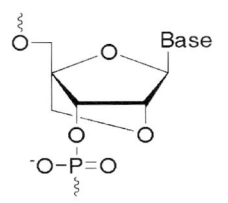

- Aqueous solubility as for DNA/RNA.
- Increased binding affinity toward complementary DNA/RNA relative to unmodified nucleic acids.
- Excellent *mismatch discrimination, particularly for short LNAs.
- Increased stability in biological media relative to unmodified nucleic acids.

Based on these basic characteristics, LNA is broadly applicable as an enhancing or even enabling technology within biotechnology, molecular biology research and drug development. LNAs can be transfected into cells and have been shown to predictably mediate *gene silencing as single-stranded *antisense agents at low nanomolar concentrations. Animal studies have confirmed the potency of single-stranded antisense LNAs and these molecules are in clinical trails as anticancer agents. Incorporation of LNA monomers into the binding arms of *DNAzymes potentiates RNA product cleavage when compared to the unmodified DNAzymes. LNA-modified *small interfering RNA duplexes mediate potent gene silencing in cells and *in vivo*, and incorporation of LNA monomers not only induces high biostability, but also reduces the number of off-target effects. The use of LNAs enables targeting of structured *RNA motifs and LNA are highly efficient as *micro-RNA-targeting probes. Other applications of LNA are as building blocks within nanotechnology (→DNA nanotechnology), and as probes and signaling molecules within biotechnology. Medically relevant *single-nucleotide polymorphisms can thus be efficiently analyzed using LNAs in solution (homogeneous assays) or array (heterogeneous) formats. Furthermore have LNA probes been demonstrated to enhance the sensitivity of *fluorescence *in situ* hybridization (FISH) and to improve RNA capture technologies. JESPER WENGEL

locus Specific site of a *gene on a *chromosome. All the *alleles of a particular gene occupy the same locus.

locus control regions (LCR) Operationally defined by their ability to enhance the expression of linked *genes to physiological levels in a tissue-specific and copy number-dependent manner at ectopic *chromatin sites. The components of an LCR commonly colocalize to sites of DNase I hypersensitivity (HS) in the *chromatin of expressing cells. The core determinants at individual HS sites are composed of arrays of multiple ubiquitous and lineage-specific *transcription factor-*binding sites. Although their composition and localization relative to their cognate genes are different, LCRs have been described in a broad spectrum of mammalian gene systems. BEATRIX SÜß

logic gate →deoxyribozyme- and ribozyme-based logic gates

long patch repair →DNA repair

long-range tertiary interaction →tertiary interactions

long terminal repeat DNA sequence repeated in direct orientation at each end of the proviral *genome or an integrated *retrotransposon.

loop →internal loop, →hairpin loop, →tetraloop

low-fidelity DNA polymerase *DNA polymerases that are believed to be involved in *DNA replication processes show low error rates (as low as only one error within one million synthesized nucleotide linkages), whereas certain enzymes that are competent to bypass *DNA lesions (caused by e.g. sun light) exhibit high error rates of up to one error within 1–10 synthesized *nucleotide linkages. These recently discovered low-fidelity DNA polymerases exhibit features like high error propensity when copying undamaged DNA or the ability to bypass DNA lesions that block the replicative enzymes. These DNA polymerases are believed to be involved in *DNA repair and the immune response. ANDREAS MARX, KARL-HEINZ JUNG

L-RNA →Spiegelmer

lysine riboswitch Conserved *RNA domain in *messenger RNAs of several bacterial species. It is found *upstream of *genes involved in lysine biosynthesis and represses their expression in the presence of elevated lysine levels by a *transcription termination mechanism (→riboswitches). The phylogeny-derived *aptamer *consensus sequence folds into a structure with five *stem–loop or hairpin domains surrounding a central core that contains most of the highly *conserved nucleotides. The conserved loop element between stems P2 and P2a conforms to the *S-turn motif consensus sequences, a structural motif that leads to a sharp bend in RNA helices (in some cases loop P2a/P2b seems to carry an additional *K-turn motif). This probably

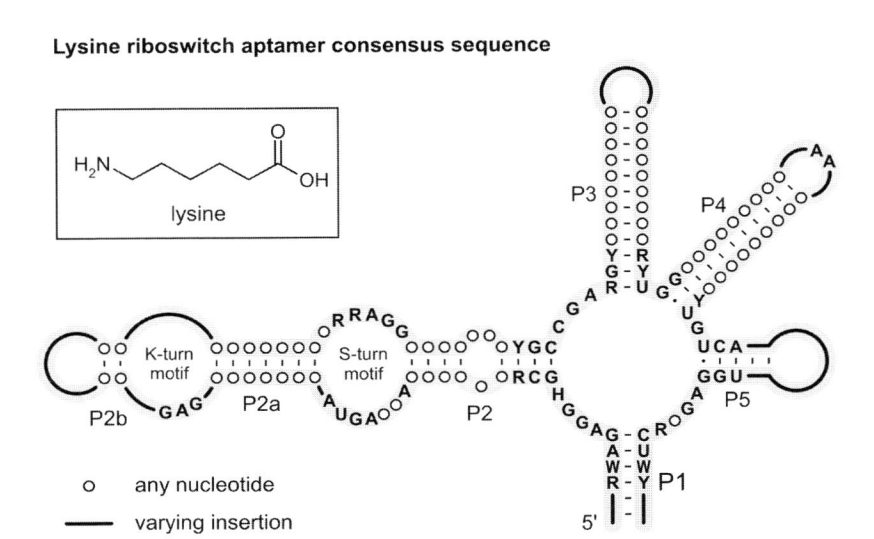

lysine riboswitch

Lysine riboswitch aptamer consensus sequence

enables the RNA to fold into a compact structure to selectively bind L-lysine over a variety of closely related natural compounds. The aptamer rejects analogs like ornithine and homolysine, which differ only in the length of the side-chain, and even discriminates against D-lysine. Interestingly, the lysine riboswitch shows high affinity to *S*-2-aminoethyl)-L-cysteine (AEC), a compound that can substitute for L-lysine in cellular reactions (a so-called antimetabolite) and is toxic to many bacteria. Consistent with this finding is the fact that *mutations leading to AEC-resistant bacterial strains could be mapped to the lysine aptamer, where they cause disruption of the conserved structure. This shows that riboswitches can be potent drug targets. →thiamine pyrophosphate riboswitch. RÜDIGER WELZ

M

macromolecule Large molecule with a large molecular mass. This term is often restricted to polymers such as *biopolymers (polypeptides, polysaccharides, nucleic acid polymers). A DNA or RNA strand is an example for a macromolecule. MATTHÄUS JANCZYK

macro satellite Variable *nucleotide *tandem repeats with a basic repeat unit that can range from half to several kilobases and that can stretch more than 100 kb in a given genomic location. MAURO SANTOS

MADS box genes Family of *genes that encode eukaryotic *transcription factors that contain a MADS domain. MADS box genes are mainly involved in developmental processes and, in particular, they are relevant for flower molecular morphogenesis in plants. MAURO SANTOS

Major groove When *base pairs are stacked into a *double helix, the *sugar–phosphate backbones of two DNA strands build the walls of a major and a *minor groove, which wind around the helix. The bottom of the major groove is paved with nitrogen and oxygen atoms that can make *hydrogen bonds with the side-chains of a protein's amino acids (and therefore have a significant role in intrinsic coding). The pattern of these hydrogen-bonding groups is different for two kinds of base pairs. Reading from *purine to *pyrimidine, an AT pair offers a nitrogen atom (a hydrogen acceptor), an amino group (a donor) and an oxygen atom (another acceptor). In contrast, the GC pair offers the same groups in a different order: first a nitrogen (an acceptor), then an oxygen (an acceptor) and finally an amino group (a donor). Since each base pair AT and GC can also be turned around (to become TA and CG), four different patterns are exhibited at each step of the helix. Therefore, the major groove carries a message about the *base sequence of DNA in a form that can be read by other large *biomolecules. The most of specific *nucleic acid–protein interactions are realized in a major groove. NINA DOLINNAYA

major tRNA *Transfer RNA (tRNA) of an *isoacceptor family which is mainly used to decode the particular amino acid (→codon usage) and is present at high cellular concentrations. The remaining, less frequently used ones are termed *minor tRNA. STEFAN VÖRTLER

mammalian artificial chromosome *Artificial chromosome containing mammalian *genomic DNA and which is designed to be episomically (→episome) maintained in mammalian cells. GEMMA MARFANY, ROSER GONZÀLEZ-DUARTE

map density Term that can be quantified using the maximum gap between adjacent genetic markers or the number of different markers in a unit of the *genome.

mapping of nucleic acid structure →chemical and enzymatic structure mapping of RNAs

marker chromosome Structurally abnormal *chromosome that is distinctive in appearance, but not fully identified. It is not a "marker" in the sense that is responsible of a specific disease, but in the

Nucleic Acids from A to Z: A Concise Encyclopedia. Edited by Sabine Müller
Copyright © 2008 WILEY-VCH Verlag GmbH & Co. KGaA, Weinheim
ISBN: 978-3-527-31211-5

sense that it can be distinguished under the microscope from the normal chromosomes. MAURO SANTOS

masked nucleoside/nucleotide →caged nucleoside/nucleotide

mass spectrometry (MS) in DNA-related research MS offers a broad range of methods relevant for DNA analysis. A schematic set-up showing the most common combination of separation techniques, ion sources and mass analyzers is presented in the figure. The most relevant abbreviations are given in *mass spectrometry: terms and definitions.

Providing the facilities for quantification (especially using MS coupled to separation techniques like those realized in *gas chromatography–mass spectrometry, *capillary electrophoresis-MS or *liquid chromatography–mass spectrometry) MS allows the simultaneous analysis of analyte mixtures (matrix-assisted laser-desorption ionization-time-of-flight and electrospray ionization-MS). Fragmentation of the analytes adds supplementary structural information. The limit of detection usually is in the medium femtomole range (10^{-15} mol L^{-1}) and thus slightly less sensitive than ^{31}P-post-labeling. This unique combination of high sensitivity with structural analysis of not separated or minor components affords assignment of unknown analytes. In particular, the latter emphasizes the unique potentials of MS in DNA research, although transferring DNA-related compounds in the gas phase is a task more demanding than ionization of peptides. Adducts formed by *methylation/alkylation, *cisplatin and other antitumor drugs or ultraviolet-induced *crosslinks and oxidation can be analyzed as well as DNA–protein complexes, DNA duplexes or whole viral *genomes.

The good linkage with database research or automated sequence analysis offers fast data analysis, although the analysis of peptides and proteins has been explored in more detail. The availability of other, very

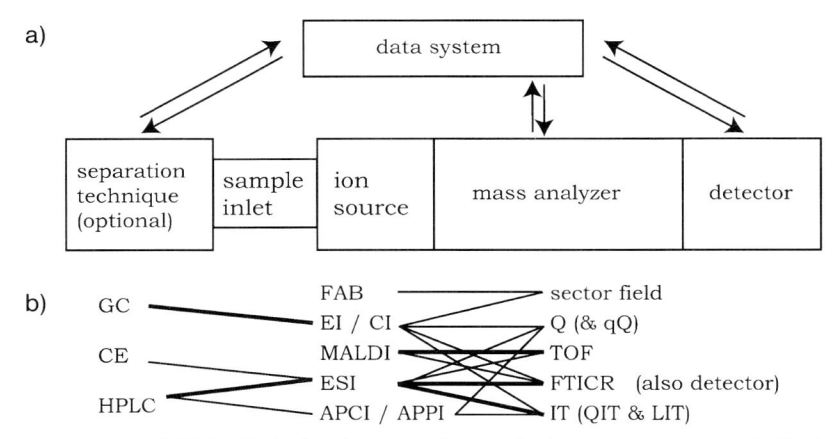

Mass spectrometry (MS) in DNA-related research: Principal set-up of mass spectrometers. An optional separation technique is followed by a sample inlet. Ion formation takes place in the subsequent ion source (atmospheric pressure or vacuum). The ions are separated in the mass analyzer and finally detected. Instrument set-up: thick lines indicate prevalent instrument set-ups, thin lines indicate common combinations.

powerful tools for *sequence analysis, the low number of repetition units and a comparable low amount of modifications in DNA still limits the application of MS, but its potential is far from being fully utilized.
ANDREAS SPRINGER

mass spectrometry (MS): ionization methods Ionization is the crucial step in all analyses by MS. Using too mild conditions, no ions may be formed and thus no information about the samples can be obtained. Too harsh ionization will lead to sample decomposition with no or poor mass spectra. However, good knowledge about ionization processes is the only way to generate mass spectra containing information about sample composition, the analyte's identity and, sometimes, quantity. All ionization methods have special requirements regarding sample composition and purity. The most important ionization methods are presented in alphabetical order:

CI (chemical ionization) uses a preliminary ionization of a reactant gas (e.g. methane) which ionizes the vaporized analyte molecules in the second step. CI mainly leads to $[M + H]^+$ and adducts of the reactant gas and the analytes as well as some fragment ions. It is used for ionization in *gas chromatography–mass spectrometry (GC-MS).

EI (electron impact ionization, also electron ionization) is a method for the ionization of small (1500 a.m.u. or less), volatile and thus relatively non-polar organic compounds. The analyte is vaporized thermally and ionized by a beam of electrons, typically with a kinetic energy of about 70 eV to guarantee high yield of ions. Typically fragment ions and the molecular ions ($M^{+\bullet}$) are formed. EI of mixtures may result in very complex mass spectra. It is widely used as an ion source in GC-MS.

ESI (electrospray ionization). The sample solution is sprayed from a narrow-bore conducting capillary on which a high electric field is applied. The enforced charge separation leads to the formation of highly charged droplets. These will loose solvent until the free charged analyte ions are formed. The desolvation can be assisted thermally or pneumatically. Depending on the polarity of the voltage, positive ($[M + n\text{Cat}]^{n+}$) or negative ions ($[M - nH]^{n-}$) can be detected. The appearance of highly charged ions ($n > 1$) is characteristic for ESI-MS spectra of large molecules. Usually multiple charge states can be observed simultaneously. This and ubiquitous proton–cation exchange can lead to very complex mass spectra for samples contaminated with salts. Contaminants like polyethylene glycols or surfactants may lead to signal suppression.

At very low flow rates (200 nL or less) the term *nanospray ionization (NSI)* is used. ESI can be easily used for ionization in *HPLC-MS. Coupling to *capillary electrophoresis is also possible (CE-MS). The very mild ionization not only offers analysis of small to medium polar or ionic compounds (e.g.(modified) nucleosides or (oligo-)nucleotides), but also of DNA–DNA, DNA–*PNA and other mixed *duplexes and the even less stable DNA multiplexes conserving their structures in the gas phase as well.

FAB (fast atom bombardment) was a revolution in the analysis of small to medium ionic or polar substances and thus *(oligo-)nucleotides when it was developed in the 1980s. It was replaced after even the milder ionization techniques, MALDI and ESI, were described few years later.

MALDI (matrix-assisted laser desorption/ionization). The analytes are embedded in a solid, rarely liquid, matrix [for oligonucleotides: often α-cyano-4-hydroxycinnamic acid (CHCA), 2,5-dihydroxybenzoic acid (DHB), dithranol (DIT) or 2,4,6-trihydroxy-acetophenone (THAP); for DNA strands/duplexes: 3-hydroxy-picolinic acid (HPA), commonly the matrix:analyte ratio is about 1000:1] which usually is highly volatile and absorbs the laser photons. The initially desorbed clusters will lose the matrix molecules and lead to the formation of $[M + nCat]^{n+}$ ions (usually $n = 1 \gg 2 \gg n$) in positive mode or $[M − H]^− (\gg [M − 2H]^{2−})$ in negative mode. With very low sample consumption the method offers brilliant limits of detection, although its use for quantification is limited. This technique is mainly used for analysis of medium to high mass polar *(bio-)molecules. MALDI is comparably robust to impurities like salts. An example for the practical use of this method can be found in the section *DNA sequence analysis by mass spectrometry. →liquid chromatography–mass spectrometry, →mass spectrometry in DNA related research, →mass spectrometry: terms and definitions. ANDREAS SPRINGER

mass spectrometry (MS): terms and definitions Generally there are a lot of unique terms and definitions in all areas of research. This makes it difficult to understand the literature for newcomers. A short overview concerning MS terms and definitions relevant in nucleic acid research is presented in the following.

AP: atmospheric pressure (approximately 1013 hPa).

FTICR (Fourier transformation ion cyclotron resonance)-MS provides a very high resolution combined with best mass precision (better than 2 ppm) guaranteeing high certainty in elemental composition determination, but is expensive and needs extensive maintenance. It is equipped with ESI or MALDI as standard ionization techniques. It offers a great variety of orthogonal MS^n techniques for structural elucidation or *sequence analysis of *oligonucleotides.

Hybrid mass spectrometers try to combine the advantages of different types of mass spectrometers or enhance the MS/MS capability, e.g. Q/TOF, TOF/TOF, qQ/FTICR or LIT/FTICR-MS instruments do exist.

IT (ion trap): facilitates the acquisition of MS^n spectra and are especially interesting for structure evaluation. QIT (quadrupole ion traps) and LIT (linear ion traps) are both very common and often used as detector in *gas chromatography–mass spectrometry (GC-MS) or equipped with ESI.

M: intact molecule, usually neutral; $M^{+\bullet}$, molecular ion (formally: $[M − e^−]^+$); $[M + nCat]^{n+}$, n-times cationized molecule, *Cat* is usually H or Na; $[M − nH]^{n−}$, n-times deprotonated molecule.

Mass units: [amu] (unified atomic mass unit), [Da] (Dalton) or [u] is defined as 1/12 of the nuclide ^{12}C or $1.66053886(28) \times 10^{−27}$ kg.

MS: mass spectrometry, mass spectrum.

MS^n: nth step of the fragmentation (e.g. isolation, fragmentation, re-isolation and again fragmentation of the ions would refer to MS^3). MS/MS or tandem MS usually refers to MS^2. Methods employed for fragmentation are base on collisions (CID/CAD, SID, PSD, SORI-CID), thermal excitation (BIRD), multi-photon processes (IRMPD) or electron capture (ECD). BIRD, ECD, IRMPD and SORI-CID are methods used in FTICR-MS exclusively (→DNA sequence analysis by mass spectrometry).

m/z (mass/charge ratio): abscissa of the mass spectra, dimensionless. Sometimes the unit *thomson* [Th] is used.

Orbitrap is a new kind of high-resolution mass analyzer, coupled to a LIT. It lacks the unique structural evaluation tools of the FTICR instruments but needs less maintenance.

Quadrupole (Q) mass spectrometers are widely used for quantification of small organic compounds. They are often a mass analyzer in GC-MS, are comparably cheap and very robust, although with low resolution. Triple quadrupole MS facilitates MS/MS (CID).

TOF (time-of-flight) analyzers have the largest mass range and are very robust. Commonly used with MALDI ionization. TOF analyzers provide a very fast analysis. rTOF (reflectron-TOF) instruments provide a better resolution but a smaller mass range and are often coupled to ESI as well. TOF mass spectrometers are mainly used for detection of oligonucleotides, *polymerase chain reaction products and for detection of *single-nucleotide polymorphisms. →liquid chromatography–mass spectrometry, →mass spectrometry in DNA related research, →mass spectrometry: ionization methods). ANDREAS SPRINGER

maturation → RNA maturation

mature mRNA → RNA maturation

Maxam–Gilbert sequencing Technique for the *sequence analysis of DNA. This approach is based on the chemical fragmentation of DNA and resembles methods for the analysis of protein primary structures. For Maxam–Gilbert sequencing, the target DNA first is labeled at its 5′-terminus with ^{32}P (→radioactive labeling) and then is submitted to four different, base-selective modification reactions that are followed by hydrolysis of the adjacent *phosphodiester bond. The reaction conditions are chosen as to modify on average one base per molecule. Thus, mixtures of *fragments that terminate at every possible occurrence of the respective base are generated. Analysis of the four reaction mixtures is achieved by electrophoretic separation on high-resolution polyacrylamide gels (→gel electrophoresis), autoradiography and reading of the sequence in a staggered ladder-like fashion. The specific reagents that are used for modification render the bases sensitive to splitting off the sugar moiety. Dimethylsulfate is used for methylation of the *purine bases, *adenine and *guanine, at positions N7 (guanine) or N3 (adenine). Methylation at these atoms facilitates hydrolysis of the *glycosidic bond leaving an *abasic site in the DNA sequence. Fragmentation of the *DNA backbone is then achieved at these positions by heating in solution. At neutral pH, the glycosidic bond of guanine nucleotides is significantly more susceptible to hydrolysis than that of adenine nucleotides; adenine–sugar bonds, however, can be split at acidic pH. The electrophoretic pattern after hydrolysis thus can be directed towards "A" or "G" fragmentations by adjusting the pH value to the respective reaction requirements. The *pyrimidines, *cytosine and *thymine, are reacted with hydrazine and subsequently with piperidine which catalyzes base elimination and *backbone hydrolysis. This type of reaction yields a mixture of fragments generated after cytosine or thymine. The two bases can be distinguished by repeating the reaction in the presence of 2 M sodium chloride which suppresses the modification of thymine nucleotides. Electrophoretic separation of the two purine reactions (acidic and neutral) and the two pyrimidine reactions (with or without sodium chloride) is performed in four distinct lanes which correspond to G, A + G, C + T and C. Sequence reading is achieved by comparison of the resulting bands starting at the shortest fragment (the one with the greatest electrophoretic mobility). Usually, a maximum of 250 bases can be determined. Meanwhile, Maxam–Gilbert sequencing has been replaced by automated sequencing approaches based on enzymatic DNA synthesis. SUSANNE BRAKMANN

maxizyme *Allosteric ribozyme that consists of two *minizyme units. One unit binds the target and thus serves as allosteric control. The other unit cleaves the substrate in response to the binding event. (The name is derived from minimized, active, x-shaped, intelligent ribozyme.) DENISE STROHBACH

megabase pairs Often used unit for the number of base pairs in a DNA double strand (1 000 000 bp).

melting →thermal melting

melting curve Curve relating some physical property of a *macromolecule solution like its viscosity, its optical absorbance or its optical rotation to temperature (also know as the thermal *denaturation profile). The shape of the curve indicates the occurrence of physical changes in the macromolecule. For example, a nucleic acid solution is slowly heated and its *absorbance at 260 nm is continuously monitored (→absorption spectra of nucleobases). Transition from double-stranded to single-stranded nucleic acids occurs over a narrow temperature range with a sigmoidal increase in absorbance at 260 nm. →thermal melting, →hyperchromic effect. STÉPHANIE VAULÉON

melting point →melting temperature

melting temperature (T_m or t_m) Temperature at which 50% of *nucleic acid *duplex molecules are dissociated into single strands. The T_m is defined as the temperature at the midpoint of the *melting curve and depends on the *base composition and sequence of the molecule. The T_m of nucleic acids can also be approximately calculated on the basis of the oligonucleotide sequence via algorithms. For DNA molecules smaller than 25 nucleotides in length the following simple equation can be used: $T_m = 4(C + G) + 2(A + T)$ in °C. →melting curve, →thermal melting. STÉPHANIE VAULÉON

messenger molecule Another name for *messenger RNA.

messenger ribonucleoprotein (mRNP) Eukaryotic *messenger RNA (mRNA) is processed by enzymes and packaged with proteins within the nucleus to create functional mRNPs before being transported to the cytoplasm. mRNPs were named *heterogeneous nuclear ribonucleoprotein, and are protein-associated complexes that contain both general nuclear proteins such as heterogeneous nuclear ribonucleoprotein A and more specific *splicing/mRNA export-associated protein factors such as the TREX complex (TRanscription/EXport). Both the RNA processing and package events interact with mRNA cotranscriptionally in order to transform *pre-mRNA to mRNP. The TREX complex is a key component in the transport of mRNA and is composed of the multi-subunit THO and the export proteins (UAP56 and Aly in human). The TREX complex is conserved from yeast to human. TINA PERSSON

messenger RNA (mRNA) Acts as a *template for protein synthesis. It is produced in the nucleus in a process called *transcription using protein-coding *genes as template and *RNA polymerase II as enzyme. The first version of the synthesized mRNA is called *pre-mRNA, which contains both protein-coding *exon fragments and non-protein-coding *intron fragments. Pre-mRNA undergoes nuclear *RNA processing events including capping (→cap structure), *splicing and *polyadenylation in order to be transformed to *mature RNA. In the splicing step the introns are removed leaving the exon fragments in the mature RNA thus only containing the protein-coding sequences. The 5′-end of mRNA is transformed by capping and the 3′-end is first modified by site-specific cleavage followed by polyadenylation. Altogether these modifications generate a translatable *open reading frame that is protected from premature degradation (→ RNA degradation). Recently, it has been established that the pre-mRNA processing steps are interconnected with each other as well as with *transcription. It is believed that the C-terminal domain of the large subunit of RNA polymerase II plays an important role in such a coupling coordination event. In addition, mRNA is packaged cotranscriptionally with proteins within the nuclei to generate functional *messenger ribonucleoprotein before being transported to the cytoplasm. TINA PERSSON

metabolite-induced riboswitching → riboswitches

metal ion–nucleic acid interaction → interaction of nucleic acids with metal ions

methionyl-tRNA *Transfer RNA (tRNA) decoding methionine. Two distinct methionine acceptors are present in all cells: one serving as *elongator tRNA during *translation and one as *initiator tRNA (in prokaryotes *formylmethionyl-tRNA) for ribosomal assembly together with initiation *translation factors.
STEFAN VÖRTLER

methyladenine (N6-methyladenine) → DNA methylation

methylase → DNA methyltransferase

methylation → DNA methylation

methylcytosine (N4-methylcytosine) → DNA methylation

5-methylcytosine → rare base, → rare nucleotide, → DNA methylation

methyltransferase → DNA methyltransferase

5-methyluracil → rare base, → rare nucleotide

MFE structure Structure of a nucleic acid molecule with minimal free energy (MFE) obtained from secondary structure prediction algorithms (→ RNA secondary structure prediction). SABINE MÜLLER

Michaelis constant → K_M value, → Michaelis–Menten model

Michaelis–Menten equation Equation used in order to describe the rate of product formation, v, in enzymatic reactions in dependence of enzyme and substrate concentrations, c_E and c_S, respectively (→ Michaelis–Menten model):

$$V = k_{cat} \cdot c_E \cdot c_S / (K_M + c_S)$$

The two parameters of this equation, the *turnover number, k_{cat}, and the *K_M value, characterize the catalytic activity and specificity of the enzyme. With the expression for the maximum rate, $v_{max} = k_{cat} \cdot c_E$, the Michaelis–Menten equation is often used in the form $v = v_{max} \cdot c_S / (K_M + c_S)$. → K_M determination. CHRISTIAN HERRMANN

Michaelis–Menten kinetics In the first place, the reaction of an enzyme with a substrate should obey second-order kinetics, meaning the observed *rate constant increases with increasing substrate concentration. In fact, this is found in a low-substrate-concentration regime, but the rate increase curves off at higher concentrations. It does not exceed an upper limiting value which is specific for the enzyme/substrate couple. The reason is

that one or more uni-molecular reaction steps of the enzyme–substrate complex set an upper limit for the overall catalytic process. This is reflected in the *Michaelis–Menten model. The term Michaelis–Menten kinetics is used jargon-like when the hyperbolic dependence of the rate of product formation on substrate concentration is observed according to the *Michaelis–Menten equation.
CHRISTIAN HERRMANN

Michaelis–Menten model Describes the kinetics of enzymatic reactions where the initial substrate binding is faster than the catalytic process which follows as a second step. This leads to the formation of a complex between enzyme and substrate. The concentration of this complex does not change with time, i.e. a steady-state situation is established. Supposed the substrate is at much higher concentration than the enzyme, this situation is stable over a reasonable amount of time and a constant rate can be measured. The rate, v, defined as the increase in product concentration, c_P, with time is described by the following relationship between the concentrations of substrate and enzyme, c_S and c_E, and the rate constants for the catalytic step, k_2, and for enzyme–substrate complex formation and dissociation, k_{+1} and k_{-1}, respectively:

$$v = dc_P/dt$$
$$= k_2 \cdot c_E \cdot c_S / \{(k_{-1} + k_2)/k_{+1} + c_S\}$$

This is the Michaelis–Menten equation and the term $(k_{-1} + k_2)/k_{+1}$ is usually replaced by K_M, the *Michaelis constant. The rate, v, shows a hyperbolic dependence on the substrate concentration, and the parameters k_2 and K_M are obtained by non-linear regression of the experimental data or from plots according to linearized forms of the equation (→Lineweaver–Burk or →Eadie–Hofstee plot). K_M has the unit mol L^{-1}, like an equilibrium *dissociation constant. Note that even in the simple two-step model K_M equals the dissociation constant of the enzyme–substrate complex only in the extreme case that k_2 is much smaller than k_{-1}.

Most enzymatic reactions are more complicated than the minimal two-step mechanism mentioned above. Reaction steps like conformational rearrangements or product release come into play and their rate constants may even govern the overall rate. Nevertheless, the rate of product formation shows the same substrate concentration dependence as described above and the *Michaelis–Menten equation can be applied for analysis. Here, K_M represents a complex expression constituted by many individual rate constants which depends on the underlying model of the mechanism. Yet, K_M can be regarded as an apparent equilibrium dissociation constant as it gives an indication at what substrate concentration the enzyme is saturated, meaning to work at maximum activity (see equation above, e.g. half maximum activity at $c_S = K_M$). Likewise, the catalytic rate constant, k_2 in the simple model above, must be replaced by a composite term of many individual *rate constants depending on the mechanism and it is usually termed k_{cat} (→turnover number). Thus, the general form of the Michaelis–Menten equation can be written:

$$v = k_{cat} \cdot c_E \cdot c_S / (K_M + c_S)$$

which is also used in the following form where the maximum rate of product formation, v_{max}, equals the product of the *turnover number and enzyme concentration:

$$v = v_{max} \cdot c_S / (K_M + c_S)$$

CHRISTIAN HERRMANN

microinjection Injection of molecules (usually DNA or RNA) into the nuclei or the cytosol of cells with the aid of a micropipette and a microinjector.

micro-RNA (miRNA) Small, around 21–26-nucleotides, RNA molecules involved in the *repression of *gene expression in a sequence-specific manner through the *RNA interference pathway. More than 2000 different miRNAs have been described in vertebrates, flies, worms, plants and even viruses. miRNAs are the product of two maturation processes. In a first step, *nascent RNA transcripts (→pri-miRNAs) are processed in the nucleus into around 70-nucleotide *hairpin structures (→pre-miRNA) via endonucleolytic cleavage by the *RNase *Drosha. Pre-miRNAs are then exported into the cytoplasm by Exportin 5 and further processed into miRNAs via a second endonucleolytic cleavage by *Dicer. The resulting short *double-stranded RNA is finally unwound and incorporated into the *RISC complex. Cleavage by Dicer and incorporation into RISC appear to be coupled to each other. NICOLAS PIGANEAU

microsatellite Identified in the early 1980s, microsatellites (or simple sequence repeats) are tandemly repetitive stretches of short (1–6 bp) motifs such as GTGTGTGT-GTGT [$(GT)_6$ for short]. Microsatellites are highly polymorphic sequences, and have become instrumental as genetic markers in areas such as forensics, genetic mapping and population genetics. These sequences have mutation rates of the order of 10^{-3} to 10^{-4}, several orders of magnitude higher than that of the bulk of DNA. These high rates of *mutations are frequently due to "replication slippage", the displacement of DNA strands thus producing a mispairing of *complementary sequences. Some human disorders are caused by the expansion of *trinucleotide repeats. For example,

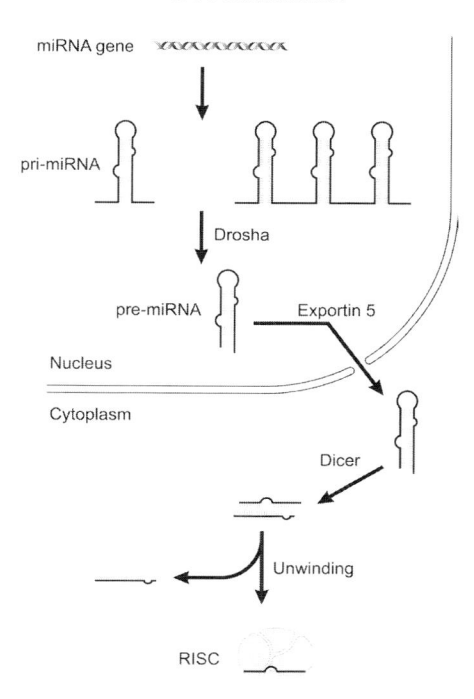

Huntington's disease (an uncommon neurological disorder) is caused by the expansion of the trinucleotide CAG: the normal *allele has between $(CAG)_{10}$ and $(CAG)_{35}$ repetitions, but in people with the disorder the number of repetitions is larger than 35. Trinucleotide repeat disorders generally show the well-known phenomenon called "genetic anticipation", a peculiar pattern of inheritance in which the symptoms appear at an earlier age and are more severe in the next generations. MAURO SANTOS

minichromosome Small *chromosome-like structure such as the package into a series of nucleosomes of the circular 5200-bp DNA of virus SV40 in both the virion and infected nucleus. MAURO SANTOS

minimal genome Smallest possible group of *genes that would be sufficient to sustain a functioning cellular life form under the presence of a full complement of essential nutrients and in the absence of environmental stress. Recent estimates suggest that the minimal genome or "minimal gene set" is about 200 genes, but this is based on a top-down approach using comparative genomics. MAURO SANTOS

minisatellite Simple sequence repeats, also named variable *nucleotide *tandem repeats, with a longer motif than *microsatellites, ranging in size from 9 to 100 bp. For most examples in the literature minisatellite motif length is 12 bp or more. The number of *tandem repeats varies (usually less than 1000). Minisatellites are less abundant than microsatellites, but share the fundamental characteristics of frequent repeat number *mutations and of repeat number influencing gene function. GEMMA MARFANY, MAURO SANTOS

minizyme *Hammerhead ribozyme with reduced structure but maintained catalytic activity.

minor base →rare base

minor groove When *base pairs are stacked into a *double helix, the *sugar–phosphate backbones of two DNA strands build the walls of a *major and a minor groove, which wind around the helix. Groove parameters (width and depth) depend on the structural form of *DNA helix. The minor groove is a poorer candidate if compare to a *major groove for information readout. The oxygen atom (O2) of pyrimidine and nitrogen (N3) of *purine faces to the minor groove. It has another important function in *B-DNA. In particular, the minor groove of the DNA helix contacts the *histone core. NINA DOLINNAYA

minor groove-binding polyamides Polyamides that may be rationally designed based on three aromatic amino acids: hydroxypyrrole (Hp), imidazole (Im) and pyrrole (Py). An Im/Py pair specifically recognizes a GC base pair, while a Py/Im pair targets a CG base pair. In a similar manner, a Hp/Py pair specifically interacts with a TA base pair and a Py/Hp pair with AT. Based on this three-letter amino acid code, pyrrole–imidazole polyamides that form a kind of hairpin structure constituting amino acid pairs can be designed for specific interaction with a desired DNA sequence. The concept has been suggested by Peter Dervan et al., and has been used for the development of strategies for molecular recognition, regulation of *gene expression and artificial *transcription activators. SABINE MÜLLER

minor spliceosome *Ribonucleoprotein (RNP) complex responsible for *atac splicing, i.e. the removal of U12-type *introns. Generally, by structure and function the minor spliceosome resembles the major *spliceosome. However, in agreement with the rare occurrence of U12-type introns the minor spliceosome is much less abundant. The active complex is also formed by five *U-snRNPs, yet composed of different *U-snRNAs and in part different particle proteins. Within the minor spliceosome, U11 and U12-snRNAs are functionally homologous to the U1- and U2-snRNAs of the major complex, respectively. Likewise, U4atac- and U6atac-snRNAs functionally replace the U4- and U6-snRNAs of the major spliceosome. However, both spliceosomes share the U5-snRNP. As to the protein content of the minor spliceosome, two groups can be distinguished: eight Sm proteins are shared by all minor snRNPs, whereas three to eight distinct proteins are particle specific.

Although generally the assembly pathway of the minor spliceosomes seems to be very similar to that of the major spliceosome, one clear difference has been observed. At the initial steps of assembly, the 5'- and 3'-*splice sites are recognized simultaneously, resulting directly in formation of the A-complex. Indeed, biochemical fractionation has verified that the U11- and U12-snRNPs exist as a stable pre-formed di-*snurp. This hetero-dimer associates via base pairing with the 5'-splice site and the branch-point region of the *pre-mRNA, respectively, in a similar way as observed with the U1 and U2 monomers of the major spliceosome. BERND-JOACHIM BENECKE

minor tRNA *Transfer RNA of a *isoacceptor family that is different to one or several other members (→major tRNA) only rarely used to decode a particular amino acid (→codon usage). Its cellular concentration is lower than that of a major tRNA. STEFAN VÖRTLER

minus strand DNA strand that is transcribed into *messenger RNA (mRNA). The resulting mRNA is complementary to the minus strand and identical in sequence to the corresponding *plus strand, with T replaced by U in the RNA. SLAWOMIR GWIAZDA

miRNA →micro-RNAs

mirror-image aptamer →Spiegelmer

mismatch Occurs when two *noncomplementary nucleotides form a *base pair. A mismatch in a *double-stranded DNA (dsDNA) is most frequently caused by a misincorporation of a *nucleotide during *DNA replication. Another frequent possibility is deamination of a *nucleobase, e.g. deamination of cytidine resulting in the change to uridine. *In vivo*, mismatches in dsDNA are recognized and repaired (→mismatch repair). SLAWOMIR GWIAZDA

mismatch repair Special case of *DNA repair in which the pre-mutagenic DNA structure does not consist of a damaged *nucleotide, but the opposition of two natural nucleotides caused during *replication by nucleotide misincorporation that escaped *proofreading by the *DNA polymerase. This nature of the structure to be repaired poses a special problem of informational ambiguity: repair will contribute to overall replicational fidelity only if it is directed with high selectivity towards the mismatched nucleotide residing in the newly synthesized DNA strand. In *Escherichia coli*, this problem is resolved by cooperation at a distance between the site of the mismatch and a nearby GATC sequence. GATC is the target of *DNA adenine-N6 methyltransferase Dam which methylates the exocyclic amino group of the *adenine moiety within the palindromic target sequence (i.e. the sequence reads GATC in *both* strands). A *replication fork passing through a GATC site leaves behind one hemimethylated site each in both of the daughter molecules. This constitutes a distinguishing chemical mark of the newly synthesized state – until Dam methyltransferase introduces the lacking methyl group. During this time window, repair endonuclease MutH can occupy the hemimethylated GATC site and cleave the unmethylated strand at the 5'-side of the G residue if activated by the presence of a DNA *mismatch. The latter information is conveyed to the MutH endonuclease by a heterotetrameric protein complex consisting of two copies of MutS and two copies of MutL. This tetramer binds to the mismatch and reels in DNA from both sides under turnover of ATP and formation of a DNA loop. Collision of the tetramer with the MutH molecule bound to the nearest hemimethylated GATC site is thought to activate its endonucleolytic activity.

Subsequently, *DNA helicase, *exonuclease and single-strand binding protein open a gap extending from the GATC sequence until beyond the location of the mismatch. This large gap (typically of the order of several hundred nucleotides length) is filled in by DNA polymerase III. Hence, DNA mismatch repair is the prototypic case of long patch repair. Organisms without DNA adenine methylation most likely use the non-continuous nature of the newly synthesized DNA strand to direct mismatch repair. HANS-JOACHIM FRITZ

missense mutation Change in coding DNA sequence, often a *point mutation, by which (at the RNA level) a *nucleotide *triplet coding for a certain amino acid is converted into one coding for a different amino acid. →nonsense mutation. HANS-JOACHIM FRITZ

missense RNA RNA that has been transcribed from a *gene containing a change in the *coding sequence and therefore encodes a non-functional gene product (→missense mutation). SABINE MÜLLER

mitochondrial DNA (mtDNA) DNA that is located in the mitochondria. Unlike nuclear DNA, mtDNA is usually maternally inherited and enables us to trace the line of descendent from a female ancestor. Mitochondria can occasionally be inherited from the father. In these cases the phenomenon of "heteroplasmy", the presence in the same individual of more than one type of mtDNA, is documented. However, if the haplotypes within each heteroplasmic individual are sufficiently closed, the variants could have arisen by *mutation rather than by exogenous introduction from the sperm. A well-known example of heteroplasmy occurs in mussels, although it has also been reported in humans. mtDNA has a relatively fast mutation rate, which renders important sequence differences between species. A 648-bp region of mtDNA, known as cytochrome oxidase I, was originally proposed as a potential DNA barcode to quickly and easily identify a particular species. This DNA-based method for species identification is called "DNA barcoding" and is becoming very popular. MAURO SANTOS

mitochondrial RNA Mitochondria have their own genetic system and have a central role in energy transduction (the powerhouse of eukaryotic cells), and also participate in several other important functions. *Mitochondrial DNA encodes a small number of proteins whose *messenger RNAs are translated by a distinctive mitochondrial protein-synthesizing system specified by the mitochondrial *genome. MAURO SANTOS

mitoribosome Mitochondrial *ribosomes. They belong to the class of prokaryotic ribosomes (70S), although they are exclusively found in eukaryotic cells. They are even smaller than bacterial ribosomes and consist of fewer proteins. DENISE STROHBACH

mobile genetic elements →transposable elements, →transposon

modified base *Nucleobase that is altered post-synthetically (→rare bases) or, alternatively, a nucleobase analog that has been synthetically incorporated into DNA or RNA (→modified DNA, →modified RNA, →oligonucleotide synthesis). SABINE MÜLLER

modified DNA DNA that deviates structurally from the nucleic acid constitution deduced from the four standard building blocks *deoxyadenosine, *deoxycytidine, *deoxyguanosine and *thymidine. Modifications may concern the *backbone, sugar or nucleobase moieties and can be naturally occurring or artificially introduced by enzymatic or chemical pathways. Modified DNA is frequently associated with *antisense oligonucleotides and *DNA damage. Automated solid-phase oligonucleotide synthesis is the method of choice for the site-specific introduction of a modification into oligodeoxynucleotides. →oligonucleotide synthesis. RONALD MICURA

modified nucleic acids →modified DNA, →modified RNA

modified RNA RNA that deviates structurally from the nucleic acid constitution deduced from the four standard building blocks *adenosine, *cytidine, *guanosine and *uridine. Modifications may concern the *backbone, sugar or *nucleobase moieties and can be naturally occurring or artificially introduced by enzymatic or chemical pathways. RNA provides a vast variety of natural modifications; more than 90 modified nucleosides are currently known, most of them encountered in *transfer RNA, *small nuclear RNA and *ribosomal RNA (→rare bases). The reasons for their modifications are far from being understood. Automated solid-phase oligonucleotide synthesis is the method of choice for the site-specific introduction of a modification into oligoribonucleotides (→oligonucleotide synthesis). RONALD MICURA

modular aptameric sensors Fluorescent sensors constructed by fusing two separate *aptamers. One of the aptamers (binding module) recognizes an analyte, while the other (fluorogenic module or fluoromodule) recognizes a *fluorescent dye that changes fluorescence upon binding; aptamers are fused through a *communication module, which allosterically transfers information about events in one module to another. For example, the malachite green aptamer can be used as a fluorogenic module as malachite green is a weakly *fluorescent dye until it binds to its aptamer. When fused to *adenosine *aptamers through a short oligonucleotide bridge, the malachite green module reports the binding of adenosine derivatives through an increase of fluorescence, because adenosine binding stabilizes the malachite green aptamer. MILAN STOJANOVIC, DARKO STEFANOVIC

molar absorption coefficient (Synonym: molar absorptivity, molar extinction coefficient) →absorptivity, →molar absorptivity

molar absorptivity (Synonym: molar absorption coefficient, molar extinction coefficient) ε (unit: $L\,mol^{-1}\,cm^{-1}$). →absorptivity

molar extinction coefficient (Synonym: molar absorption coefficient, molar absorptivity) →absorptivity, →molar absorptivity

molecular automata →deoxyribozyme-based automata, →Shapiro–Benenson–Rothemund automata

molecular beacon Molecule that is designed to bind to a target molecule and to report the binding event by emitting a light signal. Nucleic acid-based molecular beacons contain both a *fluorescent dye, which emits light upon irradiation, and a quenching molecule. Before the recognition event, the nucleic acid is in a *conformation that places the quencher in the proximity of the fluorescent dye, which effectively turns off the light signal. Upon binding to the *target molecule, the conformation of the molecular beacon changes such that the quencher is placed far away from the fluorescent dye and thereby the light signal is turned on. The simplest design for a nucleic acid molecular beacon is to place the fluorescent dye on one end of an oligomer and the quencher on the other in a sequence that favors *hairpin formation in the absence of the target molecule. This places the fluorescent dye in spatial proximity to the quencher. In the presence of a DNA or RNA sequence that is complementary to the molecular beacon sequence, a *duplex is formed between the target and the beacon, which places the quencher far away from the dye and turns on the light. This method is both sensitive and selective, and can be used to differentiate between two DNA sequences that differ by a single nucleotide. However, any event can be detected which removes the quencher from the dye, and this has, for example, been used to detect cleavage of both DNA and RNA, and is the basis for tests that detect the presence of *nucleases. *Aptamers, which are nucleic acids that bind to a variety of *ligands, have also been turned into molecular sensors using the molecular beacon strategy. Molecular beacons are the basis for the *real-time polymerase chain reaction, where *amplification of a specific sequence is quantified after each round by the light emitted by a molecular beacon that targets the sequence to be amplified. Molecular beacons have found use in biosensors that detect the genetic material of bacteria and viruses, and have even been used for monitoring expression of specific *genes in living systems. SNORRI TH. SIGURDSSON

molecular weight marker DNA, RNA or protein of known size used to identify the approximate size of DNA, RNA or protein to be analyzed on a polyacrylamide or agarose gel. If the size of the molecule is unknown, mostly a DNA, RNA or a protein ladder, which contains a pool of different fragments of known size, is used. After *gel electrophoresis the marker forms, because of the fragments differing in size, a ladder on the gel. The size of the molecule to be analyzed can then be estimated relative to the size of a fragment in the marker. IRENE DRUDE

monoisotronic mRNA Monoisotronic *messenger RNA is found in eukaryotes and carries *genetic information from only one gene product.

monomer Building block of a larger molecule (i.e. a polymer).

γ**-monomethyl phosphate cap** Modified 5′-end of certain *small nuclear RNAs synthesized by *RNA polymerase III, such as *U6-snRNA and *7S K RNA.

mononucleotide Molecule consisting of *ribose or deoxyribose ($\rightarrow$2′-deoxyribose), a phosphate attached to C5 of the ribose/deoxyribose and either a *purine or a *pyrimidine base attached to C1. Mononucleotides and desoxymononucleotides are the monomers of nucleic acids. The purine and pyrimidine bases found in natural DNA are *adenine, *cytosine, *guanine and *thymine. In RNA, the pyrimidine base thymine is replaced by *uracil. For natural enzymatic synthesis of oligomers (e.g. *DNA replication, *transcription) the monomers are activated by additional two phosphates bound via anhydride bonds to the first phosphate. During synthesis these phosphates are released as a *pyrophosphate.

Mononucleotides and the activated mononucleotide triphosphates also play an essential role in the transfer of energy in organisms, $\rightarrow$nucleoside phosphates. SLAWOMIR GWIAZDA

morpholino oligonucleotides DNA analogs where the furanose ring is replaced by a morpholine. Morpholino oligonucleotides bind to a *complementary DNA or RNA by *Watson–Crick base pairing, thereby modifying *gene expression as an

*antisense drug. Due to their *backbone, morpholino oligonucleotides are resistant against *nucleases, making them an attractive tool for therapeutic applications.
IRENE DRUDE

mosaic genes Also known as split genes or interrupted genes, consisting of alternating *exons and *introns. The introns are removed by *splicing of the primary RNA *transcript to give the *mature RNA.
MAURO SANTOS

mRNA → messenger RNA

mRNA degradation Degradation of *messenger RNA (mRNA) is an important regulatory step in the control of *gene expression. Bacterial mRNAs exhibit short half-lifes and are degraded in a highly organized cellular process by a multi-enzyme complex called the degradosome. The key enzyme for initial endonucleolytic attack and for the assembly of the degradosome is RNase E. Endonucleolytic cleavages proceed 5'–3' behind the *ribosome. The released *fragments are degraded by *exonucleases that move 3'–5'.

In eukaryotes, the path and rate of mRNA degradation is more complex and localized to specific cytoplasmic bodies, called mRNA-processing (P) bodies. Eu-karyotic mRNA has two intrinsic stability elements – the 5'-*cap structure and the *poly(A) tail, both incorporated cotranscriptionally. To initiate *degradation, either one of these two must be cleaved by an endonucleolytic attack. The majority of eukaryotic mRNAs undergo decay initiated by a deadenylation-dependent pathway. The *poly(A) tail is removed by deadenylase activity (PARN). Following deadenylation, two mechanisms can degrade the mRNA: either decapping followed by 5'–3' decay or 3'–5' decay. In the decapping pathway, the Lsm1–7 complex associates with the 3'-end of the mRNA transcript and induces decapping by the DCP1–DCP2 complex. This leaves the mRNA susceptible to decay by the 5'–3' exoribonuclease Xrn1. Alternatively, the deadenylated mRNA can be degraded in the 3'–5' direction by the exosome, with the remaining cap structure being hydrolyzed by the decapping enzyme DcpS. The pathways are not mutually exclusive and the precise pathway of mRNA decay might be flexible.

An alternative deadenylation-independent pathway in yeast requires recruitment of the decapping machinery followed by the degradation by Xrn1. Endonuclease-mediated mRNA decay initiates with internal cleavage of the mRNA which generates two fragments, each with one unprotected end degraded by Xrn1 and the exosome.
BEATRIX SÜß

mRNA processing → RNA processing

MS Abbreviation for mass spectrometry (→ DNA sequence analysis by mass spectrometry, → gas chromatography–mass spectrometry, → liquid chromatography–mass spectrometry, → mass spectrometry in DNA related research, → mass spectrometry: ionization modes, → mass spectrometry: terms and definitions).

mtDNA → mitochondrial DNA

multi-copy plasmid *Plasmids that can be maintained in prokaryotic cells in multiple copies. Most plasmids used for *cloning and *amplification of DNA or protein expression in bacteria are multi-copy plasmids. In contrast, some natural plasmids contain control elements (related to their *replication origin) to keep low copy numbers in the host cell. GEMMA MARFANY, ROSER GONZÀLEZ-DUARTE

multi-helix loop → junction loop

multiple cloning site Region of a *cloning vector containing several single *recognition sites for different *restriction endonucleases. These sites can be used to insert DNA *fragments. ANNEGRET WILDE

multiple turnover In enzymatic reactions the enzyme acts as a *catalyst for the chemical reaction of a substrate, i.e. the substrate is turned over by the enzyme molecule by molecule. Usually the concentration of the substrate is much higher than that of the enzyme which leads to multiple turnover of substrate molecules by an enzyme molecule. Such conditions are prerequisite for *Michaelis–Menten kinetics where a steady state of substrate turnover is established. Multiple turnover kinetics are governed by the enzymatic part of the reaction and/or by product release steps depending on the relative magnitude of the respective *rate constants while *single turnover kinetics are usually controlled by the binding step including the following structural rearrangements of the enzyme–substrate complex and/or by the catalytic step. CHRISTIAN HERRMANN

multiplex polymerase chain reaction Form of the *polymerase chain reaction (PCR) in which more than one set of *primer pairs are involved in a single reaction. The result is that multiple DNA products may be amplified at the same time from a single template source. Multiplex PCR has been applied to many areas of DNA testing, including analyses of deletions, *mutations and *polymorphisms as well as in *reverse transcription PCR. It is commonly used in genotyping protocols where the simultaneous analysis of multiple markers is required, for the detection of pathogens or in *microsatellite analysis. However, as multiple primer pairs are used, optimization to establish reliable and reproducible methods is required. DAVID LOAKES

multiplex sequencing Technique for the parallel *sequence analysis of multiple different DNA *fragments. These fragments are immobilized on membranes (→ immobilization of nucleic acids) and identified by *hybridization with known synthetic *oligonucleotides. Detection can be achieved by including either radioactive or fluorescent labels. SUSANNE BRAKMANN

mutagen Substance with the potential to induce *mutations. Mutagens are reactive towards DNA and produce chemically modified *nucleotide residues which are either misinterpreted during *replication or stall a moving *replication fork. HANS-JOACHIM FRITZ

mutagenesis Process generating *mutations, i.e. structural changes, in DNA (RNA) that are copied into offspring molecules. Mutagenesis can occur spontaneously or be experimentally induced by physical agents such as radiation (X-ray, ultraviolet light, etc.) or mutagenic chemicals (→ mutagen). In most cases mutagenesis proceeds via a pre-mutagenic lesion, i.e. a structure that is as such alien to DNA, but prone to be misinterpreted when serving as *template in *replication. Examples for pre-mutagenic *DNA lesions are *cis-syn* *thymine dimers resulting from ultraviolet exposure and O^6-me-G residues. Mutagenesis is also employed in genetic engineering in the form of *directed mutagenesis (precise construction of structurally predefined mutations) or *random mutagenesis (by which a DNA segment is sprinkled to a lower or higher degree with random mutations). HANS-JOACHIM FRITZ

mutation Change of a heritable trait such that the new trait is also stably heritable. Mutations have their molecular underpinning in changes of DNA (RNA) sequence; they are caused by copying errors or via structural alterations introduced into DNA (RNA) residues by the action of mutagenic agents of physical (ultraviolet light, ionizing radiation) or chemical (→mutagen) nature. Mutations can be classified either by their structural nature (→deletion mutation, →insertion mutation, →substitution mutation), by their effect on coding properties (→missense mutation, →nonsense mutation, →silent mutation, →frame shift mutation) or by their effect on a given *phenotype (→down mutation, →null mutation, →up mutation, →gene mutation).
HANS-JOACHIM FRITZ

mutational hotspot Site in a DNA sequence where *mutations occur more frequently than at other sites (→hotspot).

N

N →nucleobase

NAD →nicotineamide adenine dinucleotide

NADH Reduced form of *nicotinamide adenine dinucleotide

NADP Abbreviation for nicotineamide adenine dinucleotide phosphate (→nicotineamide adenine dinucleotide).

NADPH Reduced form of nicotineamide adenine dinucleotide phosphate (→nicotineamide adenine dinucleotide).

NAIM →nucleotide analog interference mapping

naked DNA Pure DNA which is neither attached to nor carried by proteins (e.g. *histones, virus proteins).

nanogold Labeling system for biological molecules (e.g. proteins, DNA, etc.). The technique is very similar to labeling with *fluorescent dyes. Labels, nanometer-sized gold clusters, are chemically crosslinked to the target molecule via disulfide or amide bonds. The visualization of labeled material can be achieved through electron microscopy or, after enhancement by *silver staining, the signal is visible to the naked eye, e.g. on a simple polyacrylamide gel or under a light microscope. SLAWOMIR GWIAZDA

nascent DNA DNA strand the elongation of which is in process (→DNA replication).

nascent RNA RNA strand the elongation of which is in process (→transcription).

native conformation *Conformation of a molecule (e.g. protein, nucleic acid) found under native, physiological conditions.

native DNA DNA found under natural, physiological conditions. Native DNA can be denatured by applying denaturing agents (e.g. urea) or unusual metal ion concentrations, or by dehydrating or by heating. SLAWOMIR GWIAZDA

ncDNA →non-coding DNA

N-conformation *Conformation in a five-membered furanose ring characterized by a *pseudorotation phase angle in the range $0° \leq P \leq 36°$ (northern part of the pseudorotational cycle). In double-stranded *A-form RNA or A-form DNA, the *ribose sugars are C3′-endo corresponding to an N-conformation. KLAUS WEISZ

ncRNA →non-coding RNA

NDP →nucleoside diphosphate

nearest-neighbor interaction The stability of a *DNA duplex depends on the identity and orientation of neighboring *base pairs. With the knowledge of the relative stability ($\Delta G°$) of each nearest-neighbor interaction the stability of RNA–DNA *hybrid *duplexes as well as RNA–RNA and DNA–DNA duplexes can be predicted. The nearest-neighbor interaction is also used for calculation of *oligonucleotide *melting temperatures. Physical parameters of nearest-neighbor base pairs can be used to optimize *oligonucleotide libraries of *DNA microarrays. BETTINA APPEL

negative supercoiling →supercoiled DNA

Nucleic Acids from A to Z: A Concise Encyclopedia. Edited by Sabine Müller
Copyright © 2008 WILEY-VCH Verlag GmbH & Co. KGaA, Weinheim
ISBN: 978-3-527-31211-5

neomycin Broad-spectrum *aminoglycoside antibiotic, naturally produced by the bacterium *Streptomyces fradiae*. It is effective against a wide range of Gram-negative (e.g. *Escherichia coli*) and most Gram-positive bacteria, and extremely nephrotoxic, especially compared to other aminoglycosides. Neomycin is a very active and *triplex-selective stabilization agent without affecting duplex DNA or RNA. Neomycin shows *in vitro* inhibition of Tat/TAR interaction, e.g. the inhibition of Tat-derived peptides to the *trans*-activating region (TAR) of HIV-1 RNA. Due to adverse reactions, neomycin is not given as a therapeutic in HIV therapy, but it is useful for invention of structural-based inhibitors. Neomycin was found to be a potent inhibitor of the *hammerhead ribozyme cleavage reaction. →interaction of nucleic acids with aminoglycosides.
BETTINA APPEL

nested double pseudoknot Certain type of *pseudoknot architecture appearing, for example, in the *hepatitis delta virus ribozyme or the *Diels–Alderase ribozyme. Here two helical segments form two parallel stacks that are joined by several strand-crossovers. SABINE MÜLLER

nested polymerase chain reaction Second round of *polymerase chain reaction (PCR) amplification which uses *primers located internally to those used in the first round of PCR, the first PCR product acting as a new *template for the second round. PCR is a powerful tool for amplifying small quantities of DNA from complex mixtures. However, when the desired template is at low concentration in a complex mixture of DNA then *amplification requires many cycles of PCR and the use of primers that may bind to other loci. This will often result in the amplification of other incorrect or undesired sequences. The use of a second set of PCR primers (nested primers) which bind within the first PCR product enables further amplification to give a shorter product, improving sensitivity without altering specificity. Thus, any undesired DNA sequences amplified during the first round of PCR are less likely to be amplified during the second round. DAVID LOAKES

nested primer →nested polymerase chain reaction

neurospora VS ribozyme →Varkud satellite ribozyme

neutral insertion →neutral mutation

neutral mutation Mutational change in a DNA or RNA sequence, which does not cause a change in the translated protein. The possibility of a neutral mutation is based on the fact, that there are more possible *codons (64) than coded amino acids (21), thus some amino acids are coded by multiple codons. A change of the codon does therefore not strictly change the corresponding amino acid (→wobble hypothesis). SLAWOMIR GWIAZDA

N-formylmethionine tRNA →formylmethionyl-tRNA

nick Occurs when the *phosphodiester bond between two neighboring *nucleotides breaks up in one strand of a *double-stranded DNA (dsDNA). This can be achieved by an *endonuclease or by damaging factors like simple mechanical stress. *In vivo*, a nick in a dsDNA is detected and repaired by the intracellular DNA damage repair system. SLAWOMIR GWIAZDA

nicked circular DNA *Double-stranded DNA forming a circle with a *nick in one strand. It is most commonly found during change of topology (winding or unwinding) of a circular DNA catalyzed by *topoisomerase I. SLAWOMIR GWIAZDA

nick translation Method that uses the mechanism of the intracellular DNA damage repair system in a protocol in order to create labeled DNA. The target DNA is treated with *DNase to create one strand *nicks. The cellular DNA damage repair system, in this case *DNA polymerase I, detects these nicks and starts the repair by removing some *nucleotides from the nicked strand with its *exonuclease activity. The gap is filled subsequently with specially labeled or tagged *nucleotides (e.g. radioactively). SLAWOMIR GWIAZDA

nicotineamide adenine dinucleotide (NAD) Pyridine *nucleotide coenzyme that is involved in biochemical redox processes of a large number of NAD-specific substrates. In the oxidized form, the pyridinium cation of nicotineamide is bound via an *N*-*glycosidic bond to C1 of D-*ribose, which in turn is conjugated with *adenosine via a *pyrophosphate bridge. In a redox reaction, NAD becomes reduced by reversible binding of hydrogen (which actually is a hydride transfer reaction). Due to the positive charge of the pyrimidinium cation, the oxidized form of NAD is often symbolized as NAD$^+$, while the reduced form is termed NADH. NAD occurs in living cells predominantly in the oxidized form and serves as coenzyme for oxidoreductases.
SABINE MÜLLER

nicotineamide adenine dinucleotide phosphate (NADP) Differs from *nicotineamide adenine dinucleotide (NAD) by an additional phosphate residue at the 2'-position of *adenosine. NADP functions in the same way as NAD as a coenzyme in redox reactions. However, in cells it occurs mainly in the reduced form and as such acts as a reducing agent in synthetic processes catalyzed by dehydrogenases and hydrogenases.
SABINE MÜLLER

NMP →nucleoside monophosphate

NMR studies of nucleic acids →nuclear magnetic resonance spectroscopy

non-canonical base pair All pairs of nucleobase interaction that do not fulfill the rule A pairs with T/U and C pairs with G (and *vice versa*), and thus are not *Watson–Crick base pairs. SABINE MÜLLER

non-coding DNA Describes those DNA regions that do not contain the required instructions for *translation into protein.

non-coding RNA (ncRNA) Summarizes those RNAs which are transcribed endogenously without being translated into a protein product. They are presumably biologically relevant. The most important examples for ncRNA include small RNAs involved in *messenger RNA (mRNA) processing/*splicing, *transfer

RNA (tRNA), and *ribosomal RNA (rRNA), the latter of which are involved in protein *translation. In more recent years many new small non-coding RNAs have been identified including (a) *small nuclear RNAs that are found within the nucleus of eukaryotic cells within *small nuclear ribonucleoproteins and that are involved in a variety of important processes such as *RNA splicing and the control of *telomeres, (b) *small nucleolar RNA which is involved in the chemical modification of rRNAs and contained within *small nucleolar ribonucleoprotein complexes, (c) the huge family of *micro-RNA which is thought to regulate *gene expression via interactions of their partially complementary sequences with the *5'-untranslated region of their target mRNAs, and (d) *guide RNA which functions in *RNA editing. The *gene that encodes a ncRNA is termed a ncRNA gene, sometimes also a *RNA gene, and seems to include also some of the so-called *pseudogenes. GEORG SCZAKIEL

non-coding region → non-coding DNA

non-coding sequence → non-coding DNA

non-polar nucleoside isosteres Molecules that are shaped like DNA or RNA *nucleosides, but are unable to form standard Watson–Crick *hydrogen bonds (→Watson–Crick base pair). *Pyrimidines are replaced with substituted benzenes and *purines are replaced with substituted benzimidazoles or 4-azabenzimidazoles. Exocyclic groups are also modified; amino groups are replaced with methyl groups and carbonyls are replaced with fluorines. As a result, the molecules are quite low in polarity, although they adopt structures much like the natural nucleosides in solution and in DNA helices. The *thymidine analog F and the *guanine analog H are nearly identical in size and shape to the respective natural bases. The *cytosine analog D and the *adenine analogs Z and Q are good, but imperfect isosteres because a ring nitrogen is replaced by a CH group in these compounds.

Despite their inability to form Watson–Crick hydrogen bonds, some of these molecules are able to retain normal DNA function. For example, adenine is paired opposite F by some *DNA polymerases with efficiency and fidelity nearly equal to that of natural thymidine. A series of non-polar thymidine shape analogs, based on F, with the fluorines replaced with hydrogens, chlorines, bromines or iodines, has been used to systematically probe the *active site size and tightness of polymerases in subangstrom increments. The dichloro variant of F was shown to be replicated in living bacteria.

Non-polar nucleoside isosteres have been widely used to probe the roles of *hydrogen bonding, *base pair size and base pair shape in the recognition of DNA or RNA by enzymes. These applications include probing binding and *bending of DNA by *transcription factors, recognition by *DNA repair proteins, RNA recognition in *RNA interference, and *active site sterics in *DNA polymerases. ADAM P. SILVERMAN, ERIC T. KOOL

nonsense codon →stop codon

nonsense mutation Change in coding DNA sequence, often a *point mutation by which (at the RNA level) a coding *triplet is converted into one of the three *stop codons: UAG (*amber*), UAA (*ochre*) and UGA (*opal*), Any particular nonsense mutation can, therefore, be referred to as an *amber*, *ochre* or *opal* mutation (→missense mutation). HANS-JOACHIM FRITZ

Northern blot Transfer of RNA molecules onto a carrier. In molecular biology, the Northern blot technique is used for identification and quantification of RNAs expressed in cells. In most cases a RNA population is separated by size during *gel electrophoresis followed by transfer of the RNAs onto a membrane (diazobenzyloxymethyl filter, nitrocellulose or nylon). RNA sequences to be analyzed can be detected by *hybridization with labeled RNA or DNA probes (→blotting). IRENE DRUDE

North-western technique Method that combines the *Northern blot and the Western blot techniques for identification of protein–RNA interactions. A protein population is first separated by size during *gel electrophoresis, followed by *blotting onto a membrane. The membrane is treated with buffer containing labeled RNA. Proteins that are able to bind RNA molecules can be identified based on the label (radioactive, *fluorescent dye or other specific label which can be identified by an antibody). →blotting. IRENE DRUDE

nRNA →nuclear RNA

NTP →nucleoside triphosphate

nuclear DNA DNA contained within the nucleus of eukaryotic organisms.

nuclear genes Genes located in the cell nucleus of an eukaryote.

nuclear magnetic resonance (NMR) spectroscopy Powerful method for the study of structure and dynamics of nucleic acids in solution and their interactions with *ligands such as proteins, other nucleic acids, molecules of low molecular weight, ions and solvent molecules. For NMR spectroscopic studies, an approximately 0.1–1 mM solution of the nucleic acid of interest is needed. The sample is dissolved in 0.3–0.5 mL buffer solution. NMR investigates the properties of nuclear spins. The spins of nuclei report on their chemical environment, and on the number of bound atoms, their relative distances and orientation. In most practical applications, the NMR properties of ^{1}H, ^{2}H, ^{13}C, ^{15}N, ^{31}P, and ^{19}F spins are of interest for NMR of nucleic acids.

Isotope labeling. While the natural abundance of the NMR-active but not radioactive nuclei ^{1}H, ^{31}P and ^{19}F is 100% or close to 100%, other low-abundance NMR-active nuclei have to be specifically incorporated. The availability of methods for the preparation of milligram quantities of RNA in isotope (^{13}C, ^{15}N)-labeled form has been a prerequisite for most modern NMR studies.

Information content from NMR studies. The current size limit of NMR on RNAs is at about 100 nucleotides. There is a wealth of

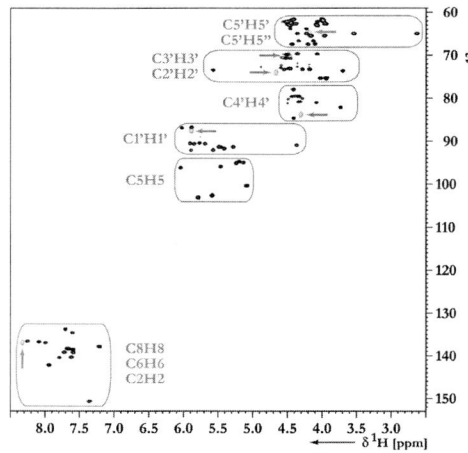

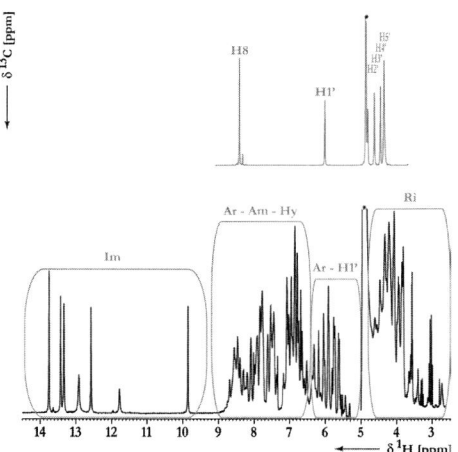

Nuclear magnetic resonance figure 1: The 1D ^{1}H- and 2D ^{13}C-HCQC spectra of the nucleotide triphosphate guanine (red) and of the RNA 14mer cUUCGg hairpin molecule (black). The typical spectral regions are assigned by grey boxes (Im: imino ^{1}H, Ar: aromatic ^{1}H, Am: amino ^{1}H, Ri: ribose ^{1}H); in the 1D ^{1}H spectrum of GTP, resonance assignment is annotated.

information that can be derived from NMR spectroscopic studies:

- *Base pairing pattern: this includes standard and non-standard *Watson–Crick-type base pairs, and allows determination of *secondary structure elements of RNA and determination of base pair dynamics.
- Information about conformational equilibria such as between *hairpin and *duplex structures.
- Site-specific information about ion binding to RNA.
- The local structure and dynamics and the overall structure of RNA derived.
- Mapping of interaction surfaces of RNA with small *ligands, other RNAs or proteins.
- Time-resolved NMR studies of *RNA folding.

Resonance assignments. The first NMR-spectroscopic step is to assign every NMR-active atom (^{1}H, ^{13}C, ^{15}N, ^{31}P) in the molecule to its respective resonance in the NMR spectra. The one-dimensional (1D) ^{1}H-NMR and the two-dimensional (2D) ^{1}H,^{13}C-HSQC spectra show that the NMR signals of specific atoms appear at distinct position within spectra; their resonances appear at characteristic chemical shifts.

NMR parameters and structure calculation. A wide range of NMR parameters can be measured that contain structural information. The following section will give a brief introduction into the origins of the respective NMR parameters and their utilization in the structural refinement of RNA molecules.

Distances from nuclear Overhauser effects (NOEs). The NOE contains information about the distance of two NMR-active nuclei. It is observed between nuclei that are less than 5 Å apart. A typical 2D NOESY experiment is shown on next page. In this experiment, each cross-peak indicates a short distance between two atoms in the 17mer RNA.

The distance restraint (r_{ij}) used in an NMR structure calculation is derived from the signal intensity of the cross-peak (I^{NOE})

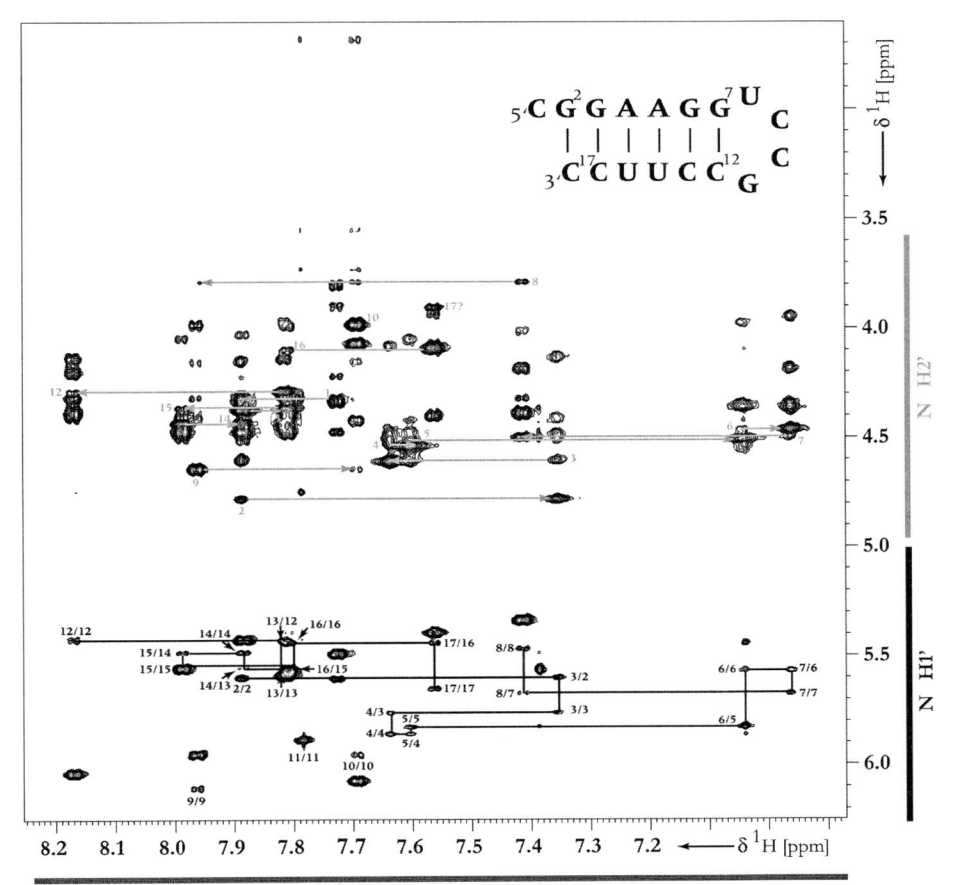

Nuclear magnetic resonance figure 2: The 2D ^{1}H-^{1}H NOESY spectrum of a RNA hairpin structure recorded at 298 K and 600 MHz on a D$_2$O sample. The connectivity walk that provides first structural information for $H_i^{ar}/H1'_{i-1} \rightarrow H_i^{ar}/H1'_i \rightarrow H_{i+1}^{ar}/H1'_i$ is shown by the black lines and for $H_i^{ar}/H2'_{i-1} \rightarrow H_i^{ar}/H2'_i \rightarrow H_{i+1}^{ar}/H2'_i$ it is shown by yellow lines.

between the two protons (i, j) and is given by:

$$r_{ij} \approx \left(c_{cal} I_{ij}^{NOE} \right)^{-\frac{1}{6}}$$

The calibration constant is defined from a reference cross-peak between nuclei with known distance, e.g. $r_{H5,H6}$ in *pyrimidine *nucleobases:

$$c_{cal} = \frac{r_{ref}^{-6}}{I_{ref}^{NOE}}$$

Torsion angles from scalar J coupling constants. Homo- and heteronuclear coupling constants provide inside into the *conformation around torsion angles. The measurement of scalar couplings is especially important for RNA because of the low proton density compared to proteins and the large number of free torsion angles. Additionally, in non-canonical regions of RNA such as *bulges and *loops, conformational heterogeneity is often observed and

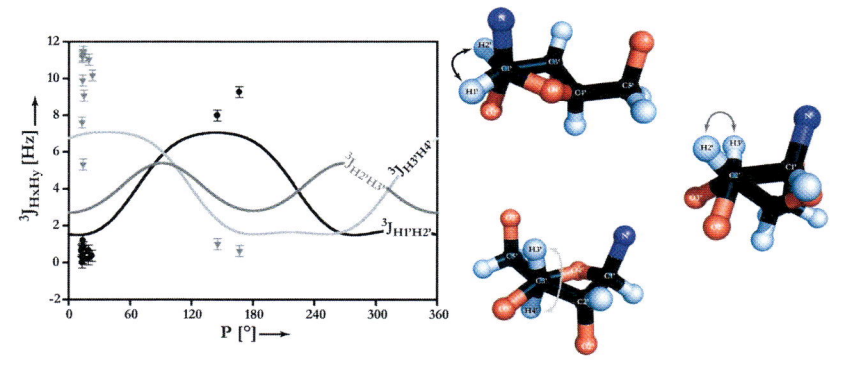

Nuclear magnetic resonance figure 3: Karplus curve for three $^3J_{HxHy}$ couplings describing the dependence on the conformation of the ribose ring in RNA (the conformation is described as the pseudorotation phase $P/[°]$ that is defined. The angles described by the displayed scalar couplings are indicated on the right side of the figure for a ribose unit in stick-and-ball representation populating the C3'-endo conformational space.

the use of NOE distance restraints is often insufficient to determine such conformational dynamics.

In particular, vicinal $^3J_{XY}$ scalar couplings between nuclei X and Y three bonds apart are used to determine torsion angles, because they directly report on the underlying torsion. The dependence of a certain coupling on a certain torsion angle is mainly derived from a parameterization, the so-called Karplus equation, of the respective J-coupling on locked nucleotides with known conformation.

Hydrogen bonding from scalar through-space J coupling constants. The electron density between hydrogen-bonded nitrogen atoms in base pairs gives rise to through-space scalar $^{2h}J_{NN}$ couplings in the range up to 7 Hz. Such information can be used to derive the geometry of hydrogen bonding in Watson–Crick, but also non-*Watson–Crick base pairs.

Long-range structural information from residual dipolar couplings (RDCs). RDCs depend on the residual orientation of the molecule relative to the external magnetic field due to an anisotropic tumbling of the RNA. A residual orientation can be achieved by the addition of a cosolvent, a so-called alignment medium, which sterically hinders the isotropic reorientation of the molecule. The widely used alignment media for nucleic acids are solutions of the filamentous bacteriophage Pf1 that has a high tolerance to the ionic strength of the solute and is robust over a wide range of temperature. Pf1 is a rod-like object orientated parallel to the external magnetic field that is substantially negatively charged and therefore induces alignment of the likewise negatively charged nucleic acids by steric and electrostatic repulsion.

NMR structure calculation of RNA. A structure calculation is defined as the minimization problem of a target function that measures the agreement between a conformation of a macromolecule and the set of restraints derived from NOEs, J couplings, RDCs, etc. Two major techniques for solving this minimization problem are (a) the metric matrix distance geometry (DG) and (b) the Cartesian or torsion angle restrained molecular dynamics simulation (rMD). The first technique is used

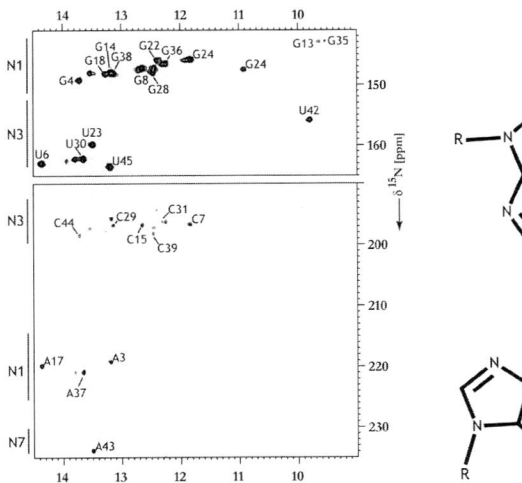

Nuclear magnetic resonance figure 4: Utilization of the scalar $^{2h}J_{NN}$ couplings to determine the base-pairing pattern in a RNA molecule (Diels–Alderase ribozyme in 4 mM Ca^{2+} and in the presence of an inhibitor). Left panel shows the diagonal and cross-peaks of an HNN-COSY experiment recorded at 800 MHz and 298 K. The upper part of the spectrum displays the peaks that arise from imino groups, the peaks for guanine and uracil are well separated due to the different chemical shift of N1 and N3, respectively. Also the cross-peaks are well dispersed due to the chemical shift differences for the nitrogen atoms in different nucleobases. Not only can the canonical Watson–Crick base pairs (AU Watson–Crick base pair illustrated in the lower part of the right panel) be identified, but due to the diverse chemical shifts also non-canonical base pairs, e.g. reverse Hoogsteen base pairs (as in the example between A43 and U23; the geometry of this base pair is displayed in the upper half of the right panel; the ellipsoids represent the electron density that procures the scalar couplings).

in programs such as DIG-II and DIANA, the latter in programs such as AMBER, CHARMM, XPLOR and CNX/CNS. In both cases, the aim is to generate an ensemble of molecular structures that fits with experimentally derived restraints. In the case of rMD, the structures are calculated using the NMR restraints and an energy minimization with a potential energy function – also called target function – that is composed of classical terms ($V_{classic}$: describing the energy of the molecule itself) and of terms describing the NMR observables (V_{NMR}):

$$V_{tot} = V_{classic} + V_{NMR}$$

with

$$V_{classic} = \omega_{bond} V_{bond} + \omega_{angle} V_{angle}$$
$$+ \omega_{dihedral} V_{dihedral}$$
$$+ \omega_{improper} V_{improper}$$
$$+ \omega_{van\,der\,waals} V_{van\,der\,waals}$$
$$+ \omega_{electrostatics} V_{electrostatic}$$

and

$$V_{NMR} = \omega_{NOE} V_{NOE} + \omega_{J\text{-coupling}} V_{J\text{-coupling}}$$
$$+ \omega_{H\text{-bond}} V_{H\text{-bond}} + \omega_{RDC} V_{RDC}$$

The ω terms describe the force constants that scale each energy contribution independently. The classical terms are fixed

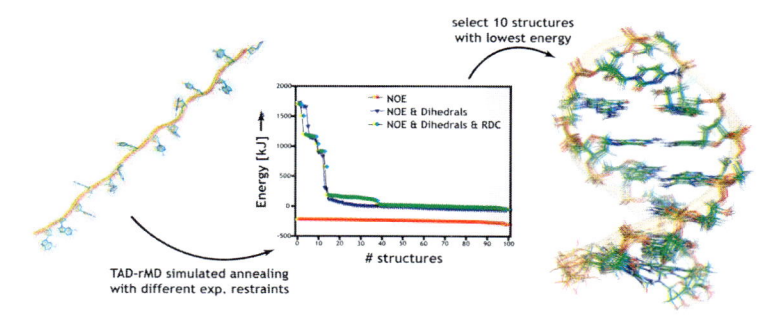

Nuclear magnetic resonance figure 5: Schematic description of the structure calculation process. Starting from a randomly elongated RNA structure, the torsion angle dynamics (TAD)-rMD calculation is conducted and yields 100 different structures that are subsequently sorted by their final energy after minimization. The 10 best structures are then assembled in a bundle of structural models representing the molecular structure best according to the input parameters.

terms of the force field; the NMR terms are heuristically chosen.

The structure calculation of the 14mer cUUCGg *tetraloop RNA is illustrated above below. Starting the simulated annealing protocol from a randomized RNA chain and with random variation of the initial conditions (=velocities), 100 structures are calculated and sorted afterwards by the total energy of the respective structure. Then the 10 structures displaying the lowest energies are selected and combined to a bundle that represents the structure best according to the input NMR structural parameter.

Three different indicators can critically evaluate the quality of an NMR structure: convergence, precision and accuracy.

- *Convergence*. How many of the 100 calculated structures exhibit a total energy in the same order as the best structures? This does not mean that an energetically converged structure calculation must have a lower structural root-mean-square deviation (RMSD) value compared to a calculation where a steep increase in energy is over the calculated structures is monitored.

- *Precision*. How large is the RMSD value between the structures of the ensemble? How large is the RMSD value between the experimentally measured input parameters and the same parameters back-calculated from the structure ensemble?

- *Accuracy*. How large is the difference between the calculated and the true structure?

Boris Fürtig, Harald Schwalbe

nuclear RNA RNA molecules that are found within the nucleus of eukaryotic cells. For instance *small nuclear RNA form a class of small RNA molecules involved in several important processes such as *RNA splicing or removal of *introns from precursor *messenger RNA transcribed from a DNA *template in the nucleus. →heterogeneous nuclear RNA.

Mauro Santos

nuclease Enzyme (protein or RNA) that cleaves nucleic acids. Nucleases are important and responsible for the turnover (processing and decomposition) of nucleic acids in all living cells, and can also be found in some (retro-) viruses. Main representatives are ribo- and deoxyribonucleases (*RNases and *DNases), and both can be divided into *endonucleases and *exonucleases. ULI HAHN

nucleic acid *Biopolymer composed of *nucleotides. Nucleic acids belong to the most important components of all living organisms. In 1869, Friedrich Miescher isolated nucleic acids for the first time from pus cells and called them "Nuclein". Due to the acidic properties of the molecule, the name "nucleic acid" was later introduced. Deoxyribonucleic acid (→DNA) and ribonucleic acid (→RNA) constitute the two major classes of nucleic acids. RNA and DNA have several similar structural properties, but clearly differ in function. While DNA is the carrier of genetic information, RNA is mainly involved in realization and regulation of gene expression (→messenger RNA, →ribosomal RNA, →transfer RNA, →ribosome, →noncoding RNA, →riboswitch), although it can also act as store of genetic information.

The molecular mass of nucleic acids varies from 20 to 10^6 kDa. They are composed of three different parts: (a) the *purine and *pyrimidine bases *adenine, *guanine, *cytosine and *thymine (in DNA) or *uracil (in RNA), and a number of *modified bases particularly in RNA (→rare bases), (b) the sugar unit being *2′-deoxy-D-ribose in DNA and *D-ribose in RNA, and (c) phosphoric acid. *Mononucleotides are connected by *phosphodiester bonds between the 3′-position of one nucleotide and the 5′-position of the following one resulting in a linear chain of nucleotides. The *nucleobases of nucleic acid strands are involved in specific *hydrogen bonds as well as *van der Waals, *hydrophobic and *electrostatic interactions allowing for the formation of a rich variety of *secondary and *tertiary structures. Nucleic acids absorb ultraviolet light with a maximum around 260 nm due to the conjugated double bonds in the heterocyclic ring systems of nucleobases (DNA structure, →RNA structure). SABINE MÜLLER

nucleic acid analogs →modified DNA, →modified RNA, →morpholino nucleic acids, →GNA, →TNA, →CeNA, →ANA, →HNA

nucleic acid backbone →sugar–phosphate backbone

nucleic acid base →nucleobases

nucleic acid-based computing →DNA computing

nucleic acid catalyst →deoxyribozyme, →ribozyme

nucleic acid chemistry Chemistry of *nucleosides, *nucleotides and *nucleic acids. This field of research includes *nucleic acid labeling.

nucleic acid circuits →deoxyribozyme and ribozyme circuits

nucleic acid library →sequence pool

nucleic acid metabolism Consists of three main processes carried out in living cells: *replication, repair and recombination. Nucleic acid synthesis is an anabolic process, whereas destruction of nucleic acids is a catabolic procedure (DNA replication, →DNA repair, →DNA recombination). BETTINA APPEL

nucleic acid probe Probes are used to detect a known nucleic acid (RNA or DNA) within a sample. Such probes in turn consist of DNA, RNA or an artificial nucleic acid motif and a reporter system. The *sequence of the probe is designed to be complementary to the target, thus enabling the probe to hybridize to the target nucleic acids. As reporter groups chromophores (→fluorescence labeling), radioactive isotopes (→radioactive labeling) or groups which enable detection by a subsequent enzyme reaction (→biotin labeling) can be used.

Nucleic acid probes can be used for heterogeneous detection of target nucleic acids. Heterogeneous assays rely on immobilization of either the target or the probe to a solid or gel phase. Commonly used formats include *Southern blotting, *Northern blotting, *dot blotting (→blotting), *DNA chips or *DNA microarrays. The heterogeneous phase is subjected to *hybridization of the probe. After hybridization, unbound excess probes are washed away. Areas in which binding had occurred are detectable by means of a reporter group that is usually appended to the soluble binder.

Nucleic acid probes are also used in homogeneous detection of target nucleic acids without removal of excess amounts of probes. Hence, the hybridization event must be accompanied by a measurable change of a certain probe property. One of the advantages of homogeneous nucleic acid detection is that nucleic acid hybridization can be monitored in real-time, if desired even within a living cell. *In vitro detection is often used in connection with the *real-time polymerase chain reaction (PCR) for amplifying target nucleic acids. Many probes for real-time PCR rely on the distance dependent interaction of two chromophores (→FRET studies). Commonly used probe designs are *molecular beacons, *TaqMan probes and adjacent probes. Chromosomal localization of *genes (→gene localization) can be achieved by treatment of fixed cells with nucleic acid probes. The predictable mutual recognition of probe and target nucleic acid allows the detection and localization of target nucleic acids even in complex biological environments.
LUCAS BETHGE, OLIVER SEITZ

nucleic acid–protein interaction →DNA-binding proteins, →protein–DNA interaction, →protein–RNA interaction

nucleic acids for crystallographic studies
Crystallization (→crystallization of nucleic acids) demands comparatively large amounts of chemically and conformationally homogeneous material. Although one technological spin-off of the structural genomics initiatives has been the availability of crystallization robotics that reduce typical sample consumption by at least an order of magnitude, a successfully completed crystallographic project normally still requires several milligrams of sample. Only few nucleic acid targets such as *transfer RNAs or *ribosomes (which can be regarded in a way as large RNAs decorated with some proteins) are available via native purification from cells. Production of DNA for crystallization usually relies on the powerful chemical synthesis using *phosphoramidite precursors. Chemical synthesis of RNA is more demanding and less efficient than DNA synthesis because an additional protection of the 2′-hydroxyl group is required. Comparatively bulky fluoride-labile 2′-silyl ether protection groups can provide synthetic RNAs suitable for crystallization. An alternative providing higher coupling yields is provided by 5′-silyl-2′-acetoxy ethyl orthoester chemistry. Still, chemical synthesis remains the method of choice only for the production of rather short *oligonucleotides (up to about 50 nucleotides). It has the advantage, however, that *modified nucleotides can be incorporated at specific positions. These include, for example, modifications at the

2′-hydroxyl group that may be useful to study the mechanism of *ribozymes or brominated and iodinated building blocks that could be useful for phase determination. For the production of RNA, an efficient enzymatic method, based on *T7 RNA polymerase (RNAP), has been developed. The rather large amounts of T7 RNAP required can be conveniently produced recombinantly. Synthetic DNA can be employed as a *template, in which only the *promoter region has to be double stranded. Alternatively, the template region for the RNA of interest can be cloned into a high-copy number *plasmid preceded by a T7 promoter and terminating in a *restriction site. After plasmid preparation and linearization, RNA can be generated by *run-off transcription. Enzymatic production poses little limits in terms of the size of the target RNA. However, efficient T7 RNAP transcription requires the transcript to start with at least one and preferably more than one guanosine imposing unwanted sequence restraints on many RNAs of interest. In addition, T7 RNAP has the tendency to add 1–3 non-encoded nucleotides at the 3′-end of the transcript, generating unwanted heterogeneity that later can hamper crystallization. These limitations have essentially been resolved by adding small *ribozyme units to the 5′- and 3′-ends of the *RNA gene. The RNA of interest can be released from the longer transcripts by induction of cleavage, e.g. by the addition of Mg^{2+} ions. Finally, chemical and enzymatic syntheses can be combined to derivatize long RNAs at specific positions. Synthetic and enzymatically produced fragments are aligned with the help of DNA splint oligos and ligated by T4 *DNA ligase or *RNA ligase.

For very short synthetic oligomers simple ethanol precipitation is sufficient as a means for purification for crystallization. Chromatographic methods provide powerful means for intermediate-size nucleic acids. The most general purification method is denaturing urea polyacrylamide gel electrophoresis, which can resolve even comparatively long molecules differing by a single nucleotide in length. The nucleic acid of interest is detected by ultraviolet shadowing, the band excised and eluted by washing or in an electric field. Affinity chromatography methods have been introduced as an alternative. The RNA of interest is produced by T7 RNAP transcription in conjunction with bordering ribozyme units. In addition, a protein-binding element is fused to one end. The product RNA can then be trapped on an affinity column of the corresponding protein and the target fragment can be selectively released from the column by sequentially inducing ribozyme cleavages. MARKUS WAHL

nucleobases Heterocyclic nitrogen bases that constitute *nucleosides. In DNA, *adenine, *cytosine, *guanine and *thymine are attached via an N-*glycosidic bond to C1 of* 2′-deoxyribose. In RNA, bases are covalently bound to *ribose with *uracil replacing thymine. In addition to these five common nucleobases, a number of modified nucleobases occur in certain *transfer

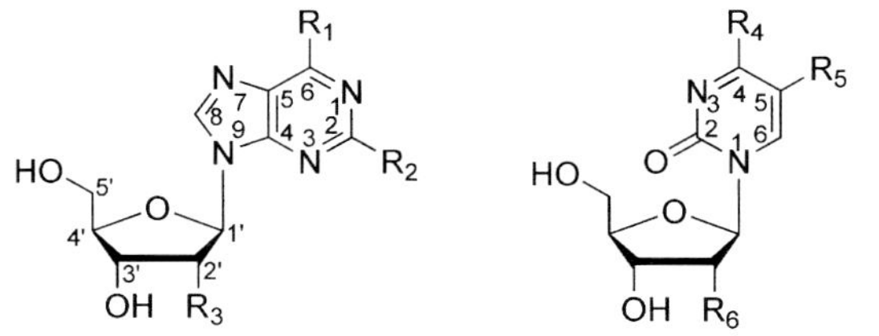

$R_1 = NH_2, R_2 = H, R_3 = H$
2'-deoxyadenosine

$R_1 = NH_2, R_2 = H, R_3 = OH$
adenosine

$R_1 = OH, R_2 = NH_2, R_3 = H$
2'-deoxyguanosine

$R_1 = OH, R_2 = NH_2, R_3 = OH$
guanosine

$R_4 = NH_2, R_5 = H, R_6 = H$
2'-deoxycytidine

$R_4 = NH_2, R_5 = H, R_6 = OH$
cytidine

$R_4 = OH, R_5 = CH_3, R_6 = H$
thymidine

$R_4 = OH, R_5 = H, R_3 = OH$
uridine

nucleosides

RNAs and *ribosomal RNAs, where they play important roles for function (→rare bases, →rare nucleotides). SABINE MÜLLER

nucleoside *N*-glycoside of heterocyclic nitrogen bases. Of particular biological importance are nucleosides composed of the *purine bases *adenine and *guanine, as well as the *pyrimidine bases *cytidine, *thymine (in DNA) or *uracil (in RNA) with pentoses. The sugar unit is either *ribose (in RNA) or *2'-deoxyribose (in DNA) in the furanose form. The C1 of the sugar residue is attached to N9 of the purine bases or to N1 of the pyrimidine bases via an *N*-*glycosidic bond. To differentiate between atoms of the sugar and the base, sugar C-atoms are termed 1' to 5'. In deoxynucleotides, ribose is replaced with 2'-deoxyribose. Some RNA molecules, in particular *transfer RNA, contain *rare nucleotides with chemically modified sugars or bases. Only the combinations of ribose or 2'-deoxyribose with the five bases as they appear in nucleic acids are properly termed nucleosides; however, often any combination of a sugar with a heterocyclic bases is called a nucleoside as well.
SABINE MÜLLER

nucleoside and nucleotide analogs Any derivative of natural *nucleosides or *nucleotides. In medicine, a nucleoside analog is a synthetic compound that is recognized as nucleoside by the cell machinery (→5-fluorouridine, →azidothymidine, →rare bases, →nucleic acid analogs).
SABINE MÜLLER

nucleoside diphosphate *Nucleoside in which a *pyrophosphate residue is attached to the 5'-hydroxyl group. Ribonucleoside and 2'-desoxyribonucleoside diphosphates are of biological relevance. SABINE MÜLLER

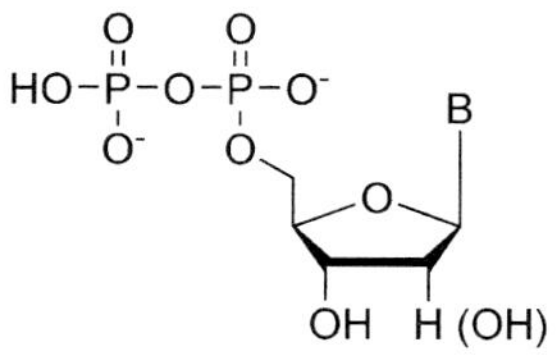

B = Adenine
Cytosine
Guanine
Thymin
Uracil

nucleoside monophosphate *Nucleoside in which a *pyrophosphate residue is attached to the 5′-hydroxyl group. Ribonucleoside and 2′-desoxyribonucleoside monophosphates are of biological relevance. SABINE MÜLLER

B = Adenine
Cytosine
Guanine
Thymin
Uracil

nucleoside phosphates →nucleotides

nucleoside triphosphate *Nucleoside in which a triphosphate is attached to the 5′-hydroxyl group. Ribonucleoside and 2′-desoxyribonucleoside triphosphates are of biological relevance. SABINE MÜLLER

B = Adenine
Cytosine
Guanine
Thymin
Uracil

nucleosome Basic repeating subunit of eukaryotic *chromatin made up of DNA and four pairs of *histones (→core DNA and →core particle).

nucleotide *Phosphoric acid monoester of a *nucleoside. Either the 5′-hydroxyl or the 3′-hydroxyl group is derivatized as a phosphoric acid ester; in case of *ribonucleosides, the 2′-hydroxyl group can also be conjugated to phosphoric acid. Furthermore, two hydroxyl groups of one nucleoside can be reacted with phosphoric acid to form *cyclic phosphates. 5′-Phosphates are of particular biological relevance. They appear as mono-, di- and triphosphorylated derivatives in which phosphoric acid, pyrophosphoric acid or triphosphoric acid is bound to the 5′-hydroxyl group. Nucleotides and 2′-deoxynucleotides are the monomeric building blocks of *oligonucleotides and nucleic acids. In particular, the 5′-nucleoside triphosphates serve as activated reactands in the synthesis of *polynucleotides (→nucleoside diphosphates, →nucleoside monophosphates, →nucleoside triphosphates, →replication, →transcription).

Cyclic 3′,5′-phosphates are of major importance in cellular metabolism (→cyclic phosphates). SABINE MÜLLER

nucleotide analog →nucleoside and nucleotide analogs

nucleotide analog interference mapping (NAIM) Chemogenetic approach to probe the contribution of a particular *functional group at every RNA *nucleotide position in a single experiment. The method utilizes a series of 5′-O-(1-thio) nucleoside analog triphosphates in a modification interference procedure. The *nucleotide analogs include a specific chemical modification to the base or sugar and γ-phosphorothioate substitution which serves as a chemical tag. The nucleotide analog triphosphate is randomly incorporated into a RNA *transcript, where the *phosphorothioate linkage can then be selectively cleaved by the addition of I_2 to produce a series of RNA cleavage products. As the phosphorothioate chemical tag is independent of the nucleotide analog, NAIM is generalizable to any analog that can be incorporated into a transcript by an *RNA polymerase. PASCALE ROMBY, PIERRE FECHTER

nucleotide composition Ratio of the four *nucleotides in an *oligonucleotide. The nucleotide composition has a strong effect on the *melting temperature of an oligonucleotide *duplex. SLAWOMIR GWIAZDA

nucleotide excision repair → DNA repair

nucleotide extrusion → base flipping

nucleotide flipping → base flipping

nucleotide incision repair → DNA repair

nucleotide overhang Protruding single-stranded ends of *double-stranded DNA or RNA, usually produced by *restriction enzymes, *nucleases or after *in vitro* annealing of *complementary strands. It is usually qualified as a 5'- or 3'-overhang, indicating which end of the nucleic acid protrudes.
GEMMA MARFANY,
ROSER GONZÀLEZ-DUARTE

nucleotide sequence For a DNA or RNA oligomer, given by a sequence of letters, which represent the *bases linked to the *phosphoribose backbone. The sequence is given by default in the 5' → 3' direction. The letters and corresponding *nucleotides are A for *adenine, C for *cytosine, G for *guanine and T for *thymine. Thymine is replaced by *uracil (U) in a RNA. It is also possible to give more choices for a specific position in the nucleotide sequence. For this purpose the alphabet was expanded by letters which represent every possible nucleotide combination. This alphabet was defined by the International Union of Pure and Applied Chemistry: A = adenine; C = cytosine; G = guanine; T = thymine; U = uracil; R = G, A; Y = T/U, C; K = G, T/U; M = A, C; S = G, C; W = A, T/U; B = G, T/U, C; D = G, A, T/U; H = A, C, T/U; V = G, C, A; N = A, G, C, T/U. SLAWOMIR GWIAZDA

nucleotide triplet → base triplet, → anticodon, → codon

null mutation *Mutation causing the complete disappearance of a *phenotype. A null phenotype may, for example, be caused by the deletion of an entire *gene (→ down mutation, → up mutation).
HANS-JOACHIM FRITZ

O

ochre codon UAA *stop codon decoded by specific ochre *suppressor *transfer RNA. Artificial name for a *nonsense mutation derived from a laboratory strain showing the ochre mutation (Oc). Following the lead of *amber codons, the artificial naming continued with the pun "ochre". STEFAN VÖRTLER

ochre mutation Special case of a *nonsense mutation.

Okazaki fragments At the *lagging strand, the synthesis of DNA is discontinuous in short pieces that are termed Okazaki fragments after their discoverer Reiji Okazaki. In eukaryotes the *fragments are 100–200 nucleotides in length, in prokaryotes they are 1000–2000 nucleotides in length. ANDREAS MARX, KARL-HEINZ JUNG

2′,5′-oligoadenosine Unusual oligonucleotide with the structure pppA-(2′p5′A)$_n$ with $n = 1$–10 that occurs as an activator of inactive RNase L, an enzyme that degrades *messenger RNA to inhibit protein biosynthesis in cells infected by viruses.

2′,5′-oligoadenosine synthetase Enzyme that supports the synthesis of *2′,5′-oligoadenosine, an oligonucleotide serving as activator for a RNA-cleaving enzyme (→2′,5′-linkage).

oligodeoxyribonucleotide →oligonucleotide

oligomer Compound (or molecule) that consists of repeating structural units.

oligonucleotide Short nucleic acid with up to about 50 *nucleotides that can be chemically synthesized (→oligonucleotide synthesis).

oligonucleotide array →DNA microarray

oligonucleotide labeling Incorporation of a label into DNA or RNA, usually for the purposes of being able to track the nucleic acid, to probe its function or to change its properties. Examples include *fluorescence labeling, *biotin labeling, *radioactive labeling and *isotope labeling. Several different approaches are available for oligonucleotide labeling using either *chemical labeling or *enzymatic labeling.

Chemical labeling enables incorporation of modified *nucleosides into specific positions of both DNA and RNA. The modifications can be chemically introduced at the 3′- or 5′-ends or at internal positions. Several atoms of a nucleoside can be modified, either at the nucleoside base or the sugar. Two different strategies have been employed for chemical labeling. One strategy is to prepare a modified nucleoside by chemical synthesis and incorporate it into the nucleic acid by the *phosphoramidite method of *oligonucleotide synthesis. The other strategy, often referred to as post-synthetic labeling, is to use the *phosphoramidite method to introduce a convertible nucleoside containing a reactive group into the nucleic acid, which is selectively modified after oligonucleotide synthesis with a labeling reagent of choice. The reactive group is usually a thiol or an amine. The main advantage of post-synthetic labeling is that once the nucleic acid containing the convertible nucleotide has been prepared, it allows for

Nucleic Acids from A to Z: A Concise Encyclopedia. Edited by Sabine Müller
Copyright © 2008 WILEY-VCH Verlag GmbH & Co. KGaA, Weinheim
ISBN: 978-3-527-31211-5

a rapid and efficient production of a variety of *modified nucleic acids. Another advantage is that sensitive labels, which would otherwise be unstable to the conditions of solid-phase oligonucleotide synthesis, can be incorporated at ease into nucleic acids.

Enzymes are also used to incorporate labels into nucleic acids and are mostly used for *end labeling, i.e. for incorporation of labels at the 3′- or 5′-end of the nucleic acid. *Kinases are used for adding a radioactive phosphate (^{32}P) to the 5′-end (→radioactive labeling) and *ligases can be used to incorporate a modified nucleotide at the 3′-end. The latter approach requires a modified nucleoside containing a 5′-phosphate. *Polymerases can also be used to incorporate labels in internal positions of DNA and RNA, and are usually employed for preparing oligomers containing isotope labels for biophysical studies of nucleic acids. This approach, however, cannot be used for a single site-specific incorporation of a label and is usually employed for uniform labeling of the oligomer. *T7 RNA polymerase can also be used to incorporate a *thiophosphate at the 5′-end of RNA during RNA synthesis from a DNA *template. The thiophosphate can subsequently be reacted with a variety of labeling reagents.
SNORRI TH. SIGURDSSON

oligonucleotide mutagenesis →sitedirected mutagenesis

oligonucleotide probe →nucleic acid probe

oligonucleotide synthesis The chemical synthesis of *oligonucleotides is most effectively achieved via solid-phase approaches based on phosphoramidite nucleoside building blocks (→phosphoramidite method). Other methods such as the *phosphodiester method, the *phosphotriester method or the *H-phosphonate method are of historical or minor importance nowadays.

The chemical synthesis of oligonucleotides is carried out on a solid phase by stepwise addition of appropriately activated and protected nucleoside building blocks until the desired sequence has been obtained. Then, all *protecting groups required during the strand elongation are removed together with the release of the oligonucleotide from the solid support. Length and integrity of the product are determined by the quality of coupling and deprotection procedures.

During the last 30 years, many research groups have contributed to make the synthesis of DNA oligonucleotides one of the most evolved chemical processes ever developed, reaching high perfection in terms of efficiency and automation. The *phosphoramidite method is now performed almost exclusively. It is based on nucleoside 3′-(2-cyanoethyl-N,N-diisopropylphosphoramidite) building blocks that are activated *in situ* by weak acids, most commonly by 1H-tetrazole and derivatives thereof. Their coupling to the 5′-hydroxyl group of the growing oligonucleotide chains results in phosphorous triesters (phosphite triesters) which are subsequently oxidized to the corresponding phosphoric acid triesters (phosphotriesters). The 5′-O-positions of the phosphoramidite building blocks are usually protected with the acid-labile 4,4′-dimethoxytrityl group that allows convenient in-line monitoring of the coupling efficiencies. After the DNA strand assembly, the release of the sequence from the solid support and the removal of the acetyl nucleobase and phosphodiester protecting groups is a one-step procedure carried out under basic conditions, commonly using concentrated ammonia in a water/ethanol.

The phosphoramidite method is also applied for the chemical synthesis of RNA. Compared to DNA, each *nucleotide unit within a RNA strand possesses a 2'-hydroxyl group which makes the RNA unstable under basic conditions. This hydroxyl group must be protected during oligoribonucleotide assembly. Since the RNA products are base labile, the removal of these supplementary 2'-O-protecting groups must be carried out separately after the basic deprotection step. As a consequence the choice of the 2'-hydroxyl protecting group is most stringent. They must be completely orthogonal to all other acid- and base-promoted operations encountered during synthesis and deprotection.

The large number of 2'-hydroxyl protecting groups reported can be divided into acid-labile, photolabile and fluorine ion-labile groups. Among them, the fluorine ion-labile *tert*-butyldimethylsilyl group has found the widest application (→TBDMS method). More recent approaches have focused very successfully on the ([triisopropylsilyl]oxy)methyl group (→TOM method) and on the acid-labile bis(acetoxyethyloxy)methyl group (→ACE method). RONALD MICURA

oligoribonucleotide →oligonucleotide

***onc* gene** →oncogene

oncogene *Gene whose product has the ability to transform the eukaryotic cell so that it will grow as a tumor cell. When the gene is endogenous, it is also called a proto-oncogene. Some viruses can be oncogene transductors, i.e. a cellular gene that has the potential to transform a cell has been incorporated into the viral *genome. In these cases, the viral and cellular genes are renamed as v-*onc* and c-*onc*, respectively. MAURO SANTOS

opal codon UGA *stop codon decoded by specific opal *suppressor *transfer RNA. Artificial name for a *nonsense mutation derived from a laboratory strain showing the opal mutation (Op). Following the lead of the *amber codon the artificial naming continued with the pun "opal".
STEFAN VÖRTLER

opal mutation Special case of a *nonsense mutation.

open reading frame (ORF) Stretch of *sense codons in a *messenger RNA embedded within the proper regulatory regions (→promoter, →terminator), having a size of larger than 100 amino acids, following a *start and ending with a *stop codon, which is likely to give rise to a polypeptide during protein biosynthesis (→translation, →reading frame) and easily identified by bioinformatics search algorithms. In eukaryotes the identification of ORFs is considerably more difficult due to interspersed insertion sequences, the *introns.
STEFAN VÖRTLER

operator →operon

operon Unit of bacterial *gene expression and regulation. It is a cluster of contiguous *genes transcribed from one *promoter and controlled as a unit that gives rise to a *polycistronic RNA. It includes one or more *structural genes and a control region. The control region contains a promoter which is the *binding site for the *RNA polymerase. Essential control elements are operators which are the binding sites for *repressor proteins and *activator binding sites. The *transcription of the structural genes takes place unless a repressor is bound to the operator or an activator is present. BEATRIX SÜß

optical density (OD) →absorbance

ORF →open reading frame

oriC *Escherichia coli* chromosomal *origin.

origin *Replication of DNA starts at a specific sequence of *nucleotides called the origin which is recognized by an origin recognition complex of proteins (ORC). The origin of the *Escherichia coli* *chromosomal DNA (oriC) is identified and its structure very well characterized. The DNA of eukaryotic *chromosomes has multiple origins.
ANDREAS MARX, KARL-HEINZ JUNG

orphan gene *Gene whose structure does not support a relationship to another gene.

outer sphere Coordination of a metal ion to a ligand via a water molecule, i.e. hydrogen bonding.

overexpression Expression level of a particular *gene or group of genes, which under certain conditions or after *transfection with a recombinant construct, is higher than usual or expected. GEMMA MARFANY, ROSER GONZÀLEZ-DUARTE

overhang Protruding end (→nucleotide overhang).

overlapping ends Protruding ends of two molecules of *double-stranded DNA, usually produced by *restriction enzymes or *nucleases, which are complementary and can anneal and be ligated. A more accurate term would be *cohesive ends.
GEMMA MARFANY, ROSER GONZÀLEZ-DUARTE

overlapping genes *Genes in which the same DNA sequence codes for two proteins using different *reading frames. They are relatively common in DNA and RNA viruses of both prokaryotes and eukaryotes. There are several examples in bacterial and eukaryote *genomes but, in general, overlapping genes are rare in these organisms. There are three classes of overlapping genes: unidirectional, convergent and divergent. In unidirectional overlapping genes the 3'-end of one overlaps with the 5'-end of the other, in convergent overlapping genes the 3'-ends overlap and in divergent overlapping genes the 5'-ends overlap. MAURO SANTOS

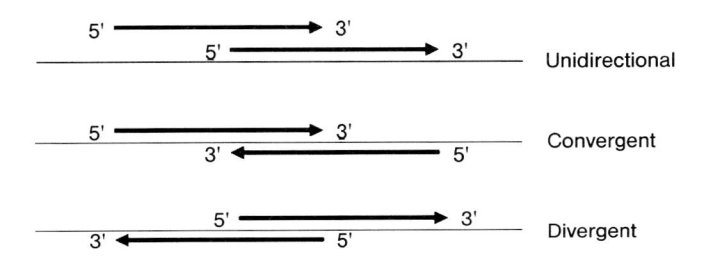

```
        5' ─────────► 3'
              5' ─────────► 3'        Unidirectional
        ─────────────────────

        5' ─────────► 3'
              3' ◄───────── 5'        Convergent
        ─────────────────────

              5' ─────────► 3'
        3' ◄───────── 5'              Divergent
        ─────────────────────
```

P

palindrome DNA carrying the same sequence symmetrically on both strands. Palindrome sequences serve as *recognition sites for many proteins such as DNA *methyltransferases or *restriction endonucleases. The figure shows two typical palindrome sites that are recognized by two different *restriction endonucleases (*Alu*I and *Eco*RI). Arrows denote the cleavage sites. →hemimethylation, →recognition site. JÖRN WOLF

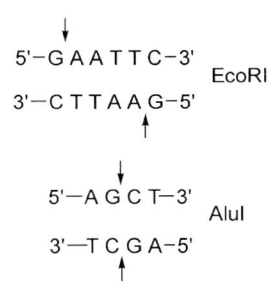

$$\begin{array}{l} \downarrow \\ 5'-\text{G A A T T C}-3' \\ 3'-\text{C T T A A G}-5' \\ \phantom{5'-\text{G A A T T }}\uparrow \end{array} \text{EcoRI}$$

$$\begin{array}{l} \downarrow \\ 5'-\text{A G C T}-3' \\ 3'-\text{T C G A}-5' \\ \phantom{5'-\text{A G }}\uparrow \end{array} \text{AluI}$$

parallel orientation Mutual arrangement of nucleic acid strands wherein their 5'–3' *phosphodiester linkages run in the same direction. *Watson–Crick base pairing forces the polynucleotide chains into *antiparallel orientation, as observed for DNA duplexes in B- and A-forms (→B-DNA, →A-DNA), and for left-handed *Z-DNA. The situation is drastically different in *duplexes with symmetrical base pairing as for a double helix formed at acidic pH by poly(A) or poly(C) and stabilized by AH^+–AH^+ or hemiprotonated C–CH^+ base pairs. Here, the dyad axis relating the glycosyl links is perpendicular to the plane of the *base pair. As a consequence, *sugar–phosphate backbones are 180° apart and in parallel orientation; this results in only one type of groove at the periphery of the *double helix. Some strands in triple and quadruple DNA helixes formed by homopurine/homopyrimidine sequences or oligoG tracts, respectively, are in parallel orientation. The parallel orientation of the strands is a characteristic feature of duplex self-structures stabilized by non-canonical AA and GG pairs. AT-containing oligonucleotides have a potential to form the parallel-stranded DNA duplexes with *reversed Watson–Crick AT pairs. These helices, whose occurrence depends on particular combinations of environmental conditions, are less stable than their antiparallel-stranded counterparts. NINA DOLINNAYA

pathogenic genes Genes associated with diseases.

PCR →polymerase chain reaction

P-element Type of *transposon (DNA) described in *Drosophila melanogaster* known to be the cause of hybrid disgenesis – a syndrome that refers to the high rate of *mutation in germ line cells of *Drosophila* strains when males carrying P-elements (P-strain or P-cytotype) are crossed to females that lack such element (M-strain or M-cytotype). MAURO SANTOS

pellet Solid residue formed after centrifugation of a suspension or solution (e.g. DNA pellet after an ethanol precipitation and centrifugation).

peptide nucleic acid →PNA

peptidyl site →P-site, →ribosome

peptidyl transferase center (PTC) →ribosome

peptidyl-tRNA *tRNA carrying a peptide rather than a single amino acid (→aminoacyl-tRNA) esterified to its terminal *adenosine. As an intermediate product of ribosomal protein synthesis (peptidyl transferase, →ribosome, →translation) it usually remains bound to the peptidyl-tRNA or *P-site of *ribosomes. However, it can be release during ribosomal stalling or premature *termination and it is a much more stable product compared to aminoacyl-tRNA as the α-amino group is part of the peptide bond. Cells contain specific "rescue" enzymes termed peptidyl-tRNA hydrolases to insure a sufficient pool of free tRNA and discard the short peptides. STEFAN VÖRTLER

periodate oxidation Periodate (IO^{4-}) is used as an oxidizing reagent for the cleavage of the C–C bond in vicinal *cis*-diols. This opens up the possibility to modify saccharide rings, as many five- and six-membered sugars have vicinal diols. For example, the *cis*-diol group at the 3′-end of RNA can be cleaved by using $NaIO_4$ to generate a 2′,3′-dialdehyde. The aldehyde then can be further reacted with amines or hydrazines, e.g. to immobilize the nucleic acid strand on a hydrazide solid phase (→immobilization of nucleic acids) or to conjugate a functional group such as a fluorophore or *spin label. BETTINA APPEL

p53 gene One of the *tumor suppressor genes of eukaryotic cells. It encodes the transcriptional activator p53 tumor suppressor protein, also known as the "Guardian of the Genome". It plays an important role in growth arrest, *DNA repair and apoptosis. In normal cells the level of p53 protein is low and *DNA damage can trigger the increased expression of the p53 gene. The ensuing growth arrest stops the progression of cell cycle, thus preventing the *replication of damaged DNA. p53 activates the *transcription of proteins involved in *DNA repair. Apoptosis (cell death) can also be triggered by p53, as a last remedy to avoid proliferation of cells containing abnormal DNA. GEMMA MARFANY, MAURO SANTOS

phenotype Physical appearance of an individual organism, which results from the interplay between its *genotype or the individual's genetic makeup and the environment where the genotype is expressed. MAURO SANTOS

Philadelphia chromosome (Ph) Discovered in Philadelphia in 1960, Ph is a *chromosome abnormality due to a reciprocal *translocation (i.e. interchange of genetic material) between chromosomes 9 and 22 in humans that explains many cases of chronic myeloid leukemia. MAURO SANTOS

phosphatase →phosphoesterase

phosphate backbone →sugar–phosphate backbone

phosphite triester →phosphoramidite method

phosphodiester Chemical structure that connects nucleotides in DNA and RNA strands (→phosphodiester method).

phosphodiesterase Enzyme that cleaves *phosphodiester bonds. Substrates might be nucleic acids (DNA, RNA) or cyclic nucleoside monophosphates like *cAMP and *cGMP. As those cyclic nucleotides are, amongst other components, involved in intracellular signal transduction as second messengers, *phosphodiesterases play an important role in cellular processes. In recent years, sildenafil, an inhibitor of one human cGMP-degrading phosphodiesterase, attained lively public interest, being the main component of Viagra, which helps to overcome male erectile dysfunction (impotence). ULI HAHN

phosphodiester backbone →sugar–phosphate backbone

phosphodiester bond Linkage of the 3'-carbon of one *nucleotide with the 5'-carbon of the neighboring nucleotide in DNA or RNA via a phosphoric acid ester structure. It is an integral part of the *sugar–phosphate backbone of nucleic acids. BETTINA APPEL

phosphodiester method First method for the synthesis of *oligonucleotides (→oligonucleotide synthesis) reported in the 1960s. The coupling step involved a 5'-O-protected deoxynucleoside with a 3'-O-protected deoxynucleoside-5'-phosphomonoester activated by triisopropylbenzenesulfonyl chloride yielding a 5'–3'-phosphodiester linkage between the two *nucleosides. RONALD MICURA

phosphoesterase Enzyme that cleaves phosphoric acid ester bonds (→DNase, →RNase, →phosphodiesterase).

phosphomonoester Chemical structure that results from reaction of phosphoric acid with an alcohol (→nucleoside phosphates).

phosphomonoesterase Enzyme that cleaves phosphoric acid esters in which only one hydroxyl group is condensed with an alcohol (→alkaline phosphatase).

phosphoramidite Nucleoside building block used during oligonucleotide synthesis following the *phosphoramidite method.

phosphoramidite method Most widely used solid-phase approach for the chemical synthesis of oligodeoxyribonucleotides (DNA) and oligoribonucleotides (RNA), named after the actual coupling step involving the ([2-cyanoethoxy]-N,N-diisopropylamino)phosphinyl or (methoxy-

Phosphodiester method

N,N-diisopropylamino)phosphinyl moiety at the 3'-oxygen of a properly protected nucleoside building block which is activated by tetrazole (or a derivative thereof) and reacted with the 5'-hydroxyl of the *nucleoside (nucleoside chain) attached at the solid support resulting in a phosphite triester linkage.

A complete cycle of nucleoside attachment requires four steps: (a) deprotection of the terminal ribose 5'-O-protecting group of the support-bound nucleoside, (b) coupling of the activated phosphoramidite building block as described above, (c) oxidation of the phosphite triester linkage to the corresponding phosphate triester and (d) capping of the unreacted nucleoside 5'-hydroxyl groups (→oligonucleotide synthesis).

For the synthesis of RNA, three specified phosphoramidite methods have found wide acceptance and are named after their protecting groups used for the 2'-hydroxyl, i.e. *TBDMS method, *TOM method and *ACE method. RONALD MICURA

phosphoribose backbone →sugar–phosphate backbone

phosphoric acid monoester →phosphomonoester

phosphoric acid diester →phosphodiester

phosphoric acid triester →phosphotriester

phosphorimaging Electronic method to detect radioactively labeled material on gel, blots, microarrays, etc. It is more sensitive and allows a far wider detection range than a classical X-ray film. A phosphorimaging plate is exposed to the radioactive material. The radioactivity induces chemical changes in the phosphorimaging plate, which can be detected by scanning the plate with a laser. Radioactively marked sites show quantitative measurable fluorescence. SLAWOMIR GWIAZDA

phosphorodithioate oligonucleotides DNA analogs where both non-bridging oxygen atoms of the internucleotide linkage are replaced by sulfur (→antisense drug). IRENE DRUDE

phosphorothioate oligonucleotides DNA analogs where a nonbridging oxygen atom of the *phosphodiester bond is replaced by sulfur (→antisense drug).

phosphorothioate sequencing Technique for the *sequence analysis of DNA that is closely related to *Sanger sequencing. The method is based on the enzymatic synthesis of DNA with a mixture of normal *nucleotides and nucleoside-γ-thiotriphosphates (dNTPγS). Starting from a *primer that is annealed to the single-stranded target DNA a complementary copy is synthesized in a *primer extension reaction. The dNTPγS also serve as substrate for the *polymerase chain reaction and are randomly incorporated. The dNTPγS positions in the newly synthesized strand can be reacted with 2-iodoethanol or 2,3-epoxyethanol to form phosphorothioate triesters, which can be more rapidly hydrolyzed than *phosphodiesters. Controlled hydrolysis releases DNA *fragments that can be analyzed exactly like the random fragments produced with chain-terminating nucleotides in the original Sanger approach. SUSANNE BRAKMANN

phosphorylation Transfer of a phosphate group onto the 5'-end of a nucleic acid strand using *polynucleotide kinase and *ATP.

phosphotriester Phosphoric acid derivative in which all three OH groups of phosphoric acid form ester bonds with alcohol groups (→phosphotriester method).

phosphotriester method Classical method for the synthesis of oligonucleotides (→oligonucleotide synthesis) applied to solid-phase synthesis in the early 1980s. The coupling step involved a 5'-O-protected deoxynucleoside 3'-O-(chlorophenyl)-phosphate with a deoxynucleoside attached at its 3'-O-position to a solid support activated by the coupling reagent mesitylenesulfonyl 3-nitro-1,2,4-triazolide yielding a 5'-3'-*phosphotriester linkage between the two *nucleosides. In recent time, the phosphotriester method has relevance in specialized applications, e.g. the solid-phase synthesis of cyclic DNA and RNA *oligonucleotides. RONALD MICURA

photo-crosslinking →ultraviolet crosslinking, →ultraviolet laser crosslinking

photolyase Monomeric protein of 450–550 amino acids and two non-covalently bound chromophore cofactors. One of the cofactors is in all enzyme isoforms the *flavin adenine dinucleotide (FAD) and the second chromophore is either methenyltetrahydrofolate (MTHF) or 8-hydroxy-7,8-didemethyl-5-deazariboflavin (8-HDF). There are two major ultraviolet-induced *DNA lesions: cyclobutane pyrimidine dimers (→thymidine dimer) and the pyrimidine–pyrimidone (6–4) photo-products. Both types of lesions are repaired in a light-dependent fashion by the DNA photolyases with similar enzyme structures and similar reaction pathways. However, a DNA photolyase that repairs one photo-product can not repair the other. Thus, the enzymes are classified in cyclobutane pyrimidine dimer (CPD) photolyases and (6–4) photolyases. Due to their light activated reactivity, the photolyases can only be found in bacteria such as *Escherichia coli* or *Anycystis nidulans*. The proteins recognize the DNA lesion substrate and form a base-flipped DNA–protein complex (→base flipping). The repair mechanism is a cascade of different kinds of photoinduced reaction steps: the MTHF or 8-HDF cofactor serve as an antenna for the absorption of a broad range within the visible light spectrum and transfer the excited state energy by a Förster-type mechanism to the FAD cofactor. The excited state of FAD represents one of the most powerful reductants in biological systems and is able to reduce the DNA lesions by an electron transfer process, e.g. to the *thymidine dimer, in the extrahelical, base-flipped conformation. Reduction of the cyclobutane ring system yields

cycloreversion. After the splitting of the dimer the electron is transferred back to the flavin cofactor.
HANS-ACHIM WAGENKNECHT

pitch of DNA helix →helix pitch

plant artificial chromosome *Artificial chromosome containing plant *genomic DNA and which is designed to be episomically maintained in plant cells.

plaque blot Technique for identification of specific recombinant phages similar to the *colony blot. Phage infection of bacteria results in lysis of the cells. On a culture plate, phage-infected colonies can be seen as small areas of clearing called plaques. For identification of phages, a copy of the culture plate is produced by stamping a filter onto the plate. Phages are then lysed and the DNA is denaturized and fixed on the filter. *Hybridization with *labeled DNA or RNA *fragments of known *sequence allows for detection of DNA fragments to be searched for. Signals on the filter allow for identification of phages containing the analyzed DNA. IRENE DRUDE

plaque hybridization →plaque blot

plasmid Circular *double-stranded DNA that can be used in prokaryotic or eukaryotic cells, and which usually contains a *multiple cloning site for DNA *cloning or an expression cassette for protein expression. Additional genetic elements, such as a *replication *origin for episomic maintenance and/or markers (i.e. antibiotic resistance or auxotrophic genes) for growth selection are generally included. Shuttle plasmids combine *genes for replication and maintenance in prokaryotic cells and cassettes for *gene expression in eukaryotic cells. GEMMA MARFANY, ROSER GONZÀLEZ-DUARTE

plasmid DNA DNA contained in a *plasmid.

plastid DNA Genetic material (DNA) of the plastids (plant cell organelles) (→chloroplast DNA).

plastome *Genome contained in the plastids responsible for photosynthesis in plants and algae.

plus strand In *double-stranded DNA, the strand that is identical in sequence to the transcribed *messenger RNA. In some cases (e.g. viral genomes) only a single-stranded genome exists. In this case the corresponding minus strand has to be synthesized prior to *transcription. SLAWOMIR GWIAZDA

PNA (peptide nucleic acid) DNA analog in which the DNA *phosphoribose backbone is replaced by an uncharged, achiral N-(2-aminoethyl)glycine polyamide structure. The *nucleobases are linked to the peptide backbone via carboxymethylene units.

Despite this rather radical change of nucleic acid structure, PNAs bind to *complementary oligonucleotides in *parallel (N-terminus of the PNA facing the 5′-end) and *antiparallel (N-terminus of the PNA facing the 3′-end) orientation forming *Watson–Crick base pairs. PNA–DNA duplexes are more stable than corresponding DNA–DNA duplexes (around 1.5 K/bp antiparallel, around 1.0 K/bp parallel). Base pair *mismatches result in a decrease of thermal stability of PNA–DNA duplexes by $\Delta T_m = 8\text{--}20$ K. In some cases this discrimination is twice as high as for DNA–DNA duplexes. Furthermore PNAs are able to form triple helical structures (→triple helix) like (PNA)$_2$–DNA and (DNA)$_2$–PNA. (PNA)$_2$–DNA triple helices are favored and sometimes even formed through strand displacement in *duplex DNA, whereas (DNA)$_2$–PNA triplexes are only formed with certain sequences. The stability of PNA against chemical and biological degradation is very high. Since PNA is neither peptide nor nucleic acid it is not degraded by *proteases or nucleases. In addition, PNAs are completely stable against acids and sufficiently stable to weak bases

B = Adenine
Cytosine
Guanine
Thymine

PNA

such as piperidine to allow the synthesis of PNAs by strategies derived from solid-phase peptide synthesis methods. A notable chemical instability results from the free N-Terminus of the PNA which can induce *N*-acyl transfer of the *nucleobase or loss of the N-terminal PNA unit by ring closure. Usually the N-terminus of a PNA is acetylated to avoid these side-reactions. The two most frequently used protecting group combinations for the synthesis of PNA are Fmoc/Bhoc (fluorenylmethyloxycarbonyl/benzhydryloxycarbonyl), and Boc/Cbz (*tert*-butyloxycarbonyl/ benzyloxycarbonyl). The described properties distinguish PNA from other DNA analogs. PNAs have become a valuable tool in DNA and RNA diagnostics as well as in molecular biology. LARS RÖGLIN, OLIVER SEITZ

point mutation Special case of a *substitution mutation in which exactly one DNA (RNA) residue has been replaced.

poly(A) →polyadenylic acid

polyadenylation Formation of the mature 3'-end of the *messenger RNA in eukaryotes involves cleavage of the nascent transcript followed by addition of around 200 *adenosines known as the *poly(A) tail.

Formation of this poly(A) tail is directed by a sequences on *pre-mRNA and the mammalian polyadenylation machinery consisting of a complex of several protein factors. The site of the cleavage is found between a highly conserved polyadenylation signal sequence, 5'-AAUAAA-3', found near the 3'-end of the pre-mRNA and a GU-rich sequence about 11–20 bases further *downstream of the polyadenylation signal.

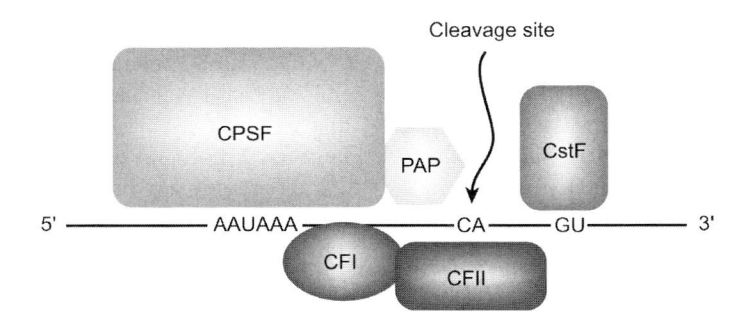

Cleavage occurs predominately at a partially conserved CA dinucleotide about 20 *nucleotides downstream of the *poly(A) signal. The polyadenylation step is initiated by recognition of the poly(A) signal sequence motif by two key multimeric mammalian factors: the cleavage and polyadenylation stimulating factor (CPSF) and the cleavage stimulation factor (CstF). Prior to cleavage two additional factors, called CF1 and CF2, are essential; cleavage of pre-mRNA is promoted after they have first associated to CPSF and CstF. The freshly formed 3'-end is then polyadenylated an addition of about 200 consequent adenines by poly(A) polymerase (PAP). PAP is usually required for the cleavage reaction and together with CPSF directs the poly(A) addition. The CPSF factor consist of four polypeptides with molecular masses of 160, 100, 73 and 30 kDa, respectively, and CstF is a heterotrimeric protein with subunits of 77, 64 and 50 kDa, respectively. The two cleavage factors CF1 and CF2 are only required for the cleavage reaction, not for the polyadenylation step. CF1 consist of three subunits with molecular masses of 68, 59 and 25 kDa, respectively. The second factor CF2 consists of the two subunits hClp1 and hPcf11. TINA PERSSON

polyadenylic acid [poly(A)] Adenylic acid is adenosine monophosphate (→polyadenylation, →poly(A) signal).

poly(A) signal Mature 3'-end of *messenger RNA in eukaryotes produced by cleavage of the nascent transcript followed by *polyadenylation at the cleavage site. The mRNA sequence itself contains the signal that determines the site of polyadenylation. The polyadenylation signal along the mRNA sequence is a hexanucleotide, 5'-AAUAAA-3' and is placed 10–30 bases *upstream of the cleavage/polyadenylation site. The 5'-AAUAAA-3' sequence is found in 90% of all polyadenylation elements and is one of the most conserved sequences known. A second recognition element is placed 20–40 bases *downstream of the cleavage site and involves a GU-rich element; in contrast to the 5'-AAUAAA-3' site, this element is present in only 70% of mammalian *pre-mRNA and is more variable in its sequence composition. TINA PERSSON

poly(A) tail In higher eukaryotes all protein-encoding *messenger RNAs (mRNAs) are modified at the 3'-end by an addition of about 200 *adenosine residues in a process called *polyadenylation. Its function is to maintain mRNA stability, promoting mRNA *translation efficiency and playing a role in transport of mRNA from the nucleus to the cytoplasm. In yeast the tail length is much shorter and consists of about 70–90 adenosine nucleotides. TINA PERSSON

polycistronic RNA *messenger RNA (mRNA) that contains the information of more than one *gene and is translated to produce more than one protein. Polycistronic mRNA is a common feature in prokaryotes, but is practically absent in eukaryotes. The only examples of polycistronic mRNA in eukaryotes are found in chloroplasts and mitochondria, which is interpreted as evidence in support of the endosymbiotic theory for the origin of these cellular organelles. According to this theory, chloroplasts and mitochondria derive from free-living prokaryotes. MAURO SANTOS

polymer High-molecular-mass molecule composed of repeating structural units, or monomers, which are connected by covalent chemical bonds. In addition to synthetic carbon-based polymers, like polyethylene, polystyrene, polyamide, polyester, etc., many naturally occurring polymers can be found. *Biopolymers like polynucleotides (DNA, RNA), polypeptides (proteins) or polysaccharides (cellulose, chitin) play important roles in biological processes. Polymer molecules can be classified into two subtypes, depending on the variety of the monomeric constituents. Polymers based on a single monomeric species, like polyethylene or polystyrene, are referred as homopolymers. In the case of polymerization of two or more monomeric species, the polymer is called a copolymer (e.g. DNA, RNA or protein). A copolymer can be subclassified as alternating, random or statistical, depending on the arrangement of the individual monomer subunits. IRENE DRUDE

polymerase Group of enzymes which catalyze a *polymerization reaction of *nucleotides into DNA or RNA. Polymerization of nucleotides occurs during *DNA replication and during *transcription, whereby polymerases in complex with other proteins catalyze the formation of a *phosphodiester bond using nucleotide *triphosphates from solution. Synthesis of DNA during replication (→DNA polymerase) as well as synthesis of RNA during transcription (→RNA polymerase) requires a DNA strand which acts as a *template. *Template-independent DNA polymerases, like terminal deoxynucleotidyl transferase, or *template-independent RNA polymerases, like poly(A) polymerase, are also described. The *reverse transcriptase that uses a RNA strand as a template for complementary DNA synthesis also belongs to the family of polymerases. IRENE DRUDE

polymerase chain reaction (PCR) First reported in 1988 by the Cetus Corporation (Saiki et al., Primer-directed enzymatic amplification of DNA with a thermostable DNA polymerase. *Science* **1988**, *239*, 487–491). This seminal article discussed the *amplification of single-copy genomic sequences by a factor of more than 10 million with very high specificity and of DNA segments up to 2000 bp. It was also shown that PCR could be used to amplify and detect a target DNA molecule present only once in a sample of 10^5 cells.

Since it was first introduced it has revolutionized all areas of molecular biology due to its ability to amplify from as little as a single copy of template DNA. An essential part of the PCR process is the use of thermostable *DNA polymerases, isolated originally from thermophilic bacteria. These organisms have evolved in hot water sources and optimally function at temperatures above 70°C. The use of thermostable polymerases allows for the use of multiple cycles of PCR, which in turn leads to an exponential amplification of the target DNA sequence. Early forms of PCR used polymerases operating at lower temperatures (37°C) which had to be replaced after each heating cycle.

PCR requires *template DNA, two *primers complementary to ends of the region of the template which is to be amplified and a thermostable DNA polymerase. The first step is a heating step, typically 90–95°C, in order to denature the *double-stranded DNA (dsDNA) template. Following this, there is an annealing step which allows the two primers to bind to their target sequences on the now single-stranded template. The primers are designed to bind to opposite strands of the template sequence with their 3'-ends oriented towards each other. Annealing temperatures will vary according to the primer sequence and length, but are typically in the region 50–60°C. The final step is an extension step where the reaction is heated to the temperature optimal for the DNA polymerase, e.g. 72°C for *Taq DNA polymerase. In this step the polymerase extends the primer along the DNA template, generating two new copies of dsDNA. Repeating this denaturing–annealing–extension cycle many times amplifies the designated sequence. The cycle is usually terminated by a final, longer extension period.

Many variants of PCR have been developed since it was first reported. There are now new polymerases capable of synthesizing longer PCR products, of replicating DNA with greater fidelity and error-prone polymerases. Error-prone polymerases are of use as they are able to copy damaged DNA regions (→competitive polymerase chain reaction, →error-prone polymerase chain reaction, →*in situ* polymerase chain reaction, →multiplex polymerase chain reaction, →nested polymerase chain reaction, →random polymerase chain reaction, →real-time polymerase chain reaction and →reverse transcription polymerase chain reaction). DAVID LOAKES

polymerization Chemical reaction in which monomeric molecules are covalently branched together to form *polymers.

polymorphism Describes the coexistence of two or more *alleles occurring at appreciable frequencies within a population.

polynucleotide →nucleic acid, →DNA, →RNA, →biopolymer

polynucleotide kinase (PNK) Enzyme that catalyzes phosphorylation of nucleic acid 5′-ends by transferring the γ-phosphate group of ATP to the 5′-hydroxyl termini. This kinase is derived from T4 bacteriophages and is therefore also called T4 PNK. PNK is often used to radiolabel nucleic acids (→phosphorylation, →radioactive labeling). VALESKA DOMBOS

polyribonucleotide →RNA, →biopolymer

polyribosome Array of *ribosomes attached to a single *messenger RNA (mRNA) molecule during the process of *translation. The ribosomes translate the mRNA simultaneously, thus allowing the synthesis of multiple protein molecules from a single mRNA strand in short time. SLAWOMIR GWIAZDA

polysome →polyribosome

polystyrene *Polymer that is used as support in *oligonucleotide synthesis.

pool Collection of nucleic acids with different sequences (→*in vitro* evolution of nucleic acids, →SELEX).

positional cloning Identification of disease-causing *genes based on their chromosomal position. It involves a previous mapping step of cosegregation and linkage analysis in affected families, followed by mutational screening of all the genes lo-cated in this mapped chromosomal region. GEMMA MARFANY, ROSER GONZÀLEZ-DUARTE

positive supercoiling →supercoiled DNA

post-transcriptional gene silencing →gene silencing, →RNA interference

post-transcriptional modification After enzymatic synthesis of RNA by *transcription within the cell, a large number of RNAs are modified. These modifications are facilitated by cellular enzymes and play a significant role in regulating the activity of RNA. For example, addition of a *poly(A) tail to the 3′-end of *messenger RNAs (mRNAs) protects the RNA from degradation by cellular nucleases (→RNA degradation) and facilitates transport of the message from the nucleus to the cytoplasm, where the message is translated. However, most post-transcriptional modifications occur at specific *nucleotides at internal sites in the RNA sequence. The modified nucleotides are derivatives of the normal nucleotides A, G, C and U. Many *nucleosides are modified by adding a single methyl group, either to the base or the sugar moiety. Others are modified in a more elaborate manner, like the conversion of uridine to *pseudouridine, which contains a C–C glycosidic bond instead of the normal C–N bond. Some nucleotides even contain more than one modification. Many of the modifications are phylogenetically conserved. The functional roles of modified nucleosides are not well understood, but are known to influence RNA *secondary and *tertiary structure. For example, an increased number of modifications in *transfer RNA (tRNA) of thermophilic bacteria indicates that modified nucleotides play a role in stabilizing the structure of tRNAs at higher temperatures. Modified nucleosides in tRNA are also known to increase the fidelity and efficiency of protein synthesis by the *ribosome. Close to 100 modified nucleosides have been isolated from cellular RNA and structurally characterized. Most of the modified nucleotides are found in tRNA and a considerable number has been identified in *ribosomal RNA and mRNA. The presence of such a large number of modified nucleosides supports the *RNA world theory because it implies that the functional repertoire of RNA was much greater when RNA performed many of the functions that are carried out by proteins today. SNORRI TH. SIGURDSSON

precipitation Separation of a solved material from a solution. It can be induced by addition of a substance to the solution, which lowers the solubility of the solved material, or by a change of the temperature of the solution. The most common use of this method in DNA/RNA chemistry is the ethanol precipitation. The addition of ethanol to a watery DNA/RNA solution causes the DNA/RNA to precipitate. The effect can be enhanced by optimizing the pH and salt conditions, and by cooling. The precipitated DNA/RNA can by obtained by centrifugation in the form of a *pellet, which can be further processed. SLAWOMIR GWIAZDA

precise substrate orientation Binding of the substrate(s) by the *catalyst and orientation of the residues that take part in a (bio)chemical reaction, resulting in optimal position and close proximity of the reacting *functional groups. In addition, the translational and rotational motions are reduced. This is one mechanistic strategy for a catalyst to decrease the activation energy of a (bio)chemical reaction. To catalyze a cleavage reaction, small *ribozymes arrange the attacking 2'-hydroxy-group, the scissile phosphate and the 5'-hydroxyl leaving group in-line, thus resulting in a perfect orientation of the reacting functional groups, following an S_N2 mechanism (→in-line attack). DENISE STROHBACH

precursor mRNA →pre-mRNA

precursor tRNA →pre-tRNA

pre-genome RNA intermediate that is reverse-transcribed into DNA, e.g. in the small enveloped DNA hepatitis B virus that replicates through *reverse transcription of its pre-genome. MAURO SANTOS

pre-miRNA Around 70-nucleotide long precursors of *micro-RNAs (miRNA) forming an imperfect *stem–loop structure. Pre-miRNAs are processed from *pri-miRNAs by *Drosha and into *miRNAs by *Dicer. NICOLAS PIGANEAU

pre-mRNA mRNA is synthesized in a process called *transcription. The sequence in the mRNA is defined either by a DNA *template or a *gene and the *genetic information is transferred to the precursor pre-mRNA using the enzyme *RNA polymerase. The pre-mRNA molecule contains both protein non-coding and protein *coding sequences called *introns and *exons. Before pre-mRNA acts a template in the translation step, pre-mRNA is transformed in several steps to create *mature mRNA. The general name for these steps are *RNA processing, and events involved are capping, *splicing and *polyadenylation. Several proteins are associated to mRNA before it is transported to the cytoplasm, producing a complex called *mRNP. TINA PERSSON

pre-tRNA Precursor of a *transfer RNA (tRNA) that is synthesized by *RNA polymerase III. Pre-tRNA are usually multicistronic *transcripts that contain several tRNAs, and additionally *messenger RNA and *ribosomal RNAs They contain

5'-*leader sequences, *spacer sequences between the tRNAs and 3'-*trailer sequences. Between 4 and 13% of all pre-tRNAs in eukaryotes contain *introns, which are 11–60 nucleotides long. The 5'-splicing site is usually located in the *anticodon loop and the 3'-splicing site is located in a *loop 3–5 nucleotides long. JÖRN WOLF

Pribnow box Also known as the Pribnow–Schaller box, a region of DNA to which *RNA polymerase binds before initiating the *transcription. The consensus sequence is TATAAT. MAURO SANTOS

primary structure For a biological *macromolecule, defined by the sequence of its monomeric units, e.g. by the sequence of amino acids in proteins or the sequence of nucleobases in nucleic acids. SLAWOMIR GWIAZDA

primary transcript *RNA produced by *transcription of *genes that code for proteins or functional RNAs. Depending on their function, primary transcripts are subjected to several post-transcriptional processes in which they are extensively modified, e.g. in *splicing the eukaryotic primary transcripts that usually consists of *exons and *introns have their *introns removed. In other cases, single *nucleotides become modified or exchanged (→editing, →maturation, →splicing). JÖRN WOLF

primase As *DNA polymerases cannot initiate DNA synthesis, but can only elongate nucleic acid polymers, a primase is required. It initiates and performs the *de novo* synthesis of *RNA primers (short RNA oligonucleotides, 10–12mers) at the *lagging strand of the *replication fork, which are then elongated by DNA polymerases. Later, the *primers are removed, e.g. by DNA polymerase I which hydrolyzes the ribonucleotides and replaces them by 2'-deoxyribonucleotides. ANDREAS MARX, KARL-HEINZ JUNG

primer Short single-stranded *oligonucleotide sequence that is required for synthesis of DNA or RNA by a *polymerase. The primer sequence is complementary to a region of the parent DNA/RNA which will act as a *template for the new strand and can be either DNA or RNA depending upon the type of polymerase involved. A primer is required because DNA and RNA polymerases can only synthesize a new *nucleic acid strand by adding to an existing strand of nucleotides. The primer anneals to the template strand to form a double-stranded *duplex which the polymerase will bind to and acts as a *recognition site for initiation of *polymerization. For enzymatic synthesis to occur it is also necessary that the primer has a free 3'-hydroxyl group for incorporation of the next *complementary nucleoside 5'-triphosphate. Synthesis of DNA by a DNA polymerase (e.g. for *polymerase chain reaction) requires primers of approximately 20–30 nucleotides. DAVID LOAKES

primer extension Frequently used to map the 5'-ends of DNA or RNA fragments. It is carried out by annealing a complementary *primer *downstream to the 5'-end of the target sequence, followed by enzymatic extension of the primer. The primer is usually labeled with a radioisotope or with a fluorophore (→oligonucleotide labeling) to aid detection and analysis following electrophoretic separation on a gel. For RNA target sequences, the polymerization enzyme is a *reverse transcriptase using a DNA primer and 2'-deoxynucleoside-5'-triphosphates (→nucleoside triphosphates). For DNA sequences, a *DNA polymerase is used – the most frequently used DNA polymerases being either *Taq DNA polymerase (a thermophilic polymerase) or the large fragment from DNA polymerase I (→Klenow fragment). Primer extension reactions are also used for *sequencing small regions of DNA, and are often used to determine *single-nucleotide polymorphisms, which may form the basis for a genetic mutation (→gene mutation).

Chemical synthesis of DNA/RNA is largely routine (→oligonucleotide synthesis) and therefore a large number of *nucleoside analogs have been incorporated into oligonucleotides (they may also be incorporated enzymatically as a 5'-triphosphate with a polymerase). A common method for studying the utility of such modified nucleosides is primer extension. This not only serves to determine how the analog interacts with the native nucleobases, but can also be used to determine the efficiency of incorporation by polymerases. DAVID LOAKES

primer extension analysis Used to map the 5'-end of *transcripts to determine the exact start site(s) for *transcription. A short *antisense 5'-end labeled oligonucleotide (either synthetic oligonucleotide or a small *restriction *fragment) is hybridized to RNA and DNA is synthesized by using a *reverse transcriptase. The reaction products are then analyzed by *electrophoresis gels along side sequencing reactions of DNA containing the same *gene and using the same *primer. The *transcription initiation site is identified as the band in the sequencing reaction parallel to the run-off reverse transcript. Multiple transcription initiation sites appear as multiple bands in the primer extension analysis. *RNA processing sites will also be observed, as will the locations of breaks or *modified bases. Primer extension analysis can also be utilized to quantify mRNA levels and to detect low abundance mRNA species. DAVID LOAKES

pri-miRNA Precursors of *micro-RNAs. Pri-miRNAs are transcribed either as independent *transcription units or as polycistrons (→polycistronic RNA). Alternatively, they may be located in *introns of protein-coding *genes. NICOLAS PIGANEAU

primosome Complex of proteins that are all involved in the synthesis of the *RNA primers which initiate *DNA replication. These proteins are responsible for the recognition of the initiation, unwinding of the *DNA double helix, the assembly of the primosome and for the elongation of the RNA primer. ANDREAS MARX, KARL-HEINZ JUNG

pRNA Abbreviation used for *pyranosyl-RNA and *promoter RNA.

promoter Non-coding sequence patch on a DNA that contains the information for activation or repression of its *gene. The promoter region defines the starting point of *transcription. In eukaryotes, the specific sequence of a promoter, which is positioned *upstream of a gene, is recognized by *transcription factors, thereby recruiting a *RNA polymerase for transcription. In prokaryotes, a promoter is recognized by the *sigma factor, which is in complex with the RNA polymerase. A prokaryotic promoter mostly consists of two short elements at the –10 (the *Pribnow box or the –10 element) and –35 (–35 element) positions. A –10 element in *Escherichia coli* is characterized by a TATAAT motif and is essential for the start of transcription. The presence of a –35 element allows a high transcription rate and consists of the 6 nucleotides TTGACA in *E. coli*. In contrast to prokaryotic promoters, eukaryotic promoters are extremely diverse. They can have regulatory elements several kilobases away from the transcription start. Many eukaryotic promoters contain a *TATA-box, about 25–35 nucleotides upstream of the transcription start site, which in turn is mostly characterized by a *pyrimidine-rich element. TATA-box binding proteins, like the transcription factor TFIID, recognize and bind to this element, which in turn assists in transcriptional complex formation. Two further basic elements which are in common are the GC-boxes and the CAAT-boxes. These elements occur in more than one copy and in different orientation in cellular and viral promoters. *Housekeeping genes show no TATA-boxes next to the transcription site, but a lot of GC-boxes, which are recognized by the transcription factor SP1. This results in recruitment of further transcription factors as well as the RNA polymerase II. CAAT-boxes are mostly positioned about −110 to −40 nucleotides from the transcription start. Transcription factors like CTF or NF1 recognize and bind to this element. IRENE DRUDE

promoter RNA (pRNA) Involved in epigenetic remodeling and *small interfering RNA (siRNA)-directed *gene silencing in which it works together with DNA *promotor regions. Those DNA promotor regions bind *antisense strands of promotor-directed siRNAs, whereas the pRNA functions as a scaffold to support correct folding of the siRNA. BETTINA APPEL

proofreading As very high fidelity is desired for the *replication of genetic material, the fidelity of several *DNA polymerases which catalyze the elongation of the *nascent DNA strain is further increased by proofreading functions. In nature, proofreading is performed by *exonuclease activities of several DNA polymerases that hydrolyze incorrect incorporated *nucleotides of the nascent strand and replace them by the correct nucleotides. ANDREAS MARX, KARL-HEINZ JUNG

protecting group Chemical moiety attached at a *functional group that prevents it exhibiting its inherent reactivity.

protein biosynthesis →translation

protein–DNA interaction Double- and single-stranded DNA is recognized in a sequence-specific or non-sequence-specific manner by a large number of cellular and viral proteins, including regulatory proteins such as enhancer proteins and transcriptional repressors (→transcription), *polymerases, *restriction enzymes, *DNA methylases, and a large number of proteins involved in transposition and DNA *recombination. Methods to study protein–DNA interactions include *DNase I footprinting, *band shift assay (also termed classically gel mobility shift assay), nitrocellulose filter binding assay, surface plasmon resonance methodology, *crosslinking, and *X-ray crystallography. GEORG SCZAKIEL

protein exit tunnel →ribosome

protein–RNA interaction →RNA–protein interaction

pseudogenes Inactive but stable components of the eukaryotic *genome. They derive of an ancestral active *gene. These copies tend to accumulate *mutations that block *transcription or *translation. MAURO SANTOS

pseudoknot Although frequently defined as *secondary structure elements, pseudoknots are typical representatives of *tertiary structure elements. A pseudoknot involves intramolecular pairing of bases in a *hairpin loop with a few bases outside the stem of the loop to form a further *stem–loop region. Consequently, a true knot is formed if each stem contains a full *helical turn.

As example, the structure of the *Diels–Alderase ribozyme is highlighted that forms a *double nested pseudoknot with a 1–2–3–1′–2′–3′ topology (equal

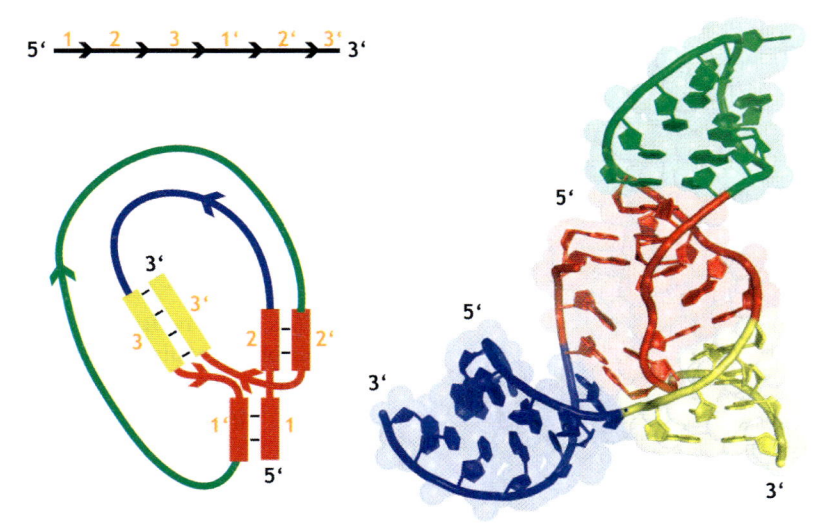

Pseudoknot: The left panel shows a schematic drawing of the RNA pseudoknot topology of the Diels–Alder ribozyme, which adopts a 1–2–3–1'–2'–3' topology.

The right panel shows the X-ray structure of the Diels–Alderase. In the same coloring scheme as the schematic drawing (PDB code: 1YLS).

numbers indicate double-helical stretches, unprimed numbers indicate the *primary structure sequential occurrence of the 5'-parts of the stretches and primed numbers indicate the structure sequential occurrence of the 3'-parts). BORIS FÜRTIG, HARALD SCHWALBE

pseudopromoter *Promoter-like sequence where *RNA polymerase can improperly bind.

pseudorotation The conformation of a five-membered ring of identical bond angles and bond lengths can be treated analytically in a form represented by the concept of pseudorotation. Accordingly, because of geometrical constraints in the ring, the five torsion angles defining the *sugar conformation can be related to only two parameters, the pseudorotation phase angle P and the maximum amplitude torsion angle Φ_m (pucker amplitude). The pseudorotation phase angle P is a positive angle and ranges from 0° and 360°. KLAUS WEISZ

pseudouridine Isomer of the nucleoside *uridine. Here, the nucleobase is linked to the *ribose unit via a C–C bond between C5 of uracil and C1 of ribose. Pseudouridine is the most common *modified nucleoside and occurs in all kinds of RNA except *messenger RNA. The isomerization of uridine to pseudouridine is catalyzed by enzymes called pseudouridine synthases (→modified base, →transfer RNA). BETTINA APPEL

P-site *Peptidyl transferase site where *peptidyl-tRNA is bound on the large subunit of *ribosomes during *translation.

PTC Abbreviation for peptidyl transferase center (→ribosome).

purine bases *Nucleobases having a *purine as basic unit (→adenine, →guanine).

purine nucleosides *N-glycosides of the *purine bases *adenine and *guanine.

purine riboswitches Regulatory *messenger RNA structures that bind to either *adenine or *guanine. They are found in several different bacterial species *upstream of *genes involved in purine metabolism. The *aptamer *consensus sequence, as derived from phylogenetic data, consists of 66 nucleotides and forms a *three-stem junction with most *conserved nucleotides residing in the terminal (L2 and L3) and internal (L1) loop regions. However, two classes have been identified that differ at one absolutely conserved nucleotide position between stems P1 and P3, which is a C residue in the case of *guanine riboswitches and a U for *adenine riboswitches. The three-dimensional structure, the first one solved for a *riboswitch aptamer by X-ray crystallography and *nuclear magnetic resonance spectroscopy, shows that the conserved residues are the key determinants for *binding pocket formation and *tertiary structure stabilization. As expected, the *ligands are recognized (and thereby selectively bound) at their Watson–Crick interface (→edge-to-edge interaction of nucleobases) by the respective conserved C or U nucleotide. In addition, the RNA encloses the ligand almost completely, which confirms biochemical evidence that any change at positions N7, C8 or N9 of the *purine leads to strong reduction in binding affinity. The overall architecture of the binding

Purine riboswitches:
Guanine aptamer consensus and tertiary structure

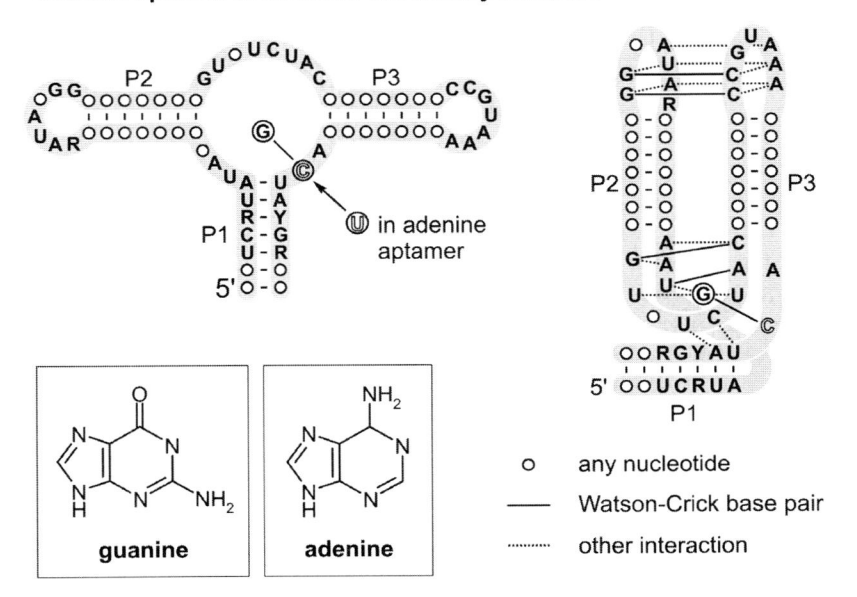

pocket allows for effective discrimination of the corresponding ligand against the diversity of other natural purine-related compounds. RÜDIGER WELZ

puromycin Aminonucleoside *antibiotic found in *Streptomyces alboniger*. The structure carries a modified *adenosine moiety linked to a tyrosine via a 3'-amino group.

Due to its close resemblance to tyrosyl-*transfer RNA, it obstructs protein biosynthesis (→translation) by binding to the *A-site of the *ribosome. In the transpeptidation step, the *ribosome transfers the nascent polypeptide chain to this aminoacylated *transfer RNA mimic, causing premature *termination of the nascent polypeptide chain. Additionally, it obstructs protein transfer into mitochondria. Gram-positive bacteria as well as several animal and insect cells are affected by puromycin, whereas Gram-negative bacteria and fungi cells display low permeability for it (→Yarus inhibitor). JÖRN WOLF

pyranosyl-RNA (pRNA) Analog of RNA where the ribofuranosyl sugar moiety is replaced by a ribopyranosyl moiety. Other oligonucleotides containing a pyranosyl-phosphate backbone are lyxopyranosyl, xylopyranosyl and arabinopyranosyl oligonucleotides. All these oligonucleotides are connected with 2',4'-*phosphodiester bonds. BETTINA APPEL

pyrimidine bases *Nucleobases having pyrimidine as the basic unit (→cytosine, →thymine, →uracil).

pyrimidine nucleosides *N-glycosides of the *pyrimidine bases *cytosine, *thymine and *uracil.

pyrophosphate (PP$_i$) Pyrophosphoric acid ($H_4P_2O_7$) is the condensation product of phosphoric acid (H_3PO_4). The esters and the salts of pyrophosphoric acid are called pyrophosphates. The anion $P_2O_7^{4-}$ is abbreviated as PP$_i$. In cells, PP$_i$ is produced by the hydrolysis of *ATP into *AMP and plays an important role in energy transfer processes in living cells. BETTINA APPEL

pyrosequencing →DNA sequence analysis

Q

QTL mapping Set of procedures for detecting *genes controlling *quantitative traits, and estimating their genetic effects and *genome locations.

Quadruplex → G-quadruplex

Quantitative trait Trait whose values are defined by a continuous distribution.

Quantitative trait loci (QTL) *Genes controlling *quantitative traits.

R

radioactive labeling Incorporation of a radiolabel into DNA and RNA (→oligonucleotide labeling, →isotope labeling). The most common radiolabel is ^{32}P, which can be introduced into phosphoesters of the *oligonucleotide. *Kinases are used to transfer a radioactive phosphate from *ATP to the 5′-end of *oligomers, and *polymerases can be used to incorporate a radiolabel into internal sites of DNA and RNA. SNORRI TH. SIGURDSSON

ram mutation Ribosomal ambiguity *mutation that allows incorrect *transfer RNAs to be incorporated into the *translation process.

random coil *Topology of a *polymer, e.g. a nucleic acid, being characterized by a statistical distribution of shapes resulting from random orientation of monomer subunits. Nucleic acid regions that lack *secondary structure are often assumed to exhibit a random coil conformation. *Denaturation of nucleic acids upon heating is considered a transition from the double-helical structure to random coil (→thermal melting) SABINE MÜLLER

random DNA DNA with non-determined sequence (→sequence pool).

random drift Change in *gene frequencies from generation to generation due to random sampling.

random mutagenesis Used for the generation of *mutations without knowing the sequence of a DNA. Random mutagenesis is used, for example, to build up a mutant *DNA library. BETTINA APPEL

random oligonucleotide *Oligonucleotide which consists of random *nucleotides.

random polymerase chain reaction (rPCR) Modification of *polymerase chain reaction which allows for the *amplification of *random DNA sequences from a *template, complex mixtures of DNA or whole cells. It is possible to generate *cDNA libraries suitable for further screening with specific DNA probes (→nucleic acid probe). There are a number of methods for conducting rPCR; it can be carried out linearly with single *random primers or exponentially under normal PCR conditions. The usual method for rPCR is the use of a *primer which has a defined sequence and a degenerate sequence. The inclusion of degenerate regions allows for multiple *binding sites along the target template (or templates). As a result of using degenerate primers, PCR products will be of varying length and sequence, and are designed to give broad coverage of the target DNA. It is commonly used to give universal *amplification of prevailing DNA or for amplification of unknown *intervening sequences which are not generally defined in length or sequence. A less common method for rPCR is to conduct the reaction at low temperatures (e.g. 30°C) where the primers will mis-prime to the template, thus generating rPCR products. DAVID LOAKES

random primer Short segments of *single-stranded DNA. These *oligonucleotides are only 8 *nucleotides long (octamers) and consist of every combination of bases

Nucleic Acids from A to Z: A Concise Encyclopedia. Edited by Sabine Müller
Copyright © 2008 WILEY-VCH Verlag GmbH & Co. KGaA, Weinheim
ISBN: 978-3-527-31211-5

($4^8 = 65\,536$ combinations are possible in the mixture). As all combinations are present, these primers can bind to any part of the DNA. MATTHÄUS JANCZYK

random RNA RNA molecules with non-determined sequence (→sequence pool).

random sequence →random DNA, →random RNA, →random oligonucleotide

rare bases *Modified bases, differing from the conventional A, C, G and U/T set of polynucleotide building blocks (→rare nucleotides).

rare nucleotides Modified nucleotides that are built by enzymatic modification of the base or the sugar of natural *nucleotides. In general, modification takes place at the level of the intact nucleic acid chain. Most common modifications are acetylation (i.e. N^4-acetylcytidine or 5-acetyluridine), glucosylation (i.e. 5-glucosylhydroxymethylcytidine), isoprenylation (i.e. N^6-isopentyladenosine), reduction (i.e. 5,6-dihydrouridine), thiolation (i.e. 2-thiouridine), cleavage of a N–C bond followed by formation of a C–C bond (i.e. pseudouridine) and methylation (i.e. 5-methyluridine or 2′-O-methyluridine). Modifications are type specific; individual nucleic acids significantly differ from each other regarding their modification level. Rare nucleotides are mainly found in *transfer RNA, located at defined positions, mostly in singe-stranded regions close to the *anticodon loop. Nucleotides that are methylated at C, N or O atoms are also found in *ribosomal RNA and in DNA. SABINE MÜLLER

rate constant Apart from the *equilibrium constants, the reaction kinetics represent important characteristics of chemical systems, particularly in the elucidation of the molecular mechanism of a reaction. Kinetics concerns the rate, v, of a reaction, which is defined by the change of the concentration, c, of one of the compounds in dependence of the time. The rate is proportional to the concentrations of the reactants and the products depending on the mechanism of the reaction. In addition, the concentrations are multiplied by proportionality constants, i.e. rate constants, k, in order to define the mathematical relationship between rate and concentrations. Each step of a more complicated reaction pathway is characterized by two individual rate constants, one for the forward and one for the backward reaction. The most simple example is a reaction of the type $A \to P$, i.e. irreversible formation of product P from compound A. Here, the rate is:

$$v = dc_A/dt = -dc_P/dt = -k \cdot c_A$$

where k is a first-order rate constant with the unit s^{-1}. Integration of this equation yields an exponential decay of the concentration of A with time.

From such exponential curves the half-life, $t_{1/2}$, or the time constant, τ, is obtained. The inverse time constant equals the observed rate constant, k_{obs}:

$$(k_{obs})^{-1} = \tau = t_{1/2}/\ln 2$$

In the example of the reaction above k_{obs} equals k, the rate constant of an individual step of a reaction. In the case that the above reaction is reversible, setting up an equilibrium between A and P, integration of the corresponding rate law $dc_A/dt = -k_f \cdot c_A + k_b \cdot c_P$ again yields an exponential equation, but here k_{obs} is composed of the sum of the rate constants of the forward and backward reaction steps $k_{obs} = k_f + k_b$.

Thus, the individual rate constants cannot be obtained directly but only by knowledge of the equilibrium constant, K, belonging to it. In this simple case the equilibrium and rate constants are related by $K = k_f/k_b$. (Note that equilibrium constants are always represented by capital K, whereas rate constants are represented by the small letter k.)

In another example two species react with each other in order to form product: A + B → P and the rate of the reaction will depend on the concentrations of both A and B:

$$dc_A/dt = dc_B/dt = -dc_P/dt$$
$$= -k \cdot c_A \cdot c_B$$

and k is therefore termed second-order rate constant. From this equation it is obvious that the unit of k is $(mol/l)^{-1} s^{-1}$. Analysis of the time course of the compounds in such a reaction is not straightforward and is often simplified by an experimental trick. One of the compounds is used in high molar excess over the other, say high excess of A over B, which results in an exponential decay of the concentration of B easy to analyze. In this case the obtained k_{obs} corresponds to $k \cdot c_A$ and it is called the pseudo-first order rate constant. Carrying out this type of experiment at various concentrations of A, in high excess over B each, allows to obtain the second order rate constant, k, from a plot of k_{obs} versus c_A.

In more complicated reactions, as mentioned above, k_{obs} is composed of a sum and/or product of individual rate constants, equilibrium constants and concentrations depending on the mechanism, experimental conditions and approximations that can be applied. CHRISTIAN HERRMANN

reading frame The flow of the *genetic information from DNA to protein requires – as described by the *central dogma of molecular biology – *transcription of the *gene into *messenger RNA (mRNA), which serves as *template for ribosomal protein synthesis (→translation). A linear chain of *nucleotides is directional due to the different sugar substituents and chemical compositions at the two ends. The letters of the *genetic code that correspond to an amino acid are made up of three consecutive nucleotides, a *codon. The linear arrangement of all consecutive, non-overlapping codons between the translational *start and *stop codons will give rise to one polypeptide chain, which is called the *open reading frame (ORF). A strand of mRNA contains three possible reading frames, due to the 3-nucleotide alphabet used for the genetic code, while in a double strand of DNA this number doubles as it could be read on the forward or reverse strands. Usually just one of them is an ORF encoding a protein of a reasonable size, while the other terminates after a few codons at a *stop codon, which is important for the interpretation of genomic sequencing data. In eukaryotes the situation is more complicated, as non-coding

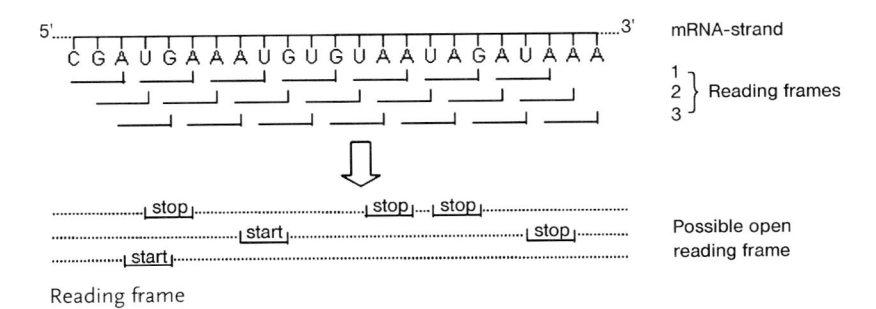

Reading frame

regions (→introns) have to be spliced out by the *spliceosome to form a continuous coding region, the situation naturally found in prokaryotes. STEFAN VÖRTLER

reading frame shift Any event in which the *reading frame is changed while *translation is ongoing and the *messenger RNA (mRNA) stays bound to the *ribosome. It results in a *recoding event of the initial *genetic information. Usually observed are the −1 or +1 frame shifts where the ribosome moves by 1 *nucleotide forward or backward and continues translation. Many viruses use a −1 frame shift to balance protein amounts while, for example, a +1 frame shift operates as feedback regulation during synthesis of many bacterial RF2 *translation factors. A premature UGA *stop codon at codon 25 in RF2 *genes terminates protein synthesis by the action of its own gene product. However, if cellular levels of RF2 are low, *termination will not occur and the pausing ribosome is rescued by the controlled frame shift which leads to completion of RF2 synthesis. Frame shifting is sequence dependent as the *A- and *P-site *transfer RNAs have both to slip along the mRNA, which requires alternative base-pairing possibilities. In case of the RF2 gene the peptidyl-tRNA^{Leu} bound to the CUU *codon 24 slips forward one position to base pair with the U in the UGA stop codon 25, therefore decoding UUU by a *wobble *base pair (→wobble hypothesis).

Alteration of the reading frame can be also caused by insertion or deletion of nucleotides in a number not divisible by 3. In this case, the reading frame shift leads to the synthesis of a *missense mRNA whose translation usually results in a non-functional protein since the message will be read in triplets specifying wrong amino acids. Reading frame shifts often create new stop codons and thus generate *nonsense mutations. STÉPHANIE VAULÉON, STEFAN VÖRTLER

reading frame shift mutation Any *mutation within the *coding region of a *gene that causes a switch during ribosomal protein synthesis from one codon reading frame to another, i.e. the insertion or deletion of 1 or 2 nucleotide residues. The properties of frame shift mutations are a consequence of the non-overlapping, comma-free nature of the *genetic code (→insertion mutation, →deletion mutation). HANS-JOACHIM FRITZ

real-time polymerase chain reaction (PCR) Developed because the *reverse transcription polymerase chain reaction is only a semiquantitative method and there is a need to quantify differences in *messenger RNA expression. The *polymerase chain reaction is the most sensitive method for detecting RNA and it can be used to discriminate between closely related mRNAs. Real-time PCR monitors the fluorescence emitted during the reaction as an indicator of *amplicon production during each PCR cycle in real-time rather than endpoint detection as used by other PCR methods. However, real-time PCR does not detect the size of the amplicon, and thus does not allow the differentiation between DNA and *cDNA amplification.

Real-time PCR is based on the detection and quantification of a fluorescent reporter (→fluorescent probe, →molecular beacon) and the fluorescence signal increases in direct proportion to the amount of PCR product in a reaction. By recording the amount of fluorescence emission at each cycle, it is possible to monitor the PCR reaction during exponential phase where the first significant increase in the amount of PCR product correlates to the initial amount of target *template. The greater the

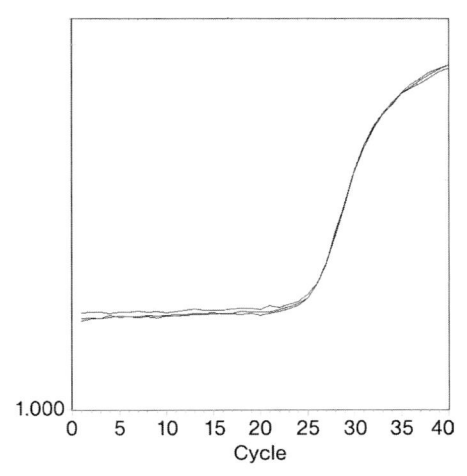

Output from a real-time PCR assay using a TaqMan probe.

starting amount of template, the sooner an increase in fluorescence is observed. Various types of *fluorescent probes or *primers are used in real-time PCR, including *TaqMan probes, *molecular beacons and *Scorpion probes. Each is based on an *oligonucleotide bearing both a fluorophore and a quencher, which are initially in close proximity and thus there is little fluorescence emission. As DNA *amplification proceeds, so the fluorophore and quencher are separated and fluorescence intensity increases in proportion to the amount of DNA produced.

The amount of DNA theoretically doubles with every cycle of PCR, so for n cycles of PCR there are 2^n times as much DNA as there was at the start of the reaction. During the early PCR cycles the increase in fluorescence is too low to register, but eventually a linear phase is reached and, as amplification cannot continue indefinitely, amplification reaches a plateau (see graph). As PCR amplification is a logarithmic reaction, if the output is analyzed on a logarithmic scale then there is a straight line relationship between the amount of DNA produced and the cycle number. It is a very important point in real-time PCR to examine the reaction while it is still in the linear phase. DAVID LOAKES

RecA Central protein of *homologous recombination in *Escherichia coli* with evolutionarily conserved homologous counterparts in all prokaryotic and eukaryotic cells. Among its unique properties, RecA monomers cooperatively form a nucleofilament on DNA, which is then able to actively search for homology in DNA and thus promote the physical exchange between DNA molecules during *recombination. Additionally, a particular form of the protein, named activated RecA (RecA$^\neq$), act as a coprotease in the autocatalytic processing of several proteins involved in the SOS response to *DNA damage in *E. coli*. GEMMA MARFANY

recoding Any event that results in an alteration of the initial *genetic information contained in the DNA *genome. This can happen on a post-transcriptional level by *RNA editing. More often recoding events are on a translational level as a result of a change of the *reading frame (termed *frame shifting) or an alternate interpretation of a particular *codon. In this way, selenocysteine and pyrrolysine are incorporated by recoding of a *stop codon (→code). Viruses use frame shifting extensively to minimize their *genome size by clustering several *open reading frames in the same sequence stretch, sometimes even in different orientations. In all cases the consequence is a protein sequence which differs at some positions to the encoding DNA. STEFAN VÖRTLER

recognition motif Local structure (often small *tertiary structure motifs) specifically recognized by a specific counterpart for interaction, e.g. a nucleic acid, protein or small-molecular (→recognition site). SABINE MÜLLER

recognition sequence →recognition site

recognition site *Nucleotide sequence (and also amino acid sequence) that is specifically recognized and bound by a protein, e.g. a *restriction endonuclease. Recognition sites formed by *nucleotides are often palindromic (→palindrome). JÖRN WOLF

recombinant DNA DNA obtained by recombinant DNA technology (can also be applied to RNA).

recombination Exchange of genetic material between two *nucleic acid strains (→homologous recombination).

recombinational repair →DNA repair

redundancy Presence in a *genome of multiple copies of the same *gene.

regulator gene →control gene

regulon Collection of *genes under the control of the same regulatory protein (generally used for prokaryote *gene expression).

relaxed DNA DNA which is not *supercoiled (→DNA structures).

renaturation Reformation of complementary *oligonucleotide sequences that were separated by thermal *denaturation.

repair enzyme Any enzyme that participates in *DNA repair. Most commonly, the term is reserved for enzymes that recognize the pre-mutagenic *DNA damage and initiate its removal from the DNA. Prototypic repair enzymes are *DNA glycosylases. In addition, there are repair *endonucleases, *exonucleases for trimming endonucleolytically cleaved DNA strands and specialized repair *DNA polymerases. Other enzymes and non-catalytic proteins, such as *DNA ligase, *DNA helicases and single-strand binding proteins, can participate in *DNA repair in addition to other functions they fulfill. HANS-JOACHIM FRITZ

repeat Motif of DNA that is repeated (→microsatellites, →minisatellites, →macrosatellites).

repetitive DNA DNA sequences that are repeated in the *genome. Repetitive DNA is a usual feature of eukaryote genomes and can be classified into two general types according to the reassociation kinetics: moderately repetitive DNA and *highly repetitive DNA. The latter consists of short sequences of about 5–100 nucleotides repeated thousands of times and includes *satellite DNA. Moderately repetitive DNA consists of longer sequences dispersed evenly throughout the genome like the retrotransposon elements SINEs (Short INerspersed Elements; <500 bp) or LINEs (Long INerspersed Elements; >5 kb). MAURO SANTOS

repetitive sequences →repetitive DNA

replacement loop *Secondary structural element in mitochondrial *transfer RNAs (tRNAs). In comparison to prokaryotic and eukaryotic cytoplasmic tRNAs, which have the classical *cloverleaf structure, *mitochondrial tRNAs have the *T-arm and the *variable loop substituted by the replacement loop. It consists of 4–12 nucleotides and is less structured. JÖRN WOLF

replacement synthesis Complementary strand synthesis that takes place after one of the *DNA strands of a *plasmid is transferred from a donor to a recipient during conjugation. JÖRN WOLF

replicase Replicase is an enzyme that catalyzes the synthesis of a *complementary strand from a *template.

replication →DNA replication

replication bubble →DNA bubble

replication fork Point where the two strands of DNA are unwound by *helicase breaking the *hydrogen bonds and forming two separated strands which are stabilized by *single-strand binding proteins to prevent rewinding, and the two strands are replicated separately. ANDREAS MARX, KARL-HEINZ JUNG

replication origin →origin

replicon Part of the DNA that contains the initiation site for *replication and undergoes replication as an autonomous unit. The *circular DNA of bacterial *chromosomes consists of one single replicon; the chromosomes of eukaryotes contain many replicons of 20–300 kb. ANDREAS MARX, KARL-HEINZ JUNG

replisome System of all protein complexes which are involved in *DNA replication. The replisome of *Escherichia coli* consists of *topoisomerases, proteins of the *primosome, *DNA polymerases and *DNA ligase. ANDREAS MARX, KARL-HEINZ JUNG

reporter gene Informative *gene that does not have any endogenous counterpart, whose easily detectable product allows us to follow the fate of another gene or assess the effect of *transcription factors. Widely used reporter genes include *lacZ* (→*lacZ* gene), which encodes the β-galactosidase enzyme, and *GFP* (green fluorescent protein) and derivatives. GEMMA MARFANY, ROSER GONZÀLEZ-DUARTE

reporter ribozyme *Ribozyme that is used in an assay to detect a special analyte. Numerous reporter ribozymes for many different molecules have been found, such as metal ions, small organic molecules, *nucleotides, amino acids, nucleic acids and even proteins. Generally, reporter ribozymes consist of an allosteric and a catalytic domain. The former mostly is obtained artificially by *in vitro* selection and binds the desired analyte. The catalytic domain consists of a *hammerhead or *hairpin ribozyme and is linked to the allosteric domain via a bridging element. It is activated or deactivated by the allosteric domain, thus producing a readout for the presence of the analyte. Combined with *immobilization strategies and *FRET, reporter ribozymes can be used as powerful tools for analyte detection, e.g. in biological samples (→RNA sensor →allosteric ribozyme). JÖRN WOLF

repression of gene expression *Gene expression is abolished by a *repressor protein that binds to a specific region of the *promoter (→operator) and decreases the rate of *transcription. Repression of gene expression is often involved in anabolic processes. Examples are enzymes for the synthesis of the amino acids arginine, tryptophan or histidine, which are solely expressed if the amino acid is not available in the environment. Gene expression is repressed until a certain threshold value is reached. BEATRIX SÜß

repressor *DNA-binding protein that decreases the rate of *transcription. Repressors bind to non-coding sequences (→non-coding DNA) on the DNA strand, impeding the progress of *RNA polymerases to progress along the strand and thus impeding *gene expression. BEATRIX SÜß

resistance gene *Gene involved in the resistance of a cell to various drugs or diseases. In bacteria, resistance can be encoded at the chromosomal or *plasmid level. Resistance mediated by plasmids is often based on genes encoding enzymes that inactivate the drug (chloramphenicol) or prevent its uptake or transport it out of the cell (tetracyclins). *Mutations in chromosomal genes often lead to an alteration of the target of the *antibiotic (streptomycin), e.g. the molecular mechanism of streptomycin resistance is mainly based on *mutations of the *rpsL* gene encoding the prokaryotic ribosomal protein S12. ANNEGRET WILDE

restriction endonuclease Recognizes a specific *nucleotide sequence in *double-stranded DNA and cleaves each strand of the *duplex at a particular site of the *sugar–phosphate backbone (also called restriction enzymes). The enzyme catalyzes the hydrolysis of the *phosphodiester bond without damaging the *bases. Restriction endonucleases are found in a wide range of bacterial species. The main function of these enzymes is to recognize and cleave *foreign DNA; such DNA is said to be *restricted*. Commonly, four types of restriction endonucleases are distinguished.

Type I consist of three different subunits, HsdM, HsdR and HsdS, that are responsible for modification, restriction and sequence recognition, respectively. The quaternary structure of the active endonuclease is $HsdM_2HsdR_2HsdS$. Type I enzymes require ATP, Mg^{2+} and S-adenosyl-L-methionine (AdoMet, →DNA methyltransferase) for activity. They interact in general with two asymmetrical bipartite *recognition sites, translocate the DNA and cleave it at random sites that can be 1000 bp or more away from the *recognition sequence (approximately half-way between two sites). Typical examples are *Eco*KI, *Eco*AI, *Eco*R124I and *Sty*SBLI.

Type II are, in most cases, simply organized homodimeric or tetrameric proteins that act independently of their companion *methyltransferases. These enzymes usually recognize rather short (4–8 bp) specific (very often palindromic) DNA sequences and cleave phosphodiester bonds at constant positions within or close to the recognition site to produce 5'-phospates and 3'-hydroxyls. They require only Mg^{2+} as cofactor. Due to the interest in type II restriction endonucleases for *recombinant DNA technology, around 4000 have been characterized. Type II enzymes are further classified according to their recognition sites and individual properties (*Nucleic Acids Res* 2003, *31*, 1805–1812).

Type III consist of two subunits, Mod (responsible for DNA recognition and modification) and Res (responsible for DNA cleavage). Active enzymes have a Mod_2Res_2 stoichiometry, require ATP and Mg^{2+} for activity, and are stimulated by AdoMet. They interact with two head-to-head arranged asymmetrical recognition sites, translocate the DNA and cleave the DNA about 25 bp from one of the recognition sites. Typical examples are *Eco*P1I and *Eco*P15I.

Type IV recognize and cleave methylated DNA. The best-studied representative is *Mcr*BC, which consists of two different subunits (B and C), responsible for DNA recognition and cleavage. *Mcr*BC recognizes in DNA at least two dinucleotides: dA or dG followed by a methylated dC, which are separated by anywhere from 40 to 3000 bp. GTP and Mg^{2+} are required for DNA cleavage; cleavage takes place about 30 bp from one of the sites. ELENA KUBAREVA

restriction enzyme →restriction endonuclease

restriction fragment Product of *double-stranded DNA cleavage by a suitable *restriction endonuclease. Each DNA molecule is cut by specific restriction endonucleases into distinct *fragments. These fragments can be separated on the basis of their size by *gel electrophoresis. The cleaved DNA is placed on top of a gel made of agarose or polyacrylamide. When an electric current is passed through the gel, each fragment moves down it at a rate that is inversely related to the log of

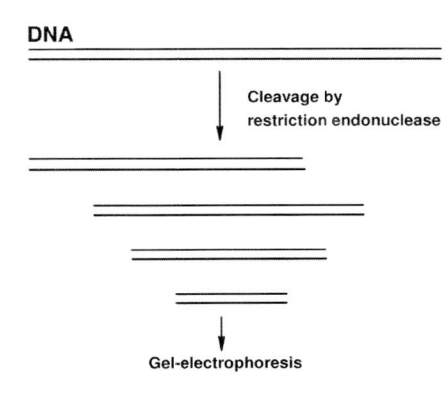

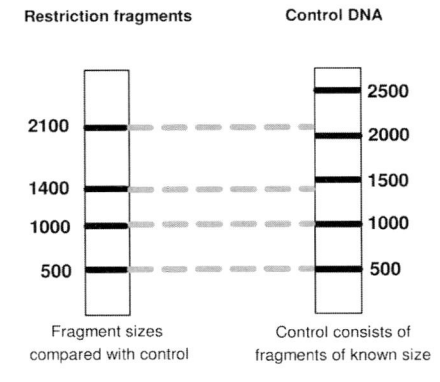

restriction map

```
DNA
═══════════════════════════════════════
            │  Cleavage by
            │  restriction endonuclease
            ▼
════════════════════
       ══════════════
           ══════════════════
              ═══════════════
            │
            │  Gel-electrophoresis
            ▼

Restriction fragments        Control DNA

                              ━━━━  2500
   ━━━━  2100
                              ━━━━  2000
   ━━━━  1400
                              ━━━━  1500
   ━━━━  1000                 ━━━━  1000
   ━━━━   500                 ━━━━   500

  Fragment sizes           Control consists of
  compared with control    fragments of known size
```

its molecular weight. This movement produces a series of bands. Each band corresponds to a restriction fragment of particular size, decreasing down the gel. The sizes of the individual fragments generated by restriction enzyme are determined by comparison with the positions of fragments of known size used as the control.

The average size of the restriction fragments depends upon the frequency with which a particular *restriction site occurs in a DNA molecule. This, in turn, depends largely on the size of the restriction enzyme recognition sequence.

Some restriction fragments have no unpaired bases on either end; these ends are often called *blunt ends. Other have 2–4 nucleotides of one strand unpaired at each end. They are referred to as *sticky ends because they can form hydrogen-bonded *base pairs with each other or with complementary sticky ends on any other DNA cut with the same enzyme. Restriction fragments with sticky ends are used in *recombinant DNA technology. ELENA KUBAREVA

restriction map Linear sequence of *restriction sites separated by defined distances on DNA. The map shows the positions at which particular restriction enzymes cut DNA and DNA is divided into a series of regions of defined lengths that lie between sites recognized by the *restriction endonucleases. An important feature is that a restriction map can be obtained for any sequence of DNA, irrespective of whether *mutations have been identified in it, or, indeed, whether we have any knowledge of its function.

The distances between the restriction sites are measured in *base pairs for short distances; longer distances are given in *kilobase pairs in DNA. At the level of the *chromosome, a map is described in *megabase pairs.

Restriction mapping is more practical for comparing smaller segments of DNA, usually a few thousand *nucleotides long. For linear DNA *fragments one can infer the position of restriction sites by carrying out a double-digestion approach. In this technique, the DNA is cleaved simultaneously with two restriction endonucleases (A and B) as well as with either one by itself. The most decisive way is to extract each *restriction fragment produced in the individual digests with either enzyme A or enzyme B and then to cleave it with the other enzyme. The restriction fragments are analyzed by *gel electrophoresis. Double digests define the cleavage positions of one enzyme with regard to the other. A restriction map can be constructed by relating the A fragments and B fragments through the overlaps seen with double digest fragments.

```
   A  B         A  B         A
 ▌  ▌▌       ▌  ▌▌       ▌
1000 200    1900  600 800   500  bp
```

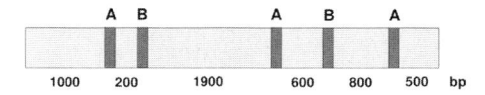

The actual construction of a restriction map usually requires recourse to several restriction endonucleases, so it becomes necessary to resolve quite a complex pattern of the overlapping fragments generated by the various enzymes. Several other techniques are used in conjunction with comparison of restriction fragments, including *end labeling in which the ends of the DNA molecule are labeled with a radioactive phosphate (→radioactive labeling). When restriction fragments are identified by their possession of a labeled end, each fragment directly shows the distance of a cutting site from the end. Successive fragments increase in length by the distance between adjacent restriction sites.
ELENA KUBAREVA

restriction-modification system Usually refers to a system containing *restriction enzyme activity and a *methyltransferase activity. It is widely believed that restriction-modification systems are used by bacteria and perhaps other prokaryotic organisms to protect themselves from viral, *plasmid and other *foreign DNAs. *Restriction endonucleases recognize and cleave incoming DNA. The cell's own DNA is not cleaved because the sequence recognized by restriction endonuclease is methylated and thereby protected by a specific *DNA methyltransferase (→DNA methylation).

Another explanation for the maintenance of restriction-modification systems is based on the observation that several restriction-modification *gene complexes in bacteria are not easily replaced by competitor genetic elements because their loss leads to cell death. This finding led to the proposal that these complexes may actually represent one of the simplest forms of life, similar to viruses, *transposons and homing endonucleases. This *selfish gene hypothesis suggested by I. Kobayashi in 1996 is now supported by many lines of evidence from *genome analysis and experimentation.

One more hypothesis that explains why restriction-modification systems are present assumes that they aid the generation of diversity. Supporting this notion is that these systems are indeed associated with *genome variation in a number of different ways.

There are four types of restriction-modification systems. They were named in the order of discovery, although the type II system is the most common.

Type I enzymes have three genes encoding a methyltransferase subunit, a restriction subunit and a specificity subunit. Type II enzymes have two separate genes, one encoding a restriction enzyme and one encoding a methyltransferase. Type III enzymes also have two genes, one encoding a methyltransferase, which can operate alone or form part of a complex with a restriction subunit. Type IV enzymes act alone and restrict DNA that is methylated.

Based on the analysis of published genome sequences a somewhat more even distribution among putative restriction-modification systems has been suggested: approximately 29% Type I, 45% Type II, 8% Type III and 18% Type IV.
ELENA KUBAREVA

restriction site Restriction sites or restriction *recognition sites are particular sequences of *nucleotides that are recognized by *restriction endonuclease as sites to cut the DNA molecule. The point in a DNA *sugar–phosphate backbone at which the cleavage of *phosphodiester bond

Restriction sites

Restriction endonuclease	Type (subtype)	Recognition sequence and cleavage site (↓)
EcoKI	I(A)	5'...AACNNNNNNGTGC...3' 3'...TTGNNNNNNCACG...5' Cleavage sites are located around 1000 bp or more from the recognition sequence
HphI	II(A)	5'...GGTGANNNNNNNN↓...3' 3'...CCACTNNNNNNN↑...5'
BplI	II(B)	5'...↓NNNNNNNNGAGNNNNNCTCNNNNNNNNNNNNN↓...3' 3'...↑NNNNNNNNNNNNNCTCNNNNNGAGNNNNNNNN↑...5'
SmaI	II(P)	5'...CCC↓GGG...3' 3'...GGG↑CCC...5'
PspALI	II(P)	5'...CCC↓GGG...3' 3'...GGG↑CCC...5'
XmaI	II(P)	5'...C↓CCGG-G...3' 3'...G-GGCC↑C...5'
MspI	II(P)	5'...C↓CG-G...3' 3'...G-GC↑C...5'
HinfI	II(P)	5'...G↓ANT-C...3' 3'...C-TNA↑G...5'
EcoP15I	III	5'...CAGCAGNNNNNNNNNNNNNNNNNNNNNNNNN↓NN...3' 3'...GTCGTCNNNNNNNNNNNNNNNNNNNNNNNNN- NN↑...5'
McrBC	IV	5'...(G/A)m^{5}C(N)$_{40\text{-}2000}$(G/A)m^{5}C...3' 3'...(C/T)—-G(N)$_{40\text{-}2000}$(C/T)—-G...5' Cleavage sites are located around 30 bp from one of the (G/A)m^{5}C dinucleotides

occurs is called the cleavage site. The length and structure of restriction recognition site as well as localization of cleavage site with respect to recognition site depend on the type or subtype of restriction endonuclease.

For example, the recognition sites of more common subtype IIP restriction endonucleases are symmetric sequences of 4–8 bp, often termed *palindromes (SmaI, MspI). Their cleavage sites have fixed symmetrical locations either within the sequence or immediately adjacent to it. Some restriction enzymes cleave DNA strands immediately opposite one another, producing *blunt ends (SmaI). Most enzymes make slightly staggered incisions, resulting in *sticky ends, out of which one strand protrudes (XmaI).

Restriction recognitions sites can be degenerated. Thus, HinfI recognizes a 5-bp sequence starting with GA, ending in TC and having *any* base between (in the table, "N" stands for any nucleotide).

The length of the *recognition sequence dictates how frequently the restriction endonuclease will cut in a *random sequence of DNA. Enzymes with a 6-bp restriction site will cut, on average, every 4096 bp; a 4-bp restriction site will be occurred roughly every 256 bp.

Different restriction enzymes can have the same recognition and cleavage sites – such enzymes are called isoschizomers (SmaI, PspALI). Neoschizomers are that subset of isoschizomers that recognize the same sequence, but cleave at different positions from the prototype (*Sma*I, *Xma*I). The recognition site for one enzyme may contain the restriction site for another (*Sma*I, *Msp*I). ELENA KUBAREVA

retrohoming Process of *intron mobility of *group II introns which involves a reverse-transcribed copy of the *intron. The retrohoming process starts with the *translation of an *intron-encoded protein which forms a RNA–protein (→ribonucleoprotein) complex with a RNA copy of the intron. This complex (a) recognizes the *insertion site in an intron-less gene, (b) inserts the RNA into one strand through *reverse splicing, (c) cuts the other strand and (d) forms a *cDNA copy of the RNA by virtue of the protein's *reverse transcriptase function. The RNA is then replaced by DNA by the cell machinery. The retrohoming process has been made use of by applying some *group II introns for *gene repair. STÉPHANIE VAULÉON

retroposon → retrotransposon

retro pseudogene Inactive *gene that lacks *introns and has originated by *reverse transcription of *messenger RNA and insertion of a *duplex copy in the *genome.

retrotransposon *Transposon that mobilizes via RNA. The DNA element is transcribed and then reverse transcribed. The *double-stranded DNA copy (→copy DNA) is inserted into a new chromosomal position. MAURO SANTOS

reversed-phase chromatography Chromatographic technique that separates molecules according to their polarity. Only unpolar compounds can be successfully separated from one another. Nucleic acids require their negatively charged phosphate groups to be masked by positively charged ammonia groups carrying alkyl residues. Usually tetrabutylammonia or triethylammonia ions are used in the buffer. Additionally, synthesized nucleic acids that are to be purified by reversed-phase chromatography are synthesized dimethoxytrityl (DMTr)-on. That means that the DMTr group on the 5'-end is not removed in the last synthesis step. The solid phases used in nucleic acid purification usually consist of C_{18}-alkyl chains that bind longer nucleic acids stronger than shorter strands. Subsequent elution is achieved by raising the fraction of the unpolar component (e.g. acetonitrile) in the buffer. →ion-exchange chromatography. JÖRN WOLF

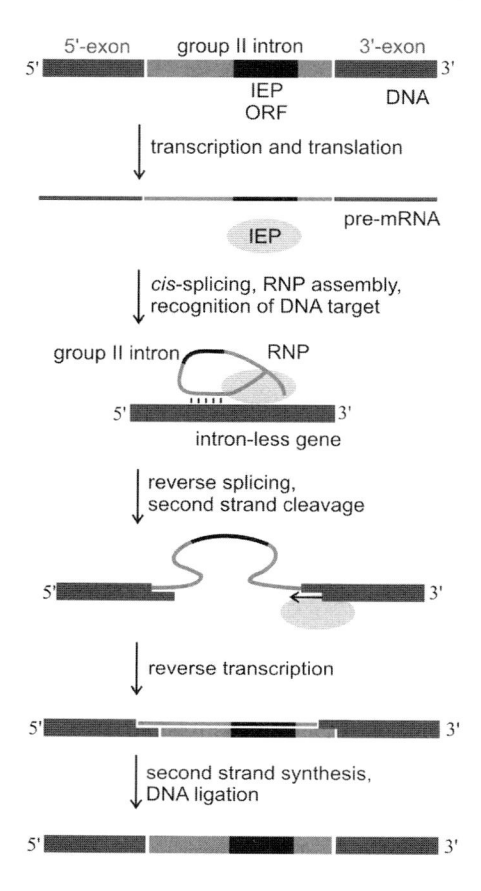

reversed Watson–Crick base pair Interaction of *nucleobases via their

*Watson–Crick edge, but with the orientation of the *glycosidic bond relative to the axis of interaction in *trans*. This requires that two nucleic acid strands, which interact via reversed Watson–Crick base pairs, are in *parallel orientation. In particular, the term "reversed Watson–Crick base pair" refers to combinations of only AT/U and CG; however, a number of base pairs interacting via the Watson–Crick edge with the glycosidic bonds being oriented in *trans* have been observed in *RNA structures (→geometric nomenclature and classification of RNA base pairs).
SABINE MÜLLER, BETTINA APPEL

reverse-joined hairpin ribozyme Structure variant of the *hairpin ribozyme where the domains A and B are joined in a reverse order: helix 4 is joined to helix 1 via a single-stranded polyadenosine *linker. Reverse hairpin ribozymes, like conventional hairpin ribozymes, can cleave and ligate specific RNA substrates through a *transesterification reaction.
STÉPHANIE VAULÉON

reverse splicing Mechanism by which *introns are transferred to new *genes (→retrohoming).

reverse transcriptase Divalent cation-dependent DNA polymerase which synthesizes DNA by using either RNA or DNA as a *template. This enzyme is the core protein in the *reverse transcription mechanism. Reverse transcriptases were found in retroviruses, in retrovirus-like particles as well as in eukaryotes. Eukaryotic cells use a reverse transcriptase with integrated RNA strands to extend G-rich *repetitive sequences at the *telomeres. The *telomerases are responsible for the maintenance of telomere length, which in turn plays an important role during ageing. Inhibitors of reverse transcriptase like Sumarin or desoxycytidine are used for treatment of retroviral infections.
IRENE DRUDE

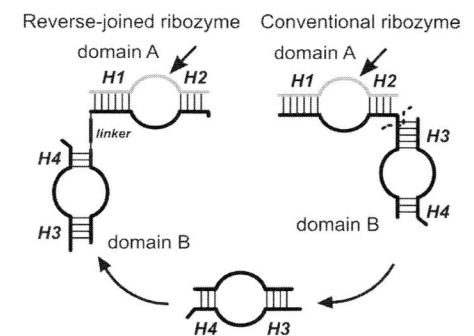

reverse transcription Mechanism for generating DNA from a RNA *template by the enzyme *reverse transcriptase. For generation of a DNA copy from RNA, the reverse transcriptase uses a short *primer to tie in *deoxynucleotides in the 5'→3' direction with the primer. An internal *RNase H domain allows the degradation of the RNA template from the *DNA–RNA hybrid directly after *DNA synthesis. The synthesized DNA strand itself acts as a template for reverse transcriptase-driven synthesis of a *double-stranded DNA (dsDNA). For catalytic activity, this enzyme requires divalent cations, like magnesium or manganese. The reverse transcriptase has no *proofreading properties, often resulting in erroneous DNA. During *polymerization, the reverse transcriptase can also exchange between the templates, which results in *recombination.

Retroviruses use reverse transcription to synthesize dsDNA from its RNA *genome, which in turn can be integrated into host genome during infection. In eukaryotic cells, reverse transcription plays an important role during ageing. The *telomerase contains an integrated RNA template for extending the repetitive G-rich sequences at the *telomeres. In genetic engineering, reverse transcriptase is used to synthesize *cDNA from mRNA for further *cloning into *vectors or for quantification of specific *messenger RNA amounts by *polymerase chain reaction. IRENE DRUDE

Reverse transcription polymerase chain reaction (RT-PCR or RNA-PCR) Powerful tool that allows for the *amplification of small quantities of RNA *templates. RT-PCR is a two step process in which RNA is first copied by a *reverse transcriptase (an *RNA-dependent DNA polymerase) into its *complementary DNA (cDNA). There are also thermostable reverse transcriptases (such as the reverse transcriptase from avian myeloblastosis virus) which are compatible with the PCR protocols. Following this, the single-stranded cDNA is exponentially amplified by a thermostable DNA polymerase using the *polymerase chain reaction to give large quantities of DNA for further analysis. It is also possible to modify the protocol to use *anchored primers with an oligo(dT)-modified primer which allows *transcription and amplification from the 3′-terminus of *poly(A) tails. As the resultant *DNA pool is derived from the original RNA sequence it is possible to sequence the DNA (after *cloning, →DNA sequence analysis) and hence determine the original RNA sequence, and this is the best method for sequencing RNA (→RNA sequencing). RT-PCR has wide applications in molecular biology as well as in clinical studies where it can be used to detect infectious agents, genetic markers and as a means for the detection of small quantities of protein. DAVID LOAKES

Rfam Abbreviation for RNA family database, is a large collection of *sequence alignments for *non-coding RNA families. It contains data about conserved natural RNAs like *ribosomal RNAs, *ribozymes, *riboswitches and *micro-RNAs, and is found online at http://www.sanger.ac.uk/Software/Rfam/. RÜDIGER WELZ

RFM element *Aptamer part of the *FMN riboswitch in the 5′-untranslated region of prokaryotic *genes.

ribonuclease →RNase

ribonuclease P (RNase P) *Ribonucleoprotein metalloenzyme that catalyzes the endonucleolytic 5′-maturation of *transfer RNA (tRNA) *primary transcripts (Fig. 1) in all three kingdoms of life (Archaea, Bacteria and Eukarya) as well as in mitochondria and chloroplasts.

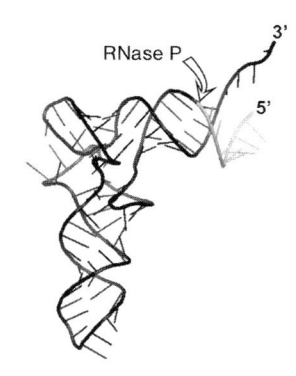

Figure 1. L-shaped 5′-precursor tRNA; the 5′-precursor segment is indicated in grey and the RNase P cleavage site is marked by the arrow.

The protein/RNA ratio increases from Bacteria over Archaea to Eukarya:

bacterial RNase P enzymes possess a single small protein subunit (around 13 kDa) that makes up 10% of the molecular mass of the holoenzyme, the archaeal number of proteins is usually four, and in eukaryotes more than 50% of the molecular mass is protein with, for example, nine polypeptides in yeast and 10 in humans. The bacterial protein subunit lacks structural similarity to any of the eukaryotic or archaeal protein subunits (with the possible exception of Pop5), whereas the archaeal proteins have homologs in eukaryotic RNase P. The bacterial protein subunit, essential *in vivo*, plays a role in binding of the 5′-precursor segment, in stabilizing local *RNA structure and in enhancing the affinity of catalytically important metal ion *binding sites.

All known RNase P RNAs share a common core structure centered around helix P4 (Fig. 2), which is also found in the eukaryotic RNase MRP RNA, indicating that they all stem from a single progenitor RNA. *In vitro*, RNA subunits of RNase P enzymes from Bacteria are highly active catalysts without the protein cofactor. Residual ribozyme activity in the absence of any protein component has also been shown for archaeal and, more recently, for eukaryal RNase P RNA. So far, RNase P (RNA) is the only naturally *trans*-acting (→*trans*-active) *ribozyme known, aside from the *ribosome. Bacterial RNase P RNA, 300–400 *nucleotides in length, has two independent folding domains (Fig. 2): the catalytic domain (*C-domain*) including all structural elements required for *catalysis and the specificity domain (*S-domain*) interacting with the *T arm of *transfer RNAs. Recent crystal structures have revealed a rather flat surface of bacterial RNase P RNA that includes the *active site. The latter is formed by a network of long-range interactions between conserved interhelical sequences. A few *functional groups of RNase P RNA are known to interact with the substrate in immediate vicinity of the cleavage site (Fig. 3a), such as the bases of residues G292 to U294 and A248 (numbering according to *Escherichia coli* RNase P RNA, Fig. 2). The *active conformation of bacterial RNase P RNA depends on complexation with the protein subunit and the substrate, both of which affect the coordination of Mg^{2+} ions crucial for catalysis.

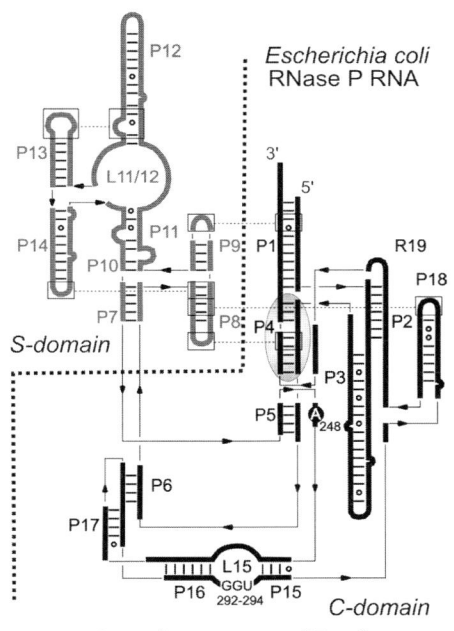

Figure 2. Secondary structure of *E. coli* RNase P RNA; the conserved P4 helix is highlighted by the grey oval; residues 248 and 292–294, relevant to Figure 3(a), are depicted.

Cleavage of precursor tRNAs by RNase P is irreversible and generates 3′-hydroxyl and 5′-phosphate termini. A solvent hydroxide or activated water molecule is

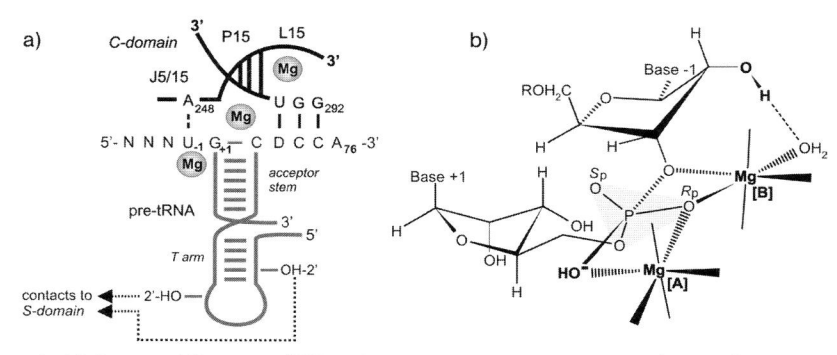

Figure 3. (a) Structural features of RNase P RNA–substrate complexes. D = G, A or U. The exact positioning and number of Mg^{2+} ions (grey spheres) is not known. (b) Proposed RNase P cleavage mechanism involving two magnesium ions. Based on nuclear magnetic resonance data, the ribose at nucleotide +1 is drawn in the A-helical C3′-*endo* and the ribose at position −1 in the C2′-*endo* conformation.

thought to act as the nucleophile in an S_N2 in-line displacement mechanism. A two-metal-ion mechanistic model for the *transition state of *phosphodiester hydrolysis by bacterial RNase P RNA has been formulated (Fig. 3b), according to which Mg^{2+} ions at sites [A] and [B] directly coordinate to the pro-Rp oxygen. Mg[A] interacts with the OH$^-$ nucleophile via *inner-sphere coordination and Mg[B] interacts with the 2′-hydroxyl at position −1 via one of its inner-sphere water molecules. Due to strong inhibition effects caused by sulfur substitutions, *active site chemistry may include additional metal ion interactions at the pro-Sp oxygen (marked Sp) and the 3′-bridging oxygen. ROLAND K. HARTMANN, MARIO MÖRL, DAGMAR K. WILLKOMM

ribonucleic acid →RNA

ribonucleoprotein (RNP) Higher-ordered complex formed of one or more RNA species and one or more proteins. Several of the crucial intracellular machineries are complex RNPs including the *ribosome, the *spliceosome, the *telomerase complex, *heterogeneous nuclear ribonucleoprotein and the *RISC complex. GEORG SCZAKIEL

ribonucleoside →nucleoside

ribonucleoside diphosphates →nucleoside phosphates

ribonucleoside monophosphates →nucleoside phosphates

ribonucleoside phosphates →nucleoside phosphates

ribonucleoside triphosphates →nucleoside phosphates

ribonucleotide →nucleotide

ribonucleotide reductase (RNR) Catalyzes the reduction of the 2′-carbons of *ribonucleotides to synthesize *2′-deoxyribonucleotides. Uniquely in enzyme reaction chemistry, the enzyme reaction involves generation of free radicals. Three classes of RNRs, differing in *primary structure and radical-generating mechanism, are known. VALESKA DOMBOS

ribo-organism Hypothetical organism (with distinct boundaries) that uses nucleic acids for inheritance and enzymatic *catalysis. Peptides and proteins may complement its function but not via *translation using the *genetic code. Ribo-organisms may have been real before the advent of the *genetic code (in the *RNA world) and they may be constructed artificially in the future, with an estimated *genome size of up to about 70–100 genes. EÖRS SZATHMÁRY

ribose Sugar unit in natural *nucleosides constituting RNA. Only the D-enantiomer occurs in natural RNA (→nucleosides, →nucleotides).

ribose zipper RNA structure motif that involves *tertiary interaction (consecutive *hydrogen bonds) between *nucleobases and 2'-hydroxyl groups of *ribose units from two regions of the RNA chain. The two interacting strands are oriented antiparallel. VALESKA DOMBOS, SABINE MÜLLER

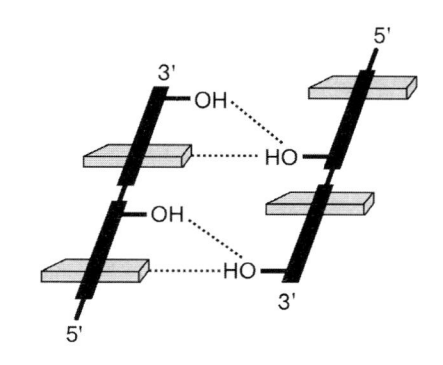

ribosomal proteins →ribosome

ribosomal RNA (rRNA) Prokaryotic rRNA includes the three species 5S rRNA, 16S rRNA and 23S rRNA; eukaryotic RNA includes the three species 5.8S rRNA, 18S rRNA and 28S rRNA (→RNA processing, →ribosome). SABINE MÜLLER

ribosomal subunits →ribosome

ribosome Universal multi-component nanomachine that translates the *genetic code into proteins. This *ribonucleoprotein cellular assembly is built of two subunits of unequal size that associate upon the initiation of protein biosynthesis to form a functional particle and dissociate once this process is terminated. In prokaryotes, the ribosomal subunits are of molecular weights of 0.85 and 1.45 MDa. The small subunit (called 30S) contains a *ribosomal RNA (rRNA) chain (called 16S) of around 1500 *nucleotides and 20−21 ribosomal proteins (r-proteins). The large subunit (called 50S) has two rRNA chains (23S and 5S RNA) of about 3000 nucleotides in total and 31−35 r-proteins. The actual biosynthetic process is performed cooperatively by the two ribosomal subunits. The small subunit provides the *decoding center and controls *translation fidelity, and the large one contains the *catalytic site, called the *peptidyl transferase center (PTC), as well as the protein exit tunnel.

Proteins are *polymers of amino acids, of which the three-dimensional structures, determined by their sequences, provide the stereochemistry required for their function. Each of the 20 natural amino acids is coded by 3 nucleotides (→codons). *Messenger RNA (mRNA) carries the *genetic code to the ribosomes and the amino acids, the nascent proteins building blocks are delivered in the form of *aminoacylated tRNA. The ribosome posses three tRNA binding sites: the *A-(aminoacyl), the *P-(peptidyl) and the *E-(exit) sites, each resides on both subunits. *transfer RNAs (tRNAs) are L-shape molecules containing an *anticodon loop that *base pairs with its *cognate codon of the mRNA and its 3'-end (almost universally composed of the *CCA-tail) carries the cognate amino acid. The *codon–anticodon interaction, called *decoding, is performed on the small subunit, whereas the tRNA *acceptor stem, together with the aminoacylated or peptidylated tRNA 3'-ends interacts with the large subunit. Hence, the tRNA molecules are the entities combining the two subunits, in addition to the intersubunit bridges, which are built of flexible components of both subunits.

Upon *initiation of protein biosynthesis, *initiation factors and mRNA bind to the small subunit, so that the *nucleotide triplet coding for the *initiator tRNA (f-Met in eubacteria, Met in eukaryotes) resides at the P-site. In bacteria, a purine-rich region, called the *Shine–Dalgarno sequence, base pairs with its mate located at the 3'-end of the 16S RNA. This binding seems to be accompanied by a "latch-like" closing and opening mechanism, performed on the other side of the small subunit by the two main regions of the small subunit. The initiator tRNA binds to the P-site within the decoding center and then the two subunits associate, while intersubunit bridges are being formed by conformational changes of their components. Within the functional assembled ribosome, the interface surfaces of both subunits are rich in RNA. Furthermore, both the decoding center and the PTC are positioned in an almost protein-free environment, indicating that the ribosome is a *ribozyme that may be assisted by the r-prokins. The selection of the correct tRNA is based on direct codon–anticodon base pairing between the tRNA *anticodon loop and its cognate mRNA. The accuracy of this selection has an error rate of 10^{-3} to 10^{-4} and the ribosome plays a major role by strictly monitoring the base pairing at the first two positions of each codon, but tolerates *non-canonical base pairs at the third position.

The *elongation cycle involves decoding, peptide bond formation, detachment of the P-site tRNA from the growing polypeptide chain and release of a deacylated tRNA molecule; all occur simultaneously with the advancement of the mRNA together with the tRNA molecules from the A- to the P- and then to the E-site. This motion is driven by GTPase activity.

Peptide bond formation is performed within the PTC by a nucleophilic attack of the primary amine of the aminoacylated A-site tRNA amino acid on the carbonyl carbon of the peptidyl-tRNA at the P-site. The PTC is built as an arched void, with shape and dimensions suitable for accommodating the 3'-ends of the aminoacylated A-site and the peptidylated P-site tRNA molecules. The PTC is situated within a universal sizable symmetry-related region (containing about 180 nucleotides), which connects all ribosomal functional centers involved in amino acid *polymerization. The linkage between the elaborate PTC architecture and the A-site tRNA position revealed that the A- to P-site passage of the tRNA 3'-end is performed by a rotatory motion. This rotatory motion leads to stereochemistry suitable for peptide bond formation as well as for amino acid polymerization and promoted substrate-mediated catalysis. Thus, the ribosome provides the frame allowing for proper and efficient elongation of nascent protein chains in addition to the formation of the peptide bonds.

The structural symmetry between the two halves of the PTC suggests that it evolved by *gene fusion. Preservation of the three-dimensional structure of the two halves of the ribosomal frame, regardless of the sequence, demonstrates the rigorous requirements of accurate substrate positioning in stereochemistry supporting peptide bond formation.

During the course of protein biosynthesis, the mRNA and the tRNAs bound to it move precisely through the ribosome, one codon at a time (→translocation). In this step, which can be described as a simple or more complicated ("hybrid states") shift, the A-site tRNA, carrying the nascent chain, passes into the P-site while creating the peptide bond. The deacylated P-site tRNA, acting as the leaving group, moves from the P-site to the E-site and eventually is ejected from the ribosome. The rotatory motion of the tRNA 3'-end is performed in concert with the mRNA/tRNA shift. Similar to all other steps of protein biosynthesis, this process, called *translocation, is assisted by non-ribosomal factors, among them EF-Tu, which delivers the aminoacylated tRNA to the A-site and EF-G, which promotes translocation. The result is a ribosome ready for the next round of elongation, with deacylated tRNA in the E-site, peptidyl-tRNA in the P-site and an empty A-site, ready to accommodate the next cognate ternary complex.

*Termination occurs when a *stop codon on mRNA is encountered. In bacteria, termination involves three release factors, RF1, RF2 and RF3, of which the first two are codon selective, whereas the third, RF3, binds to the complex of RF1/2 with the ribosome and is a GTPase. In eukaryotes, a single factor, eRF1, recognizes all three stop codons. GTP hydrolysis is required for subsequent dissociation of the release factors. After release of the peptide chain, the ribosome is left with mRNA and a deacylated tRNA in the P-site. This complex is disassembled by a factor called *ribosome recycling factor (RRF) in complex with EF-G. The latter triggers GTP hydrolysis that leads to the dissociation of ribosomes into its two subunits.

Nascent proteins emerge out of the ribosome through the exit tunnel – a universal feature of the large ribosomal subunit. The entrance to this tunnel is adjacent to the PTC and its opening is located at the other end of the subunit. It is lined primarily by rRNA, with tips of a few r-proteins reaching its walls. This tunnel plays an active role in sequence-specific gating of nascent chains and in responding to cellular signals. It also may provide services for co-translational folding of small *secondary structure elements. The encounter of the emerging nascent proteins with ribosome associated chaperones such as the trigger factor in eubacteria, which is known to prevent misfolding and aggregation, seems to cause conformational alterations which expose a sizable hydrophobic area. This area may compete with the natural aggregation of the nascent proteins. ADA YONATH

ribosome-binding site The first step in protein synthesis is binding of the small ribosomal subunit to *messenger RNA, where upon migration of the small unit in the 3'-direction is locating the *start codon AUG. In prokaryotic organisms the small 30S ribosomal subunit binds to the *purine-rich sequence called the *Shine–Dalgarno sequence (5'-GGAGG-3') in mRNA, before it starts migrating and thus locating the start codon AUG. The Shine–Dalgarno sequence is located about 10 *nucleotides *upstream of the start codon and is responsible for correct positioning of the small ribosomal subunit. The *consensus sequence in the small ribosomal subunit is found in 16S RNA. After location of the start codon the large 50S ribosomal subunit binds and the complete *ribosome complex is formed.

In eukaryotic organisms the small 40S ribosomal subunit together with the *initiator tRNA carrying the amino acid methionine recognizes and binds as a complex to the *5'-methyl guanosine cap on the mRNA. The complex then starts to move into the 3'-direction in order to locate the start codon AUG, whereupon the large ribosomal subunit 60S binds to create the complete ribosome unit. →ribosome. TINA PERSSON

ribosome interactions with antibiotics As a key player in cell vitality, the *ribosome is targeted by many *antibiotics of diverse nature. These were found to cause miscoding, to minimize essential mobility, to interfere with substrate binding at the *decoding center and at the *PTC, and to block the protein exit tunnel. As ribosomes are highly conserved, subtle sequence and/or conformational variations between ribosomes from different kingdoms of life enable drug selectivity, thus facilitating clinical usage. Specifically for macrolides/ketolides, a prominent antibiotics family that obstructs the progression of the nascent proteins within the tunnel, a single *nucleotide determines binding affinity, but the overall *binding pocket conformation governs the actual binding mode, hence indicating that there is a difference between mere binding and therapeutic effectiveness. ADA YONATH

ribosome recycling factor (RRF) Factor that mediates disassembling of the *ribosome–*messenger RNA complex after release of the terminal P-site tRNA and the nascent chain (→termination), and subsequent dissociation of the ribosome into its two subunits. ADA YONATH

ribosome subunits →ribosome

riboswitches *Gene control elements found in non-coding regions of *messenger RNAs (mRNAs) (hence *ribo-*). These structured *RNA domains undergo conformational changes (*-switch*) by binding directly to small metabolites, which leads to regulation of *gene expression in dependence of metabolite concentration or availability. Usually riboswitches are found in the *5′-untranslated region of mRNAs and consist of a highly conserved *aptamer that selectively binds to the target *ligand and a less conserved *expression platform that mediates changes in the extent of gene expression. Gene control can happen on transcriptional level, if for instance an intrinsic terminator is part of the expression platform, or on translational level, if accessibility to the *ribosome binding site is prevented by ligand-induced structural rearrangement. Depending on the organization of structural elements, riboswitches can increase or decrease the level of protein expression corresponding to "ON" or "OFF" switching, respectively.

So far 10 classes of riboswitches have been described that specifically bind coenzymes [→coenzyme B_{12} riboswitch, →thiamine pyrophosphate (TPP) riboswitch, →FMN riboswitch and →SAM riboswitch], amino acids (→lysine riboswitch, →glycine riboswitch) and nucleobases (→guanine riboswitch, →adenine riboswitch). A special case is the glucosamine-6-phosphate (→GlcN6P) riboswitch, which functions as a metabolite-sensing *ribozyme. These RNA structures are widespread in prokaryotes, where they control *genes related to the biosynthesis of the sensed metabolite. Bioinformatic analysis implies that more than 2% of the *Bacillus subtilis* genes are controlled by riboswitches. Most riboswitches are "OFF" switches, evincing negative feedback regulation by the metabolite, but examples of "ON" switches do exist, e.g. the adenine and glycine riboswitches.

Aptamer domains of riboswitches span a broad range in size and binding affinity, reflecting differing requirements for such regulatory elements in terms of specificity and metabolite concentrations to be sensed. However, *dissociation constants have been measured with complete aptamers at equilibrium, whereas genetic control usually takes place while the riboswitch

Common riboswitch mechanisms:

OFF switching

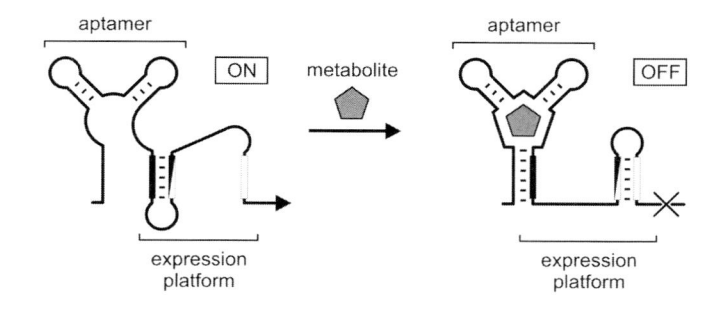

[Diagram: aptamer with ON state + metabolite → aptamer with OFF state; expression platform shown below each]

ON switching

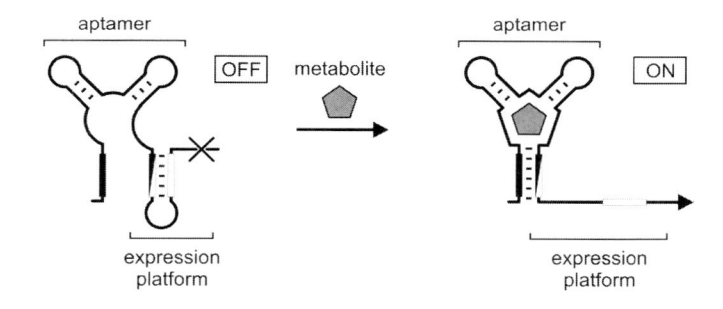

[Diagram: aptamer with OFF state + metabolite → aptamer with ON state; expression platform shown below each]

Genetic regulation

Transcription termination
by terminator formation

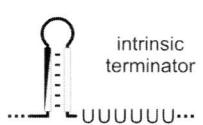

[intrinsic terminator hairpin ...UUUUUU...]

Preventing translation initiation

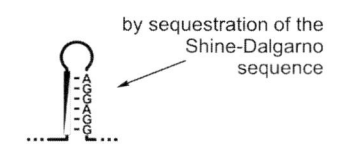

by sequestration of the
Shine-Dalgarno sequence

[hairpin with AGGAGG sequence]

is transcribed. The actual ligand concentration needed to change the amount of gene expression is therefore influenced by other factors like *transcription speed or folding rates and differs from the given dissociation constants (e.g. FMN riboswitch). Additionally, a variety of different mechanisms of riboswitch action exists. Although the majority of riboswitches have been found in prokaryotes, the TPP aptamer has been shown to occur in plants and fungi, where it is potentially involved in regulation of *splicing or mRNA stability. Riboswitches can also take on more complex arrangements. The glycine riboswitch, for example, comprises two similar aptamers that cooperatively bind two glycine molecules, which results

Riboswitch class	Gene/operon controlled		Aptamer size (nucleotides)	K_D (nM)
B$_{12}$	E. coli	btuB	202	300
TPP	E. coli	thiM	91	30
Glycine	V. cholera	gcvT	226	30000
FMN	B. subtilis	ribD	165	5
SAM I	B. subtilis	yitJ	124	4
SAM II	A. tumefaciens	metA	68	1000
Guanine	B. subtilis	Xpt	93	5
Adenine	B. subtilis	ydhL	80	300
Lysine	B. subtilis	lysC	179	1000
GlcN6P	B. subtilis	glmS	246	200000

Adapted from *The Aptamer Handbook*, Wiley-VCH, Weinheim, 2005.

in a more "digital" dependence of gene control on glycine concentration. Even alignments of two riboswitches in tandem have been found that are proposed to add an additional level of complexity to RNA-based genetic regulation. These examples support the idea that riboswitches can fulfill the same functions as protein-based regulatory systems. The existence of riboswitches also supports the *RNA world theory, adding a regulatory role to other basic RNA functions like information carriers (mRNA) and catalytic activity (ribozymes).

Comprehensive information about riboswitch sequences can be found in the *Rfam database. RÜDIGER WELZ

ribozymes RNA molecules that posses catalytic activity (*ribo-* like *ribo*nucleic acid, *-zyme* like en*zyme*). Despite having only four different chemical subunits (compare protein enzymes: 20 different amino acids) RNA molecules display an amazing structural variety. They fold into complex *tertiary structures analogous to highly structured proteins and catalyze a broad range of chemical reactions. *RNA folding is a very hierarchical process; initially formed *secondary structure elements undergo further folding into a three-dimensional complex that contains *binding pockets for *ligands or substrates and that provides *catalytic centers for chemical reactions. A number of ribozymes occur in nature: the *hammerhead ribozyme, the *hairpin ribozyme, the *hepatitis delta virus ribozyme and the *Varkud satellite ribozyme. These ribozymes are rather small and are derived from viruses, virusoides, viroids and *satellite RNAs. Apart from these small ribozymes, nature also harbors large RNAs with catalytic properties: the RNA subunit of *RNase P, the *self-splicing RNA of *group I introns and *group II introns, and the RNA subunit of the *ribosome. Furthermore, two *small nuclear RNAs (snRNAs U2 and U6) that take part in composition of the spliceosome have been shown in the absence of protein to posses activity that is related to the natural activity of the *spliceosome. More recently, two new classes of ribozymes that cleave *messenger RNA have been discovered: the *glmS ribozyme in Gram-positive bacteria and the *CoTC ribozyme responsible for cotranscriptional cleavage within human β-globin *pre-mRNA.

Ribozyme	Size (nucleotides)	Activity
Hammerhead	40	reversible self-cleavage via transesterification yielding a 2′,3′-cyclic phosphate
Hairpin	70	
Hepatitis delta virus	90	
Varkud satellite	160	
Group I intron	≥210	self-splicing via transesterification
Group II intron	≥500	
RNase P	300	cleavage of pre-tRNA (pre-tRNA processing)
Ribosome (23S rRNA)	2600	peptide bond formation
Spliceosome (U2 + U6 snRNA)	180 + 100	RNA splicing via transesterification
glmS ribozyme	246	Cleavage of the glmS gene in Gram-positive bacteria
CoTC ribozyme	200	Cleavage of human β-globin mRNA

Most of the naturally occurring ribozymes are *cis-active: they catalyze their own reversible cleavage. However, these ribozymes can be engineered to efficiently function in trans (→trans-active). The majority of ribozymes require metal ions for activity, either for direct involvement in *active site chemistry or to promote folding or both.

There are four major strategies to achieve catalysis of RNA *transesterification by ribozymes: *precise substrate orientation, preferential *transition state binding, *electrostatic catalysis and *general acid–base catalysis. Apart from this natural activity, ribozymes have been selected from large pools of RNA sequences (→artificial ribozymes) to catalyze chemical reactions such as aminoacylation, RNA *polymerization, N-*glycosidic bond formation and cleavage, *pyrophosphate bond formation and cleavage, N-alkylation, S-alkylation, porphyrin metalation, biphenyl isomerization, Diels–Alder reactions (→Diels–Alderase ribozyme) or Michael reactions. Ribozymes are paradigms of structure function relation; folding into a catalytically competent structure is mirrored in activity. Most ribozymes behave like true *catalysts; they display *multiple turnover kinetics that can be described by the *Michaelis–Menten model. The remarkable structural and functional versatility of RNA raised the question for a life form based primarily on RNA and preceding our present DNA-and protein-based life. In this *RNA world *genetic information may have resided in the sequence of RNA molecules, while the *phenotype derived from the catalytic properties of RNAs. SABINE MÜLLER.

right-handed DNA →DNA double helix, →DNA structures

RISC (RNA-induced silencing complex) Effector of the *RNA interference pathway. Its core components are an Argonaute protein and a *small interfering RNA or a *microRNA. NICOLAS PIGANEAU

RITS (RNA-induced transcription silencing complex) Effector of transcriptional *gene silencing, a variant of the *RNA interference pathway.

r-loop →replacement loop

RNA (ribonucleic acid) *Biopolymer composed of *ribonucleotides (→nucleic acid) occurring in viruses and all living organisms. Contrary to DNA, RNA does not form a contiguous double strand, but appears in a wide variety of intramolecularly folded *secondary and *tertiary structures (→RNA structures). Based on their different structure and function, traditionally, RNA molecules have been classified into *messenger RNA, *transfer RNA and *ribosomal RNA. However, over the past two decades, a number of new RNAs with new functions have been discovered (→non-coding RNA, →small nuclear RNA, →small nucleolar RNA, →small interfering RNA, →ribozymes, →riboswitches).

RNA is involved in the realization and regulation of gene expression, and also plays a major role as component of *ribonucleoprotein complexes in *RNA processing and modification. Several viruses contain RNA as the carrier of *genetic information. In eukaryotic cells, RNA occurs in the nucleus, in the cytoplasm and in the cytoplasmic organelles (→ribosome, mitochondria, chloroplasts). *In vivo*, RNA is synthesized by *transcription from DNA and subsequently undergoes further processing (→RNA processing). The discovery of RNA being capable of catalyzing a wide variety of chemical reactions as well as of RNA being significantly involved into regulation of cell metabolism has led to a boom of RNA research and to a revival of the *RNA world theory. SABINE MÜLLER

RNA alignment and structure comparison Many *non-coding RNAs evolve rapidly at sequence level while *RNA secondary structures are comparatively well conserved. A wide variety of algorithmic approaches has been explored to detect and quantify structural similarities. In the simplest case of appreciable *sequence homology, "structure-enhanced" approaches can be used, which employ simple *sequence alignment algorithms but include in their scoring function certain structural information, particularly whether a *nucleotide is unpaired or paired with an *upstream or *downstream partner. On the other hand, if the *RNA structures are known or at least reliably predicted, one typically uses tree-edit or tree-alignment approaches. Here, the *secondary structures are translated into so-called rooted planar trees. For such trees, generic edit and alignment algorithms are known that generalize the analogous problem for sequence pairs. A third category of algorithms attempts to solve the folding (structure prediction) and the alignment problem simultaneously. The exact solution to this problem, known as the Sankoff algorithm for thermodynamics-based rules and as pair-stochastic context-free grammar in the machine learning context, is computationally too expensive to be of much practical use. Practical applications thus use various heuristics approximations to reduce the search space.

As is the case with alignments in general, either global or local alignments can be computed. While global alignments are used for comparing RNAs, local alignments play a role in *RNA gene finding. PETER F. STADLER, IVO HOFACKER

RNA aptamer →aptamer

RNA backbone →sugar–phosphate backbone

RNA binding antibiotics *Antibiotics binding to DNA and RNA.

RNA binding polyamines Long-chain aliphatic amines; the most prominent members of this species are spermine, spermidine and putrescine – endogenous polycations important for cell differentiation, cell growth and cell physiology. In cells most polyamines associate with RNA, inducing structural changes in the RNA due to their positively charged backbone. Apart from RNA, these polycations can also interact with other negatively charged *macromolecules like phospholipids and proteins. VALESKA DOMBOS

RNA binding proteins (RBPs) Play a key role in *RNA processing and modification, post-transcriptional regulation, *splicing, *messenger RNA (mRNA) transport and localization, mRNA stability and *translation, and *RNA interference (RNAi). Typically, a RBP is a multi-domain protein with RNA-binding modules and other domains like an enzymatic domain or a protein–protein interaction domain. The topology of most RNA-binding domains is an antiparallel β-sheet with two or three α-helices. As compared to *DNA-binding proteins, the interaction of a protein with RNA is more complex. In addition to *minor/*major groove interactions, non-canonical RNA structures like *loops and *bulges are important for binding. Protein–RNA complex formation typically happens by a mutually induced fit. *Hydrophobic interactions are very important (contact to the *base or the *ribose) and hydrogen bonding to the RNA bases provide sequence-specific recognition. Examples of protein domains involved in RNA recognition are the RNA recognition motif (RRM), also known as RNA-binding domain (RBD) or *ribonucleoprotein domain (RNP), K homology (KH) module, double-stranded RBD, Sm domain, helix–turn–helix motif, Zinc finger/knuckle motif and Piwi/Argonaute/Zwille (PAZ) domain. Many RBPs have one or more copies of the same RNA-binding domain, whereas others have two or more distinct domains. Several RNA binding domains are suggestive for the molecular function of the RBP: the PAZ domain for short *single-stranded RNA binding in *RNAi or *micro-RNAs processes, and Sm domain for *small nuclear RNA binding in *splicing and, possibly, in *transfer *RNA processing. TOBIAS RESTLE

RNA biosynthesis →transcription

RNA blotting →blotting, →Northern blot

RNA catalysis →ribozyme

RNA chaperone Ubiquitous and abundant proteins that direct correct folding of RNA by preventing and resolving misfolded *RNA structures. Most RNA chaperones bind RNA with broad sequence specificity. They play a role in multiple and diverse cellular processes like *transcription, *translation, RNA transport, *splicing, RNA modification and decay. Examples of RNA chaperones are *cold shock proteins, ribosomal proteins and Sm-like proteins, like the bacterial Hfq protein, which facilitates RNA-RNA interactions. FRANZ NARBERHAUS

RNA computing →DNA computing

RNA degradation →alkaline hydrolysis of RNA, →mRNA degradation, →RNase

RNA-dependent DNA polymerase *DNA polymerase, which synthesizes *DNA using *RNA as a *template (→reverse transcriptase).

RNA-dependent RNA polymerase Important enzyme in all viruses containing a positive RNA strand as *genome. This enzyme catalyzes the synthesis of the complementary *minus strand, which is an intermediate during *replication, using the RNA *plus strand as *template as well as the synthesis of the plus strand using the RNA minus strand as template. The plus strand *genome serves as *genetic information for viral offspring (→rolling-circle amplification). IRENE DRUDE

RNA–DNA hybrid →DNA–RNA hybrid

RNA domain Larger part of a RNA molecule that harbors a number of structural motifs and thus can act as a platform for interaction with other molecules (proteins, nucleic acids, small *ligands) or with other parts of the same RNA. SABINE MÜLLER

RNA editing Post-transcriptional modification of RNA molecules that alters the sequence and/or structure of target RNA molecules. RNA editing can involve the modification or insertion/deletion of *nucleotides. In the case of *messenger RNA (mRNA), the post-transcriptional modification alters the *coding sequence. Three major types of mRNA editing reactions are known: *cytosine deamination of apolipoprotein B (apoB) mRNA, *adenosine deamination observed during expression of some mammalian glutamate receptors, and *uridine insertion and deletion reactions that establish the functional *open reading frame of several mRNAs in the mitochondria of trypanosomes. BERND-JOACHIM BENECKE

RNA engineering Covers all attempts to create new functional RNA. There are two basic approaches to reach this goal. First, one can rationally design sequences of a RNA for the targeted activity, either following known concepts, as in the case of *micro-RNA, or by modifying the design of an already known nucleic acid, like the adaptation of a *hairpin ribozyme or *hammerhead ribozyme to a specified substrate. Rational design of a catalytic nucleic acid with a new desired activity from scratch is very difficult. In this case, one can use the second main approach and search for an active RNA sequence in a randomized *RNA library. This approach is called *in vitro selection or *SELEX (→aptamer selection). SLAWOMIR GWIAZDA

RNA enzymes →ribozymes

RNA fold Synonym for RNA structural motif (→RNA structures).

RNA folding →RNA secondary structure prediction

RNA gene *Gene that is transcribed into *non-coding RNA.

RNA gene finding Series of recent high-throughput experiments based on various different techniques have compiled convincing evidence that a large fraction of the transcriptome of higher organisms consists of *non-coding RNA (ncRNA) and that even in prokaryotes a plethora of such non-protein-coding *genes is crucial for cellular function. The prediction and annotation of ncRNA genes has thus become an important task in bioinformatics. In contrast to protein-coding genes, however, there is no common statistical signal, analogous to *open reading frames, that could be used to find ncRNAs in *genomic DNA. Computational approaches can be classified in three main groups. (a) Methods based on homology search use sequence and structural similarity to known examples, usually the RNA families compiled in the *Rfam database. This approach underlies most of the published ncRNAs annotation, e.g. in the ENSEMBL databases. (b) The relatively small size of bacterial genes, their well-characterized *promoter structures and known terminator signals are used successfully in prokaryotic, mostly bacterial, genomes. (c) In eukaryotes, comparative genomics approaches can be used to detect evolutionarily conserved *secondary structures for which no close homologs are already known. The most widely used

programs of this type are RNAz and Evofold. Computational studies suggest that the amount of evolutionary conserved (and hence likely functional) structured RNA in mammalian *genomes is comparable to the amount of protein-coding genes.
PETER F. STADLER, IVO HOFACKER

RNAi → RNA interference

RNA-induced silencing complex → RISC

RNA-induced transcription silencing → RITS

RNA interference (RNAi) Sequence-specific *gene silencing mechanism in eukaryotic cells (also referred to as RNA silencing). RNAi is conserved from fungi to plants and mammals with the notable exception of *Saccharomyces cerevisiae*. This pathway plays important roles in protecting the cell against viruses and *repetitive DNA elements. Beyond its natural activity, it represents a promising gene silencing technology for functional genomics and a potential therapeutic strategy for a variety of diseases. Remarkably, in nematodes RNAi can be induced via feeding of bacteria producing long *double-stranded RNAs, making it the method of choice for large-scale functional genomic studies in these organisms.

Silencing of *gene expression via RNAi is mediated by small RNA molecules (around 21–26 *nucleotides), which act as specificity factors that guide silencing complexes to *complementary sequences. These RNAs are divided into two classes depending on their biogenesis. Endogenous RNAs processed from *hairpin loop structures are called *micro-RNAs (miRNAs), whereas RNAs from exogenous source and RNAs processed from long double-stranded *transcripts are called *small interfering RNAs (siRNAs).

After processing, the miRNAs and siRNAs are incorporated into a multi-

RNAi pathway

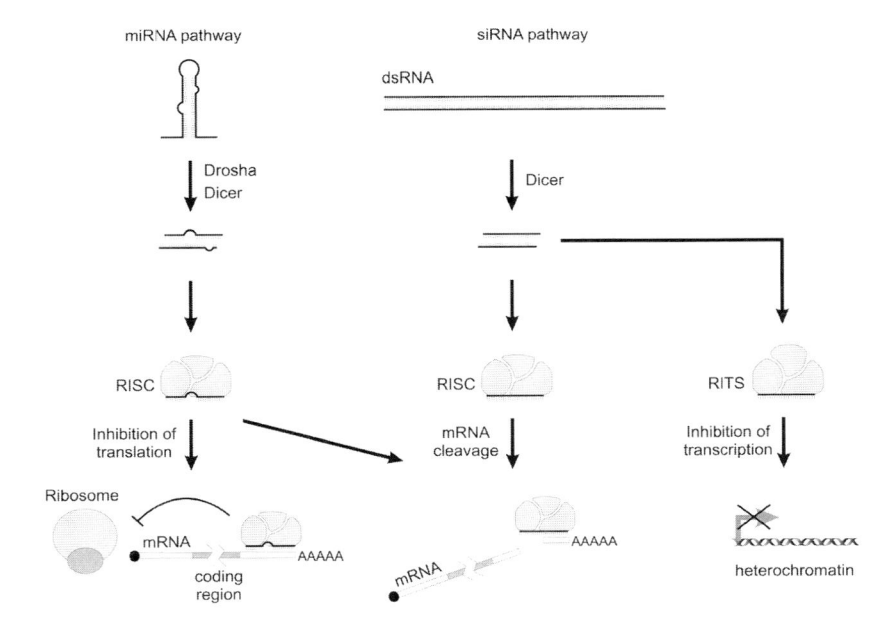

protein complex called *RISC. The mechanism of assembly of RISC is known only in the case of Drosophila melanogaster. It remains to be seen if the data gathered in the flies can be generalized to other organisms. In Drosophila there are two different forms of *Dicer process miRNAs and siRNAs, i.e. Dcr-1 and Dcr-2, respectively.

Dcr-2 is not only responsible for the cleavage of dsRNA into siRNA, but it also loads one of the two siRNA strands into RISC in combination with R2D2, a protein which contains tandem *dsRNA-binding domains. The Dcr-2/R2D2 complex seems to dictate which of the two siRNA strands is loaded into RISC. This is determined by the difference in stabilities at both ends of the siRNA duplex. Whichever strand has its 5'-end at the weaker side is incorporated into RISC. As unwinding of the siRNA strands proceeds, the Dcr-2/R2D2 complex is replaced by *Argonaute, the core component RISC.

The RISC complex targets specific *genes via base pairing of its RNA component with the target *messenger RNA (mRNA). When the mRNA and the small RNA are fully complementary, which is the case with most siRNAs and some miRNAs, the regulation of expression is caused by the cleavage of the mRNA by Argonaute and its subsequent degradation. However, most miRNAs are not perfectly complementary with their target and the regulation is rather due to inhibition of *translation.

In the case of siRNAs, it seems that in some organisms the silencing effect is amplified by a *RNA-dependent RNA polymerase, which elongates the reverse complementary strand of the cleavage products of RNAi. The resulting dsRNA is a new substrate of Dicer and leads to the production of additional siRNAs.

Regulation of *gene expression by RNAi does not only occur at the post-transcriptional level, but rather at the transcriptional level (→transcriptional gene silencing). In this case, the siRNAs are loaded in a specific complex called *RITS. This complex specifically silences *chromatin by promoting DNA or *histone modifications leading to the assembly of a repressive chromatin structure called *heterochromatin. This effect is very important for *genome stability by repressing *recombination at undesired positions and mobilization of repetitive DNA elements such as *retrotransposons. It is also required for proper *chromosome segregation. This activity of RNAi is conserved in Drosophila, plants and mammals.
NICOLAS PIGANEAU

RNA labeling →oligonucleotide labeling

RNA library →sequence pool

RNA ligase T4 RNA ligase (EC 6.5.1.3; also Rnl1 or RnlA ligase 1) is a 347-amino-acid polypeptide encoded by *gene 63 of bacteriophage T4, and which belongs to a family of *oligonucleotide end-joining enzymes involved in *RNA repair, *splicing and *editing pathways. The enzyme catalyzes formation of a *phosphodiester bond between the 5'-phosphate and 3'-hydroxyl ends of preferentially *single-stranded RNA.

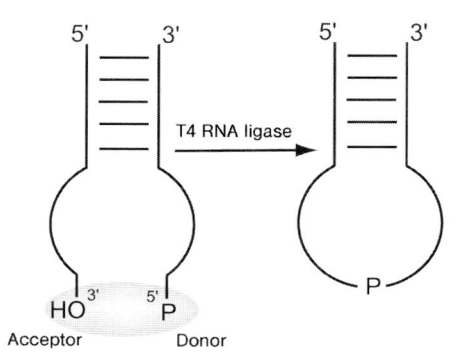

The enzyme also catalyzes the joining of single-stranded DNA ends, but less efficiently. The oligonucleotide carrying the terminal 5'-monophosphate is described as the donor substrate and the terminal carrying the 3'-hydroxyl group is termed the acceptor substrate. The biological role of T4 RNA ligase is to repair damaged bacterial tRNA^{Lys} in the *anticodon loop.
TINA PERSSON

RNA ligation Reaction catalyzed by the enzyme T4 *RNA ligase, consisting of three distinct and reversible steps. In the first step, the *ligase reacts with *ATP to form an adenylated enzyme intermediate. In the second step, a donor substrate with a 5'-terminal monophosphate bound by the enzyme-linked 5'-AMP moiety is now transferred to the 5'-phosphate of the donor substrate, yielding the adenylated donor product in which the terminal *adenosine moiety is attached via a 5',5'-*phosphodiester bridge. In the final *transesterification step, the *phosphodiester bond connecting the two 5',5'-linked phosphates is cleaved and a phosphodiester bond is created between the donor and acceptor substrate, with concomitant release of AMP. The substrate specificity of T4 RNA ligase is broad permitting to ligate essentially any RNA or DNA sequence. Substrate specificity can be summarized as follows: The minimal donor substrate is a nucleoside-3',5'-bisphosphate with a ligation efficiency decreasing in the order pCp > pUp = pAp > pGp. The deoxy-counterparts showed that except for p(dCp) they were all poorer substrates than the corresponding pNp *ribonucleosides.

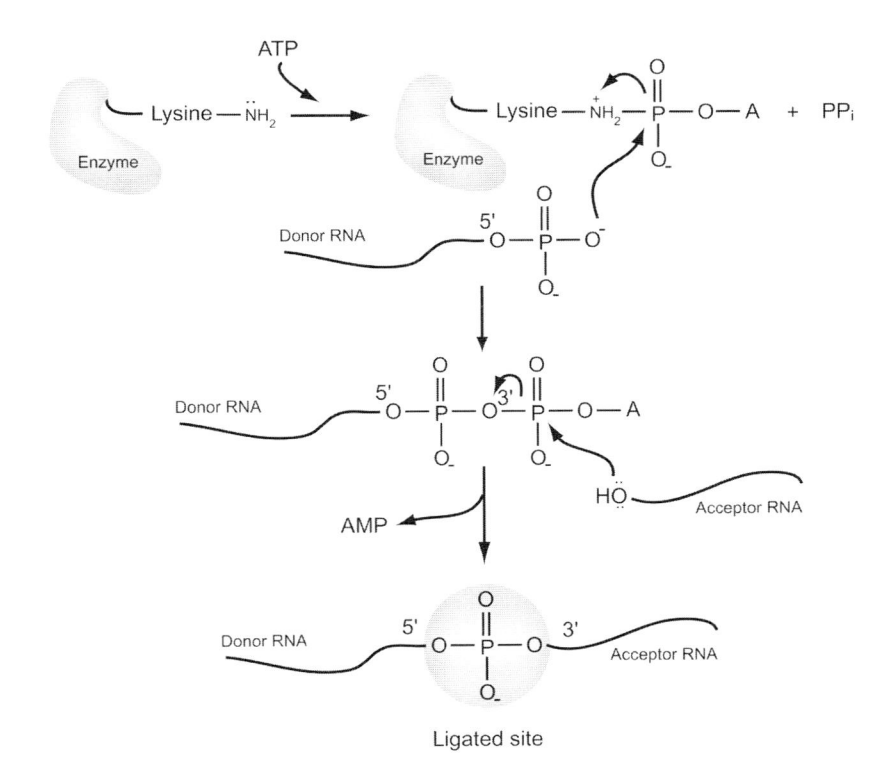

Ligated site

Isocytidine-3′,5′-bisphosphate was reactive, indicating that some base modifications are tolerated. The donor substrate seems to recognize the 5′-terminal phosphate and ribonucleoside as well as the next 3′-linked phosphate. The chain length of the donor substrate only has a marginal effect on the reaction efficiency. The 3′-terminal *ribose moiety is important for acceptor recognition and a 3′-terminal adenosine is preferred over *cytidine and *guanosine, showing intermediate reactivity, while *uridine is a poor substrate (A > C ≥ G > U). Studies on trimeric NpNpNp acceptor substrates showed that a U at any position resulted in a lower yield. A general feature of the T4 RNA ligation reaction is that DNAs are less efficient substrates than RNAs and that discrimination occurs mainly at the acceptor substrate part. Applications for the enzyme include (a) *end labeling using pCp and pCpC3′-fluorescein, (b) *circularization reactions (e.g. producing circular RNA molecules) and (c) intermolecular ligation reactions of polynucleotides (e.g. ligation of RNA oligonucleotides to the 5′-end of *messenger RNA).
TINA PERSSON

RNA maturation *Primary transcripts, i.e. the immediate products of *transcription, are not necessarily functional and may therefore have to undergo several steps of maturation. Processing of eukaryotic *messenger RNAs includes capping and addition of a *poly(A) tail as well as *splicing. Mature *ribosomal RNAs are obtained from *pre-rRNAs by endonucleolytic cleavage and methylation of specific *nucleotides. *Pre-tRNAs require extensive processing to become functional, including removal of a segment at the 5′-end and an *intron in the *anticodon loop, replacement of two *uracil residues at the 3′-end by CCA and modification of some residues to give characteristic bases. →RNA processing. JENS KURRECK

RNA melting →denaturation, →melting, →melting curve, →melting point, →melting temperature

RNA motif Local three-dimensional structure of RNA (→ RNA structure).

RNA-PCR →RT-PCR

RNA polymerase Enzyme that catalyzes *polymerization of *nucleoside triphosphates as activated RNA building blocks (→transcription).

RNA polymerization →transcription

RNA primer As *DNA polymerases cannot initiate DNA synthesis *de novo*, but can only elongate nucleic acid *polymers, *primers (short RNA oligonucleotides, 10–12mers) are required. They are synthesized by the *primosome. After elongation by DNA polymerases, the primers are removed by the *exonuclease activity of DNA polymerase I, which hydrolyzes the *ribonucleotides and replaces them by *deoxyribonucleotides.
ANDREAS MARX, KARL-HEINZ JUNG

RNA processing Maturation of the immediate products of *transcription to give functional *messenger RNAs (mRNAs), *transfer RNAs (tRNAs) or *ribosomal RNAs (rRNAs). Alterations of the *primary transcripts include removal or addition of polynucleotide segments as well as modification of specific *nucleotides.

Eukaryotic mRNAs undergo extensive post-transcriptional processing before leaving the nucleus. A 7-methylguanosine cap is linked to the first nucleotide of the *pre-mRNA by a special triphosphate linkage and a *poly(A) tail of 100 (yeast) to 250 (mammals) adenylate residues is enzymatically appended to the 3′-end. Eukaryotic pre-mRNAs consist of protein-coding sequences (→exons) interrupted by non-coding sequences (→introns). The *intervening sequences are precisely excised via *lariat intermediates and the exons are joined together by a process referred to as *splicing. This procedure is rather complex and involves five *small nuclear RNAs as well as numerous associated proteins which form a *ribonucleoprotein particle dubbed the *spliceosome. Some mRNAs are subject to further modification known as *RNA editing, which includes replacement, insertion or deletion of specific bases directed by *guide RNAs, or substitutions of bases by specialized enzymes.

Newly transcribed rRNAs are clusters of three rRNAs (16S rRNA, 23S rRNA and 5S rRNAs in prokaryotes; 18S rRNA, 5.8S rRNA and 28S rRNA in eukaryotes) that must be separated to become functional. Cleavage occurs by specific *endonucleases (RNase III, *RNase P, RNase E and RNase F). In addition, the ribosomal RNAs are methylated at specific nucleotides, yielding $O^{2'}$-methylribose and N^6,N^6-dimethyladenine (as well as 2-methylguanine in eukaryotes). Eukaryotic rRNAs are processed in special nuclear bodies known as nucleoli which are also the site of ribosomal subunit assemble. Many *uridines in the rRNAs are converted to *pseudouridines. These alterations are guided by so-called *small nucleolar RNAs.

tRNAs transport amino acids to the *ribosome for protein synthesis. They are characterized by a *cloverleaf structure. To become functional, tRNAs undergo extensive processing: a stretch of 14 nucleotides (yeast) at the 5'-end of the *primary transcript is removed by *RNase P. Furthermore, many eukaryotic pre-tRNAs contain introns adjacent to their *anticodons that have to be excised. Extra nucleotides are removed from the 3'-end and the obligatory CCA sequence is appended to the immature tRNAs by the enzyme *tRNA nucleotidyl transferase. Maturation of tRNAs also includes modification of some residues to give uncommon bases like inosine, dihydrouridine and pseudouridine (→rare nucleotides). JENS KURRECK

RNA–protein interaction Structural and functional interplay between RNA and proteins is necessary for the controlled execution of crucial biological processes such as *RNA processing, including *splicing, *translation, and the control of mRNA stability and degradation (→RNA degradation). RNA–protein interactions may be specific for *nucleotide sequences (→recognition motifs) or unspecific which is true for many cellular *RNA-binding proteins that, for example, serve as *RNA chaperones. Experimentally, RNA–protein interactions can be identified and characterized by various means, e.g. *band shifting, *crosslinking (including *photocrosslinking), filter binding assays, *FRET studies and *footprinting. GEORG SCZAKIEL

RNA repair Therapeutic modification of the *genetic information at the RNA level. In contrast to conventional *gene therapy, where correct copies of abnormal *genes are added to the *genome, RNA repair conserves endogenous spatial and temporal *gene regulation, and simultaneously reduces the expression of the mutant gene. A possibility to reprogram *transcript sequences is to artificially induce a *trans-splicing event, thus using the cells own *spliceosome machinery or a *trans-splicing ribozyme like the *Tetrahymena ribozyme. Both methods were shown to correct genetic information at the transcript level in therapeutic cell and animal models. Alteration of *pre-mRNA *splicing can be also achieved by using

Ribonucleases from *E. coli*

Enzyme	Substrate/involved in
Endoribonucleases	
RNase I, I*, M, R, IV, F, N	nearly all RNAs
RNase III	double-stranded RNA/rRNA, mRNA processing
RNase P	tRNA precursors/tRNA processing
RNase HI, HII	DNA–RNA hybrids/DNA replication and repair
RNase P2, O, PC	multimeric tRNAs
3′,5′-Exoribonucleases	
RNase II	non-structured RNAs/mRNA decomposition, tRNA processing
PNPase	non-structured RNAs/mRNA decomposition
RNase D	denatured tRNAs, tRNA precursors/tRNA processing
RNase BN, PH	tRNA precursors/tRNA processing
RNase R	mRNA, rRNA

mRNA, messenger RNA; rRNA, ribosomal RNA; tRNA, transfer RNA.

*antisense RNAs to block undesired *splice sites. STÉPHANIE VAULÉON

RNA–RNA interaction →RNA structure, →RNA motifs, →RNA secondary structure prediction

RNase Enzyme (protein or ribonucleic acid) that catalyzes the cleavage of RNA chains. RNases are *endonucleases or *exoribonucleases and can be single- (→RNase A, T1 etc.) or double-strand specific (→Dicer, RNase III) or even specific for DNA–RNA hybrids (→RNase H). ULI HAHN

RNase A *RNase from the bovine pancreas, belongs to the most thoroughly investigated enzymes (first determined protein primary structure, first crystallized protein, third protein X-ray structure solved). RNase A is an *endonuclease which cleaves *single-stranded RNA specifically after *pyrimidines (C and U) in two steps, producing *mono- or *oligonucleotides with a 3′-phosphate group. The cleavage reaction of RNases A (as well as of RNase T1) can be divided into two independent non-consecutive or non-processive steps. (a) Chain cleavage takes place during a *transesterification leading to a mono- or oligonucleotide with a terminal 2′,3′-cyclic phosphate which dissociates from the enzyme. (b) This *cyclic phosphate is then hydrolyzed to the final product with a 3′-phosphate group. The product of the first step is 2′,3′-cyclic C or U. RNase A, among others, may also be used for RNA synthesis if it is forced, under appropriate reaction conditions, to catalyze the first step's back reaction, i.e. the addition of a 3′-nucleotide phosphate to a 2′,3′-cyclic C or U terminus. Enzymatically synthesized poly(U), for example, served Nirenberg and Mathei as the first artificial *messenger RNA to be translated into polyphenylalanine when they started to solve the *genetic code. RNase A may also be used for enzymatic *RNA sequencing or RNA structure probing (→chemical and enzymatic structure mapping of RNAs). ULI HAHN

RNA secondary structure In contrast to *tertiary structures, which refer to the three-dimensional structure of the molecule, *secondary structures of nucleic acids are defined only by the base-pairing pattern (→base pair). Secondary structures therefore consist of two distinct classes of residues: those that are incorporated in double-helical regions (→stems) and those that are not part of helices. For RNA, the double-helical regions consist almost exclusively of Watson–Crick CG and AU pairs as well as the slightly weaker GU wobble pairs. All other combinations of pairing nucleotides, called *non-canonical pairs, are usually neglected in secondary structure prediction, although they do occur, particularly in tertiary structure motifs.

A secondary structure is the list of *(canonical) base pairs. In the strict sense, a secondary structure satisfies three conditions. (a) A base cannot participate in more than one *base pair. (b) Bases that are paired with each other must be separated by at least three (unpaired) bases. (c) No two base pairs (i,j) and (k,l) "cross" in the sense that $i < k < j < l$. The last condition excludes *pseudoknots. The term secondary structure is often used in a looser sense that allows pseudoknots.

Secondary structures in the strict sense can be uniquely decomposed into *loops. These are simply the faces of the planar drawing of the structure. A loop is characterized by its length, i.e. the number of unpaired *nucleotides in the loop, and its degree, i.e. the number of helices that emanate from the loop. Loops of degree 1 are called *hairpin loops, interior loops (→internal loops) have degree 2 and loops with a degree larger than 2 are called multiloops. *Bulge loops are a special cases of interior loops in which there are unpaired bases only on one side, while stacked pairs correspond to an interior loop of size zero.

The thermodynamic stability of a RNA secondary structure can be described by the *nearest-neighbor model. In this standard energy model the total free energy of a structure is modeled as the sum of energy contributions of its constituent loops. These energy contributions depend on size, degree and the nucleotides in the loops. Qualitatively, the major energy contributions are *base stacking, *hydrogen bonds and loop entropies. The secondary structure model considers free energy differences between folded and unfolded states

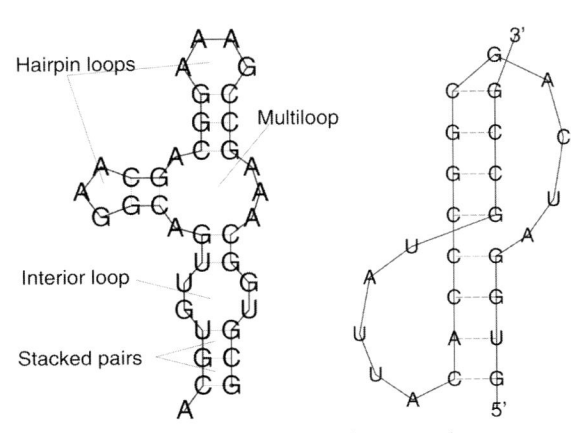

Loop types in RNA secondary structures: pseudoknot-free secondary structures (left) and an H-type pseudoknot (right).

in an aqueous solution with rather high salt concentrations. As a consequence one has to rely on empirical energy parameters which are determined from large sets of *melting curves measured for small RNA molecules.

A good energy model for the thermodynamics of pseudoknoted structures is still missing because steric effects are hard to incorporate into the secondary structure model. PETER F. STADLER, IVO HOFACKER

RNA secondary structure prediction Predicting the *secondary structure of a RNA molecule from the nucleic acid sequence is the most basic problem in RNA bioinformatics. The most common approach is folding by energy minimization, which computes the secondary structures with the minimal free energy (MFE), called the MFE structure. For *pseudoknot-free *RNA secondary structures this problem can be solved efficiently by a dynamic programming algorithm whose run-time scales as the cube of the sequence length. The algorithm is based on the observation that any *base pair divides the structure into two independent parts with additive energies. This allows us to compute the optimal structure of a larger RNA recursively from smaller substructures. In essence the same types of algorithms can be used for linear and circular RNA molecules, as well as for handling two or more interacting RNAs.

Due to inaccuracy of the parameters as well as limitations imposed by the model assumptions, predicted structures will contain errors and accuracies, and vary widely from sequence to sequence. A serious limitation of MFE folding is that it returns only a single structure and therefore gives no indication as to the reliability of the prediction. Several variants of the basic folding algorithm address this problem by producing a list of suboptimal structures. Alternatively, one may compute the probability of each possible *base pair in order to capture the ensemble of thermodynamically plausible structures.

Another school of thought prefers algorithms based on the stochastic context-free grammar (SCFG) formalism. Despite the differences in vocabulary, the two approaches are algorithmically very similar. The most important distinction to thermodynamic folding, where parameters are derived from experiment, is that in SCFGs all parameters are estimated from data, i.e. sequences with known structure. While thermodynamic folding has been traditionally preferred at least for single sequence prediction, recent work has demonstrated that methods based on trained parameters can indeed reach the same level of accuracy.

In some cases the native structure of RNA may differ from the thermodynamic ground state. This can happen especially

RNA folding software

Name	Download	Web service	URL
mfold / unafold	source code	yes	http://www.bioinfo.rpi.edu/applications/mfold/
Vienna RNA	source code	yes	http://www.tbi.univie.ac.at/RNA/
RNAstructure	windows exe	no	http://rna.urmc.rochester.edu/rnastructure.html
RNAshapes	source code	yes	http://bibiserv.techfak.uni-bielefeld.de/rnashapes/
Sfold	–	yes	http://sfold.wadsworth.org/
CONTRAfold	source code	yes	http://contra.stanford.edu/contrafold/

for longer RNAs where the molecule does not reach thermodynamic equilibrium on biologically relevant time scales. In such cases one has to resort to kinetic folding methods that attempt to simulate the folding process itself. Typically, these methods treat folding as a stochastic process using insertion and deletion of either single base pairs or of entire helices as the move set. Compared to the dynamic programming algorithms above this is usually quite slow. However, it is comparatively easy to include *pseudoknots and more complicated energy evaluation in these methods. PETER F. STADLER, IVO HOFACKER

RNase H *RNase that cleaves the RNA chain of a *DNA–RNA hybrid; it can be found in every living cell and even in retroviruses as it catalyzes reactions involved in important turnover processes like *replication. ULI HAHN

RNA sensor Any RNA molecule (in many cases a *ribozyme) that undergoes a reaction upon the presence of a certain analyte resulting in a detectable readout (→reporter ribozyme). JÖRN WOLF

RNA sequence →nucleotide sequence

RNA sequencing The sequence of *monomers (bases) forming a RNA molecule is determined using RNA sequence analysis. Modern RNA sequencing is based on the *reverse transcription of *template RNA and analysis of the *cDNA using the *Sanger technique. First RNA sequence analyses elucidated the *primary structure of alanine and serine *transfer RNA (tRNA) using gram amounts of RNA and a technique that was based on specific RNA *fragmentation with different *ribonucleases and combination of the resulting fragment sequence information using overlapping regions. Further developments for RNA sequencing included the use of *radioactive labeling, specific fragmentation of the RNA by *RNase A and *RNase T1, two-dimensional *electrophoresis and isolation of the fragments, their degradation by *exonucleases, and another electrophoresis. These tedious approaches were successfully employed for analyzing the sequences of tRNAs, *ribosomal RNAs (5S and 16S) and of bacteriophage MS2. Currently, the sequencing by fragmentation is used only if *rare nucleosides need to be determined in tRNA. SUSANNE BRAKMANN

RNase T1 initially isolated from Takadiastase, a growth supernatant of the slime mold *Aspergillus oryzae*, RNase T1is the key member of a family of microbial *RNases of prokaryotic and eukaryotic origin. This enzyme, which is also well-characterized, cleaves *single-stranded RNA with high specificity at the 3'-site of guanylyl residues in a two-step mechanism. The first one is a *transesterification of the *phosphodiester bond yielding a 2',3'-cyclic phosphate. The *cyclic phosphate is then hydrolyzed in a second and rate-limiting step yielding the 3'-phosphate product. RNase T1 may also be used for enzymatic *RNA sequencing, *RNA structure probing or RNA synthesis (→RNase A, →chemical and enzymatic mapping of ribonucleic acid structure). ULI HAHN

RNA silencing →RNA interference

Rnasin Originally isolated from human placenta, this *ribonuclease inhibitor reduces the activity of *RNase A-type enzymes. The 51-kDa acidic protein forms an enzymatically inactive complex by binding to RNases non-covalently, probably via leucine-rich repeats. DENISE STROHBACH

RNA software →RNA secondary structure prediction

RNA splicing Processing of non-functional *primary transcripts by removal of intervening sequences (→introns). Typically, the term refers to the removal of introns from *heterogeneous nuclear RNA (hnRNA) (or *pre-mRNA), although splicing of other RNA species has been demonstrated, such as *transfer RNA (tRNA splicing) or *ribosomal precursor RNA (→group I introns). Following *transcription, hnRNA molecules are associated with specific proteins. The resulting *ribonucleoprotein (→heterogeneous nuclear ribonucleoprotein) complex forms the substrate for the splicing reaction. Basically, the splicing reaction is characterized by two subsequent and independent *transesterification steps, resulting in removal of the intron and ligation of *exons. For the sake of *reading frame integrity splicing must be absolutely precise. That precision depends on short conserved *consensus sequences defining the 5'- and 3'-*splice sites. Most pre-mRNA molecules follow the GT–AG rule, meaning that the first two nucleotides of the intron usually are G and T, while the 3'-end of the intron is constituted by two A and G residues. A third element is the intronic *branch point, almost invariantly an A residue located near (about 30 nucleotides) and separated from the 3'-splice site by a poly-*pyrimidine track. The mechanism of the splicing reaction can be summarized as follows. The first step of splicing involves a nucleophilic attack by the 2'-hydroxyl group of the branch point A on the *phosphodiester bond at the 5'-splice site. In a second step, the now free 3'-hydroxyl group of the preceding exon attacks the phosphodiester bond at the 3'-splice site. As a result, both exons are ligated via a new phosphodiester bond. The intron is released in form of a *lariat, due to the bond between the 5′-phosphate (G) and the 2′-hydroxyl of the branch point A. The two transesterification reactions themselves do not require energy; however, the catalytic compound of the splicing reaction the *spliceosome strictly depends on ATP. BERND-JOACHIM BENECKE

RNA structures RNA double-stranded structures occur almost exclusively as *A-form RNA. These double-helical stems are limited to lengths of several standard *Watson–Crick base pairs connected by single-stranded regions that include a wide variety of *secondary and *tertiary structure elements, some of which have been classified and named. RNA structure determination at the atomic level (→crystallization of nucleic acids, →nuclear magnetic resonance spectroscopy) has allowed for elucidation of the structure of a large number of functionally diverse RNA molecules including the *ribosomal RNAs, many *transfer RNAs, a variety of *ribozymes, the RNA components of *ribonucleoproteins and *aptamers bound to their *ligands. Based on these analyses, a number of RNA structural motifs have been defined (e.g. →A-platform, →A-minor motif →hairpin loop, →internal loop, →Junction loop, →pseudoknot, →tetraloop, →U-turn, →K-turn, →S-turn) and a nomenclature based on the geometry of base pair interactions has been suggested (→A-form RNA, →geometric nomenclature and classification of RNA base pairs, →RNA secondary structures, →RNA secondary structure prediction, →SCOR). SABINE MÜLLER

RNA switch →riboswitch

RNA synthesis →oligonucleotide synthesis, →transcription

RNA synthesizer →DNA synthesizer

RNA thermometer Like *riboswitches, RNA thermometers are *cis*-acting (→*cis*-active) sensory RNA elements usually located in the *5′-untranslated region (5′-UTR) of *messenger RNAs (mRNA). RNA thermometers *upstream of bacterial *heat shock genes and *cold shock genes reduce expression at low and high temperatures, respectively. In pathogenic bacteria like *Listeria* and *Yersinia*, the expression of a virulence *gene activator is controlled by a RNA thermometer such that virulence genes are only expressed at 37°C, i.e. when a warm-blooded host has been infected. An RNA thermometer in the 5′-UTR of the bacteriophage λ cIII gene controls the lysis/lysogeny decision. Under severe heat shock conditions, expression of cIII is downregulated and the lytic cycle is induced allowing the phage to escape from its host *Escherichia coli*.

All known RNA thermometers control *translation by a temperature-responsive mRNA structure that controls access of the *ribosome to the *Shine–Dalgarno sequence. RNA thermometers found in the 5′-UTR of heat shock and virulence genes fold into a complex structure at low temperatures. The *ribosome-binding site, including the Shine–Dalgarno sequence, and in some case the AUG *start codon, is masked. *Melting of the structure at increasing temperatures permits ribosome binding and *translation initiation. FRANZ NARBERHAUS

RNA topology →RNA structures

RNA world Hypothetical phase in early evolution where RNA functioned in genetic as well as enzymatic roles. Conceptually, it goes back to suggestions in three overlapping papers by Woese (1967), Crick (1968) and Orgel (1968). The strongest experimental support for the idea comes from *in vitro* selection of RNA *aptamers and *ribozymes, and the existence of a few catalytic RNAs in extant organisms. The *nucleotide coenzymes (such as *ATP, *NAD, etc.) are arguably remnants of this evolutionary stage; their existence is consonant with the view that the RNA world was metabolically complex. The RNA world may have started on mineral surfaces or in compartmentalized systems (→ribo-organisms). EÖRS SZATHMÁRY

RNP →ribonucleoprotein

rolling-circle amplification Sensitive technique for amplification of *plasmid DNAs or other *circular DNAs in various surroundings such as on a *solid support or in cells. In the linear mode, a complementary *primer is hybridized with the target *circular DNA and a strand-displacing *DNA polymerase generates hundreds of concatermerized copies of the target per minute. Using two *primers, whereas one hybridizes to the *minus and the other to the *plus strand of the DNA, an exponential growth of copies of the target can be achieved (10^9 copies in 90 min, →rolling circle replication). JÖRN WOLF

rolling circle replication Process in bacteria or viroids in which multiple copies of a *circular DNA or RNA are generated. These nucleic acid species can be *plasmids and bacteriophage or viroid *genomes. In contrast to linear *DNA replication, the two DNA strands of circular DNA are replicated successively during *rolling-circle amplification. Small ribozymes like the *hairpin or *hammerhead ribozyme are important in the rolling-circle amplification mechanism of circular *satellite RNA of certain viroids. JÖRN WOLF

rPCR →random polymerase chain reaction

r-Proteins Ribosomal proteins (→ribosome).

RRE (Rev response element) RNA *RNA structural motif in HIV which is approximately 250 *nucleotides long. Among its hairpin motifs, the IIB hairpin provides a *binding site that binds HIV Rev protein with high affinity. As in the case of *TAR RNA, the RRE RNA–protein interaction depends strongly on an arginine-rich region. This interaction is essential for the HIV life cycle, since it allows for export of viral *messenger RNAs to the cytoplasm, which are responsible for structural and accessory proteins. JÖRN WOLF

RRF →ribosome recycling factor

rRNA →ribosomal RNA

rRNA genes *Ribosomal RNA (rRNA) is the major product of *transcription and amounts to 80–90% of the total mass of RNA in both prokaryotes and eukaryotes. rRNA genes can vary in number, from seven in *Escherichia coli* to several hundred in higher eukaryotes. rRNA genes in eukaryotes are generally organized in *tandem repeats. In humans, about 300–300 rRNA genes are present in five clusters on different *chromosomes (→RNA processing). MAURO SANTOS

rRNA processing →RNA processing

RS Abbreviation of *aminoacyl-tRNA synthetase used to specify the specific enzyme, e.g. Ala-RS = alanyl-tRNA synthetase.

RT-PCR →reverse transcription polymerase chain reaction

run-off transcription *In vitro* *transcription where *RNA polymerase synthesizes the *transcript until it falls off the DNA *template, so that the transcript ends at the DNA template end. STÉPHANIE VAULÉON

S

S-adenosylmethionine riboswitch → SAM riboswitch

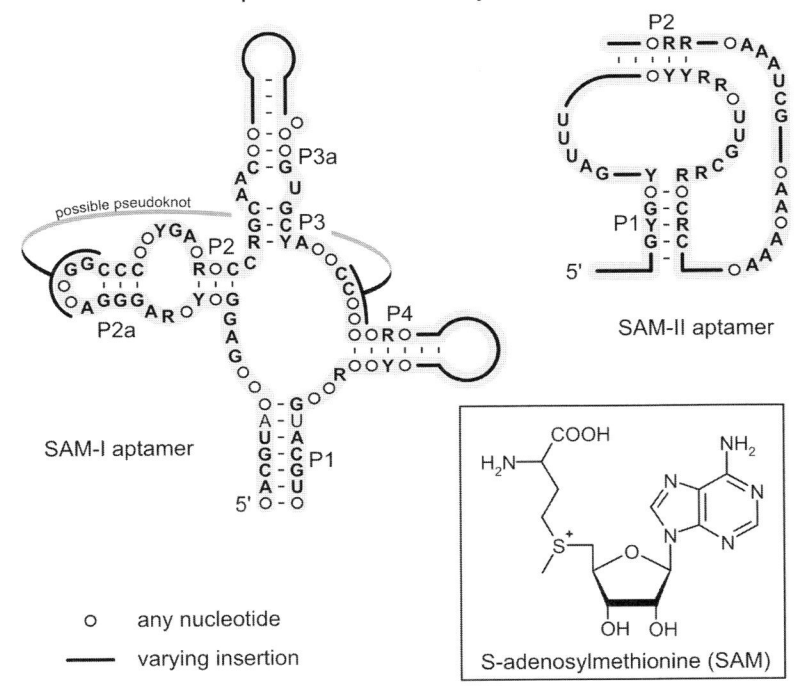

SAM (S-adenosylmethionine) riboswitch A highly conserved *RNA domain called the S-box is found frequently in the *5'-untranslated region of *messenger RNAs in Gram-positive bacteria. This domain was identified as the *aptamer part of a SAM riboswitch, which regulates expression of *genes involved in sulfur metabolism and biosynthesis of methionine, cysteine and SAM. Bioinformatic screening (→ Rfam) for other conserved RNA elements revealed that a different SAM riboswitch exists in Gram-negative bacteria, which was named SAM-II. Similar to SAM-I, it is found *upstream of genes related to SAM and methionine biosynthesis, and regulates their expression at the transcriptional level. Two distinctive RNA structures can be derived from phylogenetic data for these two aptamer types. The larger SAM-I aptamer forms a *four-way junction with putative *K-turn and *pseudoknot elements, whereas the smaller SAM-II structure is based on a simple *hairpin and a pseudoknot interaction. The differing affinities for SAM are consistent with the size and complexity of the aptamers, with *dissociation constants in the low nanomolar range for SAM-I and low micromolar range for SAM-II aptamers. However, both aptamers show strong selectivity for SAM and discriminate more than 100-fold against closely related metabolites like S-adenosylhomocysteine and S-adenosylcysteine, which differ from SAM by only a single methyl or methylene group, respectively.

Different conclusions can be drawn from the fact that multiple regulatory RNA structures have evolved that sense the same metabolite in different organisms. This suggests, for instance, that RNA has a high level of complexity and chemical sophistication and, similar to proteins, can adapt to perform diverse cellular functions. Additionally it could imply that SAM riboswitches have evolved independently and are therefore not as old as the first cellular life forms. This would mean that riboswitches might not be simple remains of a hypothetical *RNA world, but are still used as efficient gene control elements in modern life. RÜDIGER WELZ

Sanger sequencing Technique for the *sequence analysis of DNA. This method is based on the enzymatic synthesis of DNA with a mixture of normal and chain terminating *2′-3′-dideoxynucleotides (ddNTPs). Starting from a *primer that is annealed to the single-stranded target DNA, a complementary copy is synthesized in a *primer extension reaction. The chain-terminating *nucleotides used in Sanger sequencing lack the 3′-hydroxyl group necessary for elongation of the *sugar–phosphate backbone with another nucleotide. If a complementary ddNTP is incorporated instead of the corresponding dNTP, the chain elongation terminates.

Sanger sequencing requires the set-up of four reactions, each including a proportion of one of the respective ddNTP (ddATP, ddGTP, ddCTP, or ddTTP) as well as a radiolabeled nucleotide (usually [α-^{32}P] dATP or [α-^{35}S]dATP). These reactions result in mixtures of DNA *fragments terminating randomly at positions filled with a dideoxy analog. In the case of the *ddATP reaction, for example, chains are terminated at every possible A occurrence in the sequence. Separation of the reaction mixture using a high-resolution polyacrylamide gel and autoradiography ($\rightarrow$ gel electrophoresis) will result in a ladder representing all chain lengths from the sequencing primer to each A base in the sequence.

Combination with the other three base type reactions in adjacent gel lanes yields autoradiographs that can be read off as the DNA sequence. The resulting sequence is complementary ($\rightarrow$ complementary sequence) to that of the targeted *template DNA and can be converted into the original sequence according to the *base-pairing rules.

Sanger sequencing has been greatly improved and accelerated by the substitution of radiolabels with fluorescent labels ($\rightarrow$ fluorescence marker) that can be detected with comparable sensitivity ($\leq 10^{-15}$ mol DNA/band). Different from radioactively labeled samples, fluorescent gel bands can be detected directly during electrophoretic separation by using machines that combine electrophoresis units with fluorescence detection set-ups. In principle, the fluorescent label can be attached to the primer (dye-primer method) or to the terminating nucleotides (dye-terminator method). If distinctly emitting dyes four each of the bases (A, G, C and T) are attached to the terminators the reactions with all four ddNTPs can be combined in a single reaction because the fluorescent products can be determined in parallel.

Originally, Sanger sequencing was performed with single-stranded template DNA which was generated using bacteriophage M13 *cloning. This approach was substituted by *plasmid vector cloning of target DNA, which is easier and enables the sequence analysis of both directions using primers that flank the region of interest. Furthermore, the incorporation of terminating nucleotides can be combined with *amplification in a linear *polymerase chain reaction (cycle sequencing). SUSANNE BRAKMANN

Sankoff algorithm → RNA alignment and structure comparison

SAT chromosome Also referred to as satellite *chromosome, has a *bulge on the telomeric end and contains the enzyme *sine acido nucleinico*. Plays a vital role in the formation of the nucleolus after division is completed. MAURO SANTOS

satellite chromosome → SAT chromosome

satellite DNA Consists of many *tandem repeats of short DNA sequences often located in the telomeric (at the end of a linear *chromosome, → telomere) and centromeric regions (→ centromere).
MAURO SANTOS

satellite RNA Subviral agents composed of RNA. If it encodes the viral coat protein it is referred to as satellite virus or single-stranded RNA satellite virus.

S-box → SAM riboswitch

scanning force microscopy (SFM) Technique where surfaces are physically scanned in order to generate topographic images of molecules [also called atomic force microscopy (AFM) or scanning probe microscopy (SPM)]. At best, atomic resolution can be achieved, but this is usually not the case with biological material. The sample is mounted on a stage that is controlled along the X-, Y- and Z-axis by three piezo elements. An elastic arm (cantilever) with a fine tip (3–10 nm in diameter) is brought into contact with the surface and a laser beam is adjusted to the back of the cantilever. The laser beam is reflected to a four-quadrant photodiode that registers the position of the beam.

When the sample stage is moved in the X-direction, the tip registers height differences in that the cantilever arm is bent when the tip encounters an obstacle. Consequently, the position of the laser beam changes. In constant height contact mode, topography of the sample is simply registered by deflection of the laser beam that in turn depends on the bending of the cantilever arm. For biological samples, constant force contact mode better conserves the object. Here, the deflection of the laser beam is also registered by the photodiode, but directly feeds back to the cantilever arm via the Z-piezo and keeps the force between the tip and sample constant, i.e. bending of the arm is immediately compensated by movement of the stage in the Z-direction.

At the end of an X-scan, the stage moves one unit along the Y-axis and the next X-line is scanned. Images are composed of, for example, 512 X-lines with 512 image points each. With conventional instruments, scanning is done at a velocity of 1–2 Hz (lines per second). In constant force and constant height modes the tip is always in contact with the sample and may cause damage. In fact, high forces are now being used to dissect samples, e.g. to cut sections out of *chromosomes.

For imaging of biological materials, the tapping mode is the least destructive method. The cantilever is externally excited to oscillate close to its resonance frequency (depending on the cantilever, resonance frequencies range between 1 kHz and several hundred kHz). Direct contact to the sample is thus reduced to a minimum. Cantilevers oscillate with a free amplitude of approximately 100 nm. Similar to contact modes, the deflection of the laser beam registers changes in topography and provides a height image of the sample. With materials of different properties, e.g. softness, the phase of the oscillations may change, thus allowing us to differentiate between materials of similar height but distinct make-ups. With some materials the amplitude may be damped, this is usually adjusted by feeding information from the photodiode back to the Z-piezo so that the measurement is carried out at constant amplitude.

In the most simple applications, *biomolecules or reactions between different biomolecules are prepared in the appropriate buffer, if possible diluted in order to reduce salt concentrations, spread on a flat surface, washed with water and dried. Surfaces of sufficient flatness are freshly cleaved mica, silanized glass or gold-coated glass. Samples can be directly observed in the scanning force microscope without further treatment.

A substantial advantage of SFM is the possibility to observe single molecules under close to physiological conditions. The sample is applied to a fluid chamber mounted on the piezo stage, the cantilever is immersed into the liquid and imaging is carried out as described above. In liquid, special care is required to fix molecules properly to the surface, otherwise they will be shifted by the cantilever even in tapping mode. Plasma-activated mica in combination with appropriate divalent cations is suitable for nucleic acids and proteins. Silanized glass is also a good substrate for nucleic acids, but may bind proteins too strongly.

Images of dried samples and samples in liquid can define *binding sites of proteins to nucleic acids, and, to a limited extent, structures of protein–nucleic acid and protein–protein complexes.

Measurements in liquid may eventually allow for imaging enzymatic reactions in real-time; however, the scanning speed of cantilevers is currently limiting for such investigations.

Combinations of conventional light microscopy or fluorescence microscopy can be set up and allow us to address a specific position in a sample by SFM.

WOLFGANG NELLEN

scanning force spectroscopy (SFS) Special application of *scanning force microscopy (SFM) to measure intramolecular (e.g. unfolding of a protein) and intermolecular (e.g. protein–protein or protein–nucleic acid interaction) forces. Similar to SFM, bending of a small cantilever arm with a fine tip is monitored by a laser beam reflected from the cantilever to a four-quadrant photodiode. In contrast to SFM, the cantilever is only moved in the Z-direction. For the analysis of intermolecular interactions, the tip is derivatized with one reaction partner while the other one is attached to the surface mounted on the piezo stage. The tip is approached to the surface until contact is indicated by bending of the arm and, consequently, deflection of the laser beam that is focused on the back of the cantilever. The tip is then retracted and, if partner molecules on the surface and the tip have interacted, an increasing force will be elicited on the arm and bend it. Disruption of the interaction results in a release of the force and the cantilever jumps back to its original position, the zero line. Unbinding may occur in multiple steps indicated by multiple increases in force followed by a partial release.

For intramolecular forces, the target molecule is usually modified with different tags at both ends to allow for an oriented application on the surface and a defined attachment site for the cantilever tip. The tip is derivatized with an appropriate receptor to specifically bind the modified target (e.g. *biotin–*streptavidin). Upon approach to the surface, a molecule is bound to the tip. When the cantilever is then retracted from the sample, the molecule is first lifted up and stretched. Intramolecular forces then bend the arm until the retracting force equals and finally exceeds the structural forces that form the most unstable part of the molecule. Forced unfolding of this domain allows for further stretching without application of external forces by the cantilever until the next stable part/domain of the molecule has to be unfolded. Eventually, the substrate is entirely stretched out and finally the most stable bond to the cantilever or the surface is disrupted. The forces required for unfolding or disruption of binding are calculated from the recorded bending of the cantilever and the spring constant (a value describing the flexibility of an individual cantilever). Forces as low as approximately 15 pN can be registered.

The measured unfolding or disruption forces depend on the velocity of cantilever retraction (loading rate = retraction velocity × elasticity of molecule). With low loading rates, thermal fluctuation contributes significantly to overcoming the energy barrier to separate molecules, therefore lower forces are measured. With higher loading rates, higher forces are usually required to achieve unbinding or unfolding. Different protein binding characteristics to target molecules may only be detected in the high loading rate regime.

Instruments that combine SFM and SFS are available. These allow us to first establish a topographic image and then to address a specific molecule on the surface with high precision. WOLFGANG NELLEN

S-conformation Conformation in a five-membered furanose ring characterized by a *pseudorotation phase angle P with $144° \leq P \leq 190°$ (southern part of the pseudorotational cycle). In double-stranded B-form DNA ($\rightarrow$ B-DNA), the *deoxyribose sugars are in C2'-*endo* corresponding to an S-conformation ($\rightarrow$ pseudorotation). KLAUS WEISZ

SCOR Database of RNA structure, function and tertiary interactions, and their relationships: http://scor.lbl.gov/scor.html.

Scorpion primers Contain a self-complementary *stem–loop probe structure containing a fluorophore ($\rightarrow$ fluorescent primer, $\rightarrow$ fluorescence labeling) at one end and a quencher at the other. The stem–loop structure has the same sequence as a region of the *template close to the *primer *binding site of the probe. Attached to the stem–loop structure is a primer complementary to the template sequence desired for *amplification. The primer is often attached to the probe by a non-hydrogen bonding linker such as hexaethylene glycol. During the *polymerase chain reaction the primer is extended and in the presence of this extended strand the probe, which is now complementary to the new strand, hybridizes to the new DNA strand. As the probe hybridizes to the target the fluorophores and quencher separate, and there is therefore an increase in fluorescence emitted ($\rightarrow$ FRET). DAVID LOAKES

scRNA $\rightarrow$ small cytoplasmic RNA

secondary structure Any regular, stable structure taken up by some or all of the nucleotides in a nucleic acid. Different structural forms of DNA fall under the heading of secondary structure. NINA DOLINNAYA

second genetic code →aminoacyl-tRNA synthetase

second-site mutation →suppressor mutation

second strand DNA strand (→complementary DNA) which is complementary to a single-stranded RNA is called a second strand.

second-strand synthesis Synthesis of a *complementary DNA strand, e.g. in the *replication process of the retroviral *genome. ANDREAS MARX, KARL-HEINZ JUNG

segmentation genes *Genes that divide the early embryo into a repeating series of segmental primordial along the anterior–posterior axis. *Mutations in these genes cause the embryo to lack certain segments or parts of segments. MAURO SANTOS

selection in vitro →in vitro selection, →SELEX

selenium modified RNA RNAs with selenium modifications (e.g. 2′-methylseleno groups or selenophosphates) represent useful derivatives with anomalous scattering propensity to facilitate phasing of nucleic acid crystallographic data (→crystallization of nucleic acids). 2′-Se-methyl modified *nucleosides can be site-specifically introduced by chemical oligonucleotide synthesis (→oligonucleotide synthesis). RONALD MICURA

SELEX (Acronym for Systematic Evolution of Ligands by EXponential enrichment.) SELEX is a variant of *in vitro selection and constitutes an iterative technique that allows the isolation of *aptamers from combinatorial *nucleic acid libraries (→aptamer selection). The SELEX process selects for aptamers on the basis of binding between a *target and *nucleic acid molecules, and relies on standard molecular biology techniques. Steps of the SELEX cycle are: (a) library preparation, (b) selection, (c) *amplification and (d) aptamer isolation. In step (a), a large library (or pool) is synthesized. Each molecule in the library (typical complexity: up to 10^{15} different compounds) contains a unique *nucleotide sequence that can, in principle, adopt a unique three-dimensional shape. Only very few of these molecules – the

Selenium modified RNA

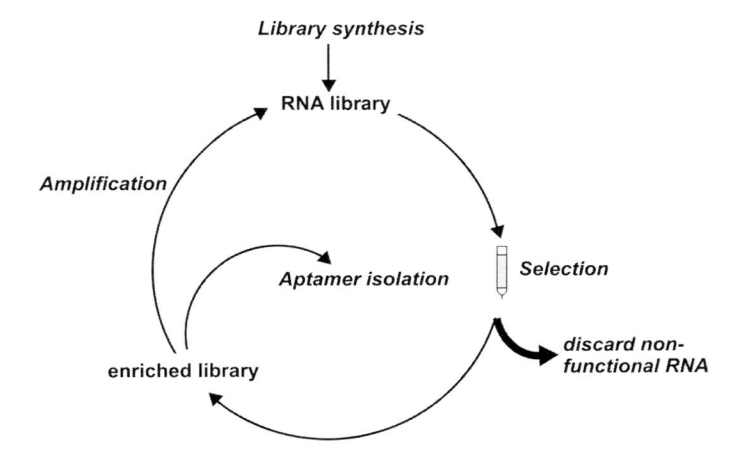

Library synthesis → RNA library → *Selection* → discard non-functional RNA → *Aptamer isolation* → enriched library → *Amplification* → RNA library

aptamers – present a shape that is complementary to the target molecule. The selection step is designed to find those molecules with the greatest affinity for the target of interest. The library is incubated with the target and the library members will either associate with the target or remain free in solution. Several methods exist to physically separate the aptamer target complexes from the unbound molecules in the mixture. The unbound molecules are discarded, and the target-bound aptamers released and copied enzymatically, to generate a new library of molecules that is enriched for those species that can bind the target. This enriched library is used to initiate a new cycle of selection and binding. Typically, after 5–15 cycles, the library is reduced from the 10^{15} unique sequences to a small number of tight-binding molecules. Individual members of this final pool are then isolated, their sequence determined, and their properties with respect to *binding affinity and specificity are measured and compared.

SELEX can be applied to RNA, DNA or modified nucleic acid libraries, as long as these can be enzymatically copied. Several variations of the SELEX technique exist that also allow the isolation of *ribozymes and allosterically regulated ribozymes (→allosteric ribozymes). Using SELEX, one can find aptamers with high affinity to virtually any target, from small organic molecules to peptides, proteins, cell surfaces and tissue sections. *In vivo* SELEX has been carried out in living animals. SELEX can be performed either manually or in an automated fashion. ANDRES JÄSCHKE

self-assembly →DNA self-assembly

self-cleaving RNA *Ribozyme that catalyzes an intramolecular cleavage reaction. Naturally occurring ribozymes are mostly acting in *cis* (→*cis*-active), mediating viral *replication during the *rolling-circle amplification mechanism. DENISE STROHBACH

selfish DNA Stretches of DNA that replicate out of phase with other parts of the *genome, even if it is harmful to the organism: it generates intragenomic conflict. Conceptually the notion is linked to that of the selfish *gene (Dawkins, 1976), but the relation is not one-to-one: in evolutionary terms all genes are selfish, but selfish DNA is more selfish than well-behaved genes (following Mendelian inheritance rules), e.g. *transposons can jump around and multiply within the genome, sometimes causing a lot of trouble. Selfish DNA in an organism is parasitic. EÖRS SZATHMÁRY

self-organization Leads to an increase in complexity without evolution by natural selection. Good examples are stars, Benard cells in fluids heated from below, vortices, etc. There is a lot of self-organization in living organisms as well, since *genes do not have to encode for the laws of physics and chemistry. Pure self-organized systems do not build up adaptations in a cumulative manner; they just exist. EÖRS SZATHMÁRY

self-priming Process of *base pairing in which the 3'- end of a *DNA folds back and creates intramolecular *hairpin *loops that can serve as *primers for *reverse *transcriptases or *DNA polymerases. Retroviruses like HIV have developed complicated strategies to prevent self-priming of synthesized DNA *in vivo*. JÖRN WOLF

self-replication In the process of self-replication, agents (replicators) make copies of themselves, with or without the aid of others agents. The structure of the replicator is passed on largely intact. Chemically, autocatalysis always results, in some sense, in replication; conversely, replication of chemically based agents always requires autocatalysis. Replicators have limited or unlimited hereditary potential. In the case of the former, the number of individuals is commensurate with the number of possible types (e.g. sequences); for the latter, there are many more possible types than individuals. *Genes today have unlimited hereditary potential; this must have been different early in evolution. The mode of *replication is either holistic (as for the intermediates in the reductive citric acid cycle or the Calvin cycle) or modular (as in nucleic acid replication). Prions are chemical, but protein-based replicators, memes replicate in the cultural domain and computer viruses replicate in cyberspace. EÖRS SZATHMÁRY

self-splicing Self-extraction of a RNA *intron out of a *pre-mRNA molecule. This reaction is performed by the *group I and *group II introns (→self-splicing RNA) through two *transesterification reactions. STÉPHANIE VAULÉON

self-splicing RNA RNA intron that can splice itself out of its *pre-mRNA molecule (→self-splicing). Self-splicing RNAs belong to the *group I and *group II introns.

semiconservative replication After *replication, the *DNA double helix consists of a newly synthesized strand and a conserved parental strand. Thus, this process is called semiconservative replication. ANDREAS MARX, KARL-HEINZ JUNG

semidiscontinuous replication The synthesis of DNA is continuous at one strand, the *leading strand, and discontinuous in short pieces (→Okazaki fragments) at the other strand, the *lagging strand. ANDREAS MARX, KARL-HEINZ JUNG

sense codon All *codons in an open *reading frame of a *messenger RNA which lead to the incorporation of amino acids into a polypeptide by the *ribosome (→translation) and are not *stop codons (which are also termed nonsense codons if they occur inside an open *reading frame). STEFAN VÖRTLER

sense strand DNA strand that has the same *nucleotide sequence as the transcribed *messenger RNA. As it contains the sequence of *codons (nucleotide triplets) which interact with the *anticodons of the *transfer RNAs during *translation leading to the primary sequence of the resulting proteins, the sense strand is also called the *coding strand. ANDREAS MARX, KARL-HEINZ JUNG

Sequenase Enzyme used for *Sanger sequencing with radioactive or fluorescent *dideoxynucleotides. The enzyme is an engineered variant of bacteriophage T7 *DNA polymerase that shows 5′–3′ polymerase activity, but no 3′–5′ *exonuclease activity. Sequenase exhibits a high rate of *nucleotide incorporation as well as a high processivity, i.e. forms stable polymerization complexes and dissociates from the *primer *template after incorporation of large numbers of nucleotides. SUSANNE BRAKMANN

sequence →nucleotide sequence

sequence alignment Comparison of nucleic acid sequence screening for identical regions/areas (→RNA alignment and structure comparison, →Rfam).

sequence analysis →Human Genome Project, →DNA sequencing, →Maxam–Gilbert sequencing, →Sanger sequencing

sequence homology Describes similarities between DNA sequences and structures to measure possible evolutionary origins. DNA sequences that share a high sequence similarity obviously share a common ancestor and also indicate a common function. In general, if two *genes have almost the same sequence it is clear that they are homologous. MATTHÄUS JANCZYK

sequence library →sequence pool

sequence pool (or library) Collection of nucleic acid molecules that is characterized by its diversity (number of different sequences). The library can consist either of DNA or RNA or *modified nucleic acids. Nucleic acid sequence pools can have a diversity of up to 10^{16} different sequences. Most libraries are generated by chemical synthesis of a DNA library (→oligonucleotide synthesis), which – if desired – can be transcribed into a RNA library (→*in vitro* transcription). The most common approach to library synthesis is use of monomer (→phosphoramidite) mixtures in solid-phase DNA synthesis. Such a mixture of the four standard phosphoramidites, coupled N times, gives a library of 4^N different sequences, the positions where N was coupled being referred to as "randomized". This leads to two different cases. (a) Complete libraries: up to $N = 25$, the library will contain at least one molecule of all theoretically possible sequence (4^{25} or around 10^{15} molecules). (b) At $N > 25$, coverage of sequence space becomes incomplete and only a fraction of the theoretically possible sequences are physically present. A third approach uses focused (or biased or doped) libraries in which a defined "wild-type" DNA sequence is synthesized using monomers that have intentionally been contaminated with low amounts of other monomers (doped), so that all members of the synthesized pool will resemble the "wild-type", but have a small number of randomly scattered mutations. All three strategies can be combined in one library. The diversity of a library can be further increased by *random mutagenesis techniques (e.g. mutagenic *polymerase chain reaction). Long libraries (over 120 nucleotides) can be assembled from smaller precursors using *ligation or other standard techniques. For use in *SELEX experiments, the variable part(s) of the library must be flanked by two constant-sequence *primer *binding sites. ANDRES JÄSCHKE

sequence tagged sites (STS) Short unique *fragment of DNA, usually shorter than 1 kb.

sequencing →Human Genome Project, →DNA sequencing, →Maxam–Gilbert sequencing, →Sanger sequencing

SFM →scanning force microscopy

SFS →scanning force spectroscopy

Shapiro–Benenson–Rothemund automata
Such automata embody a mathematical model of a restricted two-state finite automaton: a simple "computer" that switches between two states. It processes sequences of symbols (strings) one symbol at a time with the next state determined by the current state and the incoming symbol. A *double-stranded DNA (dsDNA) "input molecule" encodes a string. The incoming (left-most) dsDNA symbol is partially converted to a *sticky end whose offset relative to the symbol reflects the state of the automaton. Cascades of cleavages performed on this molecule by *restriction enzymes (e.g. *Fok*I) and directed by "cutting" (transition) rule molecules accomplish a computation. A cutting molecule comprises a *Fok*I *binding site appropriately distanced from a sticky end. The latter binds to a sticky end of the input molecule and guides the enzyme to expose a sticky end within the next symbol, thereby switching the state. Given that every state–symbol combination encountered in a computation has an appropriate cutting molecule, the process is repeated until all input symbols are processed. The cascade generates a molecule that uniquely encodes the final state of the automaton. More advanced automata can analyze *oligonucleotide inputs that can interact (block or activate) with cutting rules and can produce new oligonucleotides by cleavage, performing arbitrary Boolean calculations. YAAKOV BENENSON

sheared base pair Base-pairing mode involving the *Hoogsteen edge of one *purine nucleoside with the *sugar edge of another, with the *glycosidic bonds being oriented in *trans* relative to the axis of interaction. The term sheared base pair is mostly used for AA and GA interactions corresponding to these rules. SABINE MÜLLER

Shine–Dalgarno sequence Sequence 5'-AGGAGGU-3' that is found at the 5'-end of prokaryotic *messenger RNA and that defines the binding platform for the small subunit of the *ribosome before *translation is initiated. The *consensus sequence 3'-UCCUCCA-5' in the small ribosomal subunit is found at the 3'-end of the 16S RNA. After localization of the Shine–Dalgarno sequence the ribosome starts to migrate in the 3'-direction and, as next step, identifies the *start codon (AUG) usually 10 nucleotides *downstream of the platform sequence. In eukaryotes, the small ribosomal subunit recognizes the *cap structure at the 5'-end before localizing the start codon. TINA PERSSON

short interspersed repetitive element (SINE)
Group of short, dispersed and related

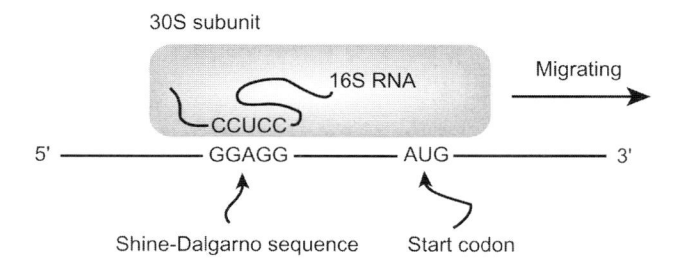

sequences frequently found in mammalian *genomes. The Alu elements belong to this family, each is approximately 300 bp long and they are highly represented in the human genome. MAURO SANTOS

short patch repair →DNA repair

shotgun cloning Collection of randomly cloned DNA *fragments of organisms. The DNA of the respective organism is digested with *restriction endonucleases and DNA fragments of a certain size are ligated into a *cloning vector. This method is used to create *genomic libraries. ANNEGRET WILDE

shotgun sequencing Technique for (large-scale *genome) *sequencing that is based on the random (i.e. shotgun) subcloning of *template DNA. The DNA to be sequenced is randomly broken down to smaller *fragments and cloned into a suitable *vector producing a subclone library. Random breakage is achieved either by disruption in a French press, passing the DNA through a fine-gauge needle, enzymatically by *digestion with *DNase I or shearing by sonication. Recombinants from the subclone library are chosen at random and sequenced. Using computers with appropriate software, individual sequence reads are compared and overlaps are combined to reconstruct the original sequence. Shotgun sequencing produces significant redundancy that depends on the proportion of the sequence determined, total fragment length, average read length and number of *clones sequenced (→sequence analysis of DNA). SUSANNE BRAKMANN

shRNA (small hairpin RNA molecules) Precursors of *small interfering RNAs mimicking *pri-miRNAs.

shuffling Altering the order of a given nucleic acid sequence, although maintaining the overall *nucleotide composition. *Exon shuffling means sorting or rearranging the given array of *exons of a gene or *genes. GEMMA MARFANY, ROSER GONZÀLEZ-DUARTE

shuttle transfer Transfer of a shuttle DNA between two different hosts, usually between a prokaryotic and a eukaryotic cell. The shuttle DNA usually contains genetic elements that allow maintenance or expression in both types of host. In a different context, it may refer to endogenous proteins being mobilized from cytosol to nucleus and *vice versa* (→shuttle vector). GEMMA MARFANY, ROSER GONZÀLEZ-DUARTE

shuttle vector *Vector that replicates in at least two different host species (e.g. yeast and bacteria). They have been constructed by the fusion of two vectors and contain different *origins of *replication for the respective host. Shuttle vectors are often used to clone *genes in the easily transformable *Escherichia coli* host that are subsequently analyzed in another cell type. Instead of shuttle vectors, broad host range *plasmids (e.g. RP4) can be used. Plasmids of the IncP1 incompatibility replicate in many Gram-negative bacteria. ANNEGRET WILDE

sigma factor Subunit of the bacterial *RNA polymerase needed for *promoter recognition. *Escherichia coli* contain seven different sigma factors. The housekeeping sigma factor σ^{70} (RpoD) directs the RNA polymerase to promoters which carry characteristic *recognition sequences in the −35 and −10 region (TTGACA and TATAAT, respectively). Under specialized conditions, alternative sigma factors control the expression of a subset of genes by binding to promoter sequences that deviate from the σ^{70} consensus. The cellular level of alternative sigma factors depends on the environmental conditions, e.g. the starvation sigma factor σ^S (RpoS) increases in stationary phase when nutrients become limiting and the heat shock *sigma factor σ^{32} (RpoH) is induced under *heat shock conditions. FRANZ NARBERHAUS

signal recognition particle (SRP)
Protein–RNA complex which recognizes and transports specific proteins to the endoplasmic reticulum (ER) (eukaryotes) or to the plasma membrane (prokaryotes). The eukaryotic complex is composed of six protein subunits and RNA. During *translation, the *GTP-bound SRP binds to a signal peptide element of eight or more non-polar amino acids immediately after its ribosomal synthesis, as well as to the large ribosomal subunit, thus inducing a translational delay. The whole complex migrates to the ER and is recognized by a SRP-specific receptor. This gives *ribosomes the opportunity to attach onto the ER membrane. After dissociation of the SRP, which is driven by GTP hydrolysis, protein synthesis is continued thereby transporting the growing polypeptide chain into the ER lumen. Binding of the SRP and the resulting delay of protein synthesis is important to circumvent the early release of immature proteins to the cytosol and to facilitate protein maturation in the ER lumen.
IRENE DRUDE

signal sequence Sequence of amino acids (in proteins) or *nucleotides (in DNA or RNA) that is recognized by cell proteins or other *ligands. It only affects molecules it contacts directly. A typical example of a signal sequence in DNA from some bacteria is the uptake signal sequence (USS), a short DNA sequence (9–10 nucleotides) that is recognized by certain bacteria. The DNA uptake machinery of those bacteria specifically recognizes the USS during natural *competence. The *genomes of those species carry multiple copies of their preferred USS (→ cis-active, →competence, →signal structure). JÖRN WOLF

signal structure *Oligonucleotide structure that is used as characteristic feature in many processes such as *replication, *transcription and *translation. Signal structures can mark the *origin of replication where *DNA replication is initiated; they can be involved as *promoters, *operators and *terminators in *transcription mechanism; and they also play an important role in *translation (→signal sequence). JÖRN WOLF

silencer →RNA interference, →small interfering RNA

silent mutation Structural change in a DNA (RNA) sequence without consequences for the coding properties of the affected nucleic acid. Within a *coding sequence, a silent mutation is typically a *substitution mutation of a third *codon letter. HANS-JOACHIM FRITZ

silver staining Histochemical staining method that was first applied to proteins and polypeptides, and later also to nucleic acids. The samples are separated by *gel electrophoresis and treated with a solution of silver ions that bind to the *bases and are reduced chemically or upon light exposure. Silver staining methods are 100 times more sensitive than other stains (Coomassie) and comparable to radioisotopic methods. JÖRN WOLF

simple sequence repeats (SSRs)
→microsatellite, →minisatellite

SINE →short interspersed repetitive element

single-base mutation →point mutation

single-copy genes Active *gene copies without close relatives in the *genome.

single-copy sequence *Coding or *non-coding sequence present only once in the *genome.

single-molecule sequencing →DNA sequence analysis

single-nucleotide polymorphisms (SNPs) Inherited *nucleotide substitutions (→point mutation) between individuals of a species, occurring at a rate greater than 1% in a population. SNPs are the most common type of human *mutations, comprising approximately 0.1% of the average human *genome. To date, there are more than 1 million validated human SNPs. They are classified by their genomic location:

genome SNP (gSNP)	located within the non-coding region of the *genome
coding SNP (cSNP)	located within the coding regions of genes
regulatory SNP (rSNP)	located within regulatory regions of genes (→regulator gene)
intronic SNP (iSNP)	located within *introns

It has been suggested that most SNPs are located within the 95% non-coding region (→non-coding DNA) of the *genome. SNPs can alter the function of DNA, RNA and proteins. cSNPs effect the protein sequence by means of a non-synonymous exchange (→genetic code). SNPs with a known *phenotype can be discovered from association studies, while phenotypically unknown SNPs can be discovered by SNP screening or *genome analysis. SNPs are responsible for diseases, such as sickle cell anemia, cystic fibrosis, phenylketonuria and familial hypercholesterolemia. In addition, the combination of several SNPs can be associated with an increased risk for cancer and other diseases. TOM N. GROSSMANN, OLIVER SEITZ

single-strand assimilation Displacement of one strand of a *DNA duplex by a homologous external *single-stranded DNA molecule. The assimilation of single-stranded DNA into an existing duplex has importance in *DNA repair and *recombination functions. This process is usually supported by various proteins. SLAWOMIR GWIAZDA

single-strand-binding proteins Proteins that specifically recognize and bind single-stranded regions of DNA and RNA (→DNA-binding proteins, →RNA-binding proteins).

single-strand break (SSB) Break of only one strand within a *double-stranded DNA. Chromosomal SSBs are the common lesions in cells where the DNA was attacked by endogenous free radicals and *alkylating agents. In addition SSBs can also be induced by some environmental genotoxins, ionizing radiation, anticancer drugs or *topoisomerases. MATTHÄUS JANCZYK

single-strand exchange →homologous recombination

single turnover Enzymatic reactions follow a more or less complicated mechanism. Different experimental conditions allow focusing on the kinetics of earlier or later phases of the reaction. Single turnover means mixing a substrate with a large molar excess of the enzyme which leads to a second order binding process and which can be analyzed correspondingly (→rate constant). Binding is followed by the enzymatic process (one or more steps) where single turnover guarantees that an enzyme molecule reacts only with one substrate molecule. Therefore, single turnover conditions ensure that product release steps have no impact on the observed rate of the enzymatic reaction (→multiple turnover). CHRISTIAN HERRMANN

siRNA →small interfering RNA

sister chromatids Two identical DNA strands of a *chromosome.

site-directed mutagenesis Technique that allows specific *mutation of DNA. Defined *nucleobases of a *gene are changed or completely removed. In 1978, Michael Smith received the Nobel Prize in Chemistry for site-directed mutagenesis using *oligonucleotides for *in vitro* synthesis of mutant DNA. BETTINA APPEL

site-specific recombination Specialized reversible *recombination mediated by site-specific recombinases, which recognize, cut and ligate two specific DNA targets. Depending on the biological effects, the site-specific recombinases may be classified as integrases, invertases or resolvases if they promote integration, excision or inversion of the recombinogenic DNA molecules, respectively. GEMMA MARFANY, ROSER GONZÀLEZ-DUARTE

7S K RNA *Small nuclear RNA of 331 nucleotides in length. 7S K RNA is synthesized by *RNA polymerase III under control of an exclusively gene-external *promoter. The human *genome contains only one functional *gene per haploid *genome, but several thousand non-functional more or less divergent *pseudogene sequences of unknown significance are present. Mammalian cells contain about 100 copies of the RNA. Although the RNA is highly conserved among vertebrates, significantly diverged 7S K RNA sequences are found in higher invertebrates; however, 7S K RNA is not detectable in and most likely absent from lower eukaryotes. In contrast to other small stable RNA molecules, 7S K RNA is not found in the form of a stable *ribonucleoprotein complex. Rather, 7S K RNA has been found to be transiently associated with the positive *transcription elongation factor b (pTEFb) via a bridging protein hexim. That protein is also absent from lower eukaryotes. It appears that 7S K RNA functions as a negative regulator of transcription elongation in vertebrates and higher invertebrates.
BERND-JOACHIM BENECKE

Slicer Putative protein carrying the endonucleolytic activity of *RNA interference-mediated *messenger RNA cleavage. It is now known that members of the Argonaute family carry out Slicer activity.
NICOLAS PIGANEAU

slot blot Technique by which RNA or DNA molecules can be directly applied onto a membrane as a slot. The *oligonucleotide is fixed on the membrane by ultraviolet light or heat. *Hybridization of labeled RNA or DNA (→labeling of DNA and RNA) probes complementary to the target oligonucleotide allows for detection of the RNA or DNA to be analyzed (→blotting).
IRENE DRUDE

7S L RNA Forms the RNA constituent of the *signal recognition particle (SRP). That particle specifically interacts with the mostly hydrophobic residues that constitute the amino-terminal signal sequence of secretory and lysosomal proteins. Via interaction with the docking protein, the SRP mediates *translocation of the nascent polypeptide chain into the lumen of the endoplasmic reticulum. The mammalian 7S L RNA is 299 nucleotides in length and is synthesized by *RNA polymerase III. The only one functional *gene per haploid *genome is under control of a split *promoter, localized in part *upstream and *downstream of the *transcription start site. About 50% of the RNA is highly homologous or even identical to the highly repetitive mammalian Alu *RNA gene sequences. In addition to 7S L RNA, the SRP contains six specific polypeptides (p9, p14, p19, p54, p68 and p72), some of which (p9/p14 and p68/p72) bind to the RNA as *heterodimers. BERND-JOACHIM BENECKE

small cytoplasmic RNA (scRNA) Any one of the many classes of scRNAs present predominantly or exclusively in the cytoplasm of eukaryotic cells. Conversely, *small nuclear RNA are exclusively located in the nucleus of cells. GEORG SCZAKIEL

small hairpin RNA →shRNA

small interfering RNAs (siRNAs) Around 21- to 26-nucleotide long RNA molecules involved in the *repression of *gene expression in a sequence-specific manner through the *RNA interference (RNAi) pathway in eukaryotic cells. The natural role of siRNAs is the repression of expression of viral and *repetitive DNA elements, thus providing a protection against viruses and increasing *genome stability. However, the celebrity of siRNAs is due to their employment as specificity determinants for the knock-out of *genes of interest. This activity has made them an essential tool in functional genomics, and a potential therapeutic component in the treatment of various viral and genetic diseases. The mode of action of siRNAs is similar to that of *micro-RNAs, but their biogenesis is different. Endogenous siRNAs are processed from long *double-stranded RNA precursors via successive cleavages by the *RNase III-like *endonuclease *Dicer. siRNAs are delivered via different means depending on the organism under study. Nematodes can be immerged in a siRNA solution or fed with bacteria producing *double-stranded RNA. In human cells two different strategies are used. On one hand, *short hairpin RNA can be transcribed by *polymerase III after *transfection of appropriate *vectors. On the other hand, siRNA *duplexes mimicking the product of Dicer cleavage can be transfected directly. (a) Important factors must be taken into account when designing siRNAs. First, from the double-stranded siRNA product of Dicer cleavage, only one strand is incorporated into the RNAi machinery. (b) It should be kept in mind that several genes can be targeted by one siRNA even with less than perfect complementarity, so-called "off-target" effects.
NICOLAS PIGANEAU

small nuclear ribonucleoprotein (snRNP) Stable complexes formed between *small nuclear RNA molecules and several proteins. Proteins of these complexes may be distinct for a specific snRNP or, as in the case of the spliceosomal RNPs (→spliceosome), may be shared among several snRNP complexes.
BERND-JOACHIM BENECKE

small nuclear RNA (snRNA) Small stable RNA molecules predominantly located within the nucleus. Quite often "snRNA" is used to refer exclusively to the spliceosomal RNAs (→spliceosome), although other species, such as *7S K RNA, also clearly belong to the group snRNAs. snRNAs can be synthesized by either *RNA polymerase II or III. In some cases, snRNAs are synthesized in the form of a precursor molecule which has to be processed to mature snRNA (→RNA processing). This maturation may take place in the cytoplasm. Consequently, those snRNAs shuttle between the nucleus and the cytoplasm, although the majority of molecules are localized within the nucleus. With the exception of 7S K RNA, snRNA molecules in their functional form exist in form of stable *small nuclear ribonucleoprotein complexes. BERND-JOACHIM BENECKE

small nucleolar ribonucleoprotein (snoRNP) Protein–*small nucleolar RNA (snoRNA) complex in the nucleolus, responsible for modification of *ribosomal RNA (rRNA). snoRNPs containing C/D box snoRNA catalyze 2'-O-methylation of rRNA. snoRNPs containing H/ACA box snoRNA catalyze rRNA pseudouridylation.
VALESKA DOMBOS

small nucleolar RNA (snoRNA) Growing number (more than 150) of *small nuclear RNAs localized within the nucleolus. snoRNAs are involved in processing and modification of *ribosomal RNA (rRNA) precursor (→pre-rRNA, →RNA processing). They are transcribed by *RNA polymerase II or III and some are encoded in the *introns of other *genes. Non-intronic polymerase II transcribed snoRNAs are capped with a *trimethylguanosine cap. The box C/D family of snoRNAs functions as *guide RNA for 2'-O-methylation of *ribosomal RNA, whereas snoRNAs of the box H/ACA family guide rRNA pseudouridylation (→pseudouridine). Certain members of both families are required for cleavage of the pre-rRNA molecules.
BERND-JOACHIM BENECKE

small nucleolytic ribozymes Class of *ribozymes to which belong the *hairpin ribozyme, the *hammerhead ribozyme, the *hepatitis delta virus ribozyme and the *Varkud satellite ribozyme. These ribozymes are small RNA structures of about 40–160 nucleotides that catalyze the site-specific reversible cleavage of RNA. They are found in viral, virusoid or *satellite RNAs, where they take part in processing the multimeric products of *rolling-circle amplification into *genome-length strands. More recently, two more catalytic RNA motifs, catalyzing the same type of cleavage reaction, have been identified in mRNAs of several Gram-positive bacteria (→glmS ribozyme) as well as in the *transcript of the human β-globin gene (→CoTC ribozyme). The cleavage reaction proceeds via *in-line attack of the 2'-hydroxyl group on the adjacent phosphorous atom, leading to the departure of the 5'-oxygen atom on the adjacent *ribose and to generation of a *2',3'-cyclic phosphate. The reaction mechanism is a S_N2 reaction which is accompanied by an inversion of configuration at the phosphorous. The *transition state is an oxyphosphorane intermediate with a trigonal bipyramidal structure containing the attacking 2'-O and departing 5'-O atoms at apical positions. The reverse reaction may also be efficiently catalyzed involving attack of the oxygen atom of the free 5'-hydroxyl group of one RNA fragment on the phosphorous atom of the *cyclic 2',3'-phosphate group of another, which results, through ring opening and release of the 2'-hydroxyl unit, in ligating the two fragments.

The small nucleolytic ribozymes catalyze the same reaction as many protein *RNases do. However, they act only at specific *phosphodiester bonds. Mainly *base pairing and in some cases *tertiary interactions are used to align the cleavage site within the ribozyme *active site. The catalytic strategies used by these ribozymes involve *general acid–base catalysis, charge stabilization and conformational effects like proximity and orientation for alignment of the attacking nucleophile.
SABINE MÜLLER

small ribosomal subunit →ribosome

S1 mapping Biochemical method based on *hybridization in which a special RNA sequence in a mixture of RNAs (e.g. from a biological sample) can be analyzed. A DNA segment that is complementary to the target RNA sequence is added to the mixture and left to hybridize. All unpaired regions are then digested by *S1 nuclease, a *nuclease that is specific for single-stranded segments. The *RNA–DNA hybrids that are left after the *incubation can be separated, visualized, and analyzed.
JÖRN WOLF

snake venom phosphodiesterase Phosphodiesterase isolated from snake venom (→phosphodiesterase).

snoRNA →small nucleolar RNA

snoRNP →small nucleolar ribonucleoprotein

SNP →single-nucleotide polymorphism

snRNA →small nuclear RNA

snRNP →small nuclear ribonucleoprotein, →spliceosomal RNP

S1 nuclease Single-strand specific *nuclease isolated from the slime mold *Aspergillus oryzae*. It cleaves RNA as well as DNA both endo- and exonucleolytically; often used to trim the ends of a *cDNA resulting from a *reverse transcription.
ULI HAHN

snurp Casual term for small nuclear uridylic acid-rich ribonucleoprotein as an abbreviation for spliceosomal *small nuclear ribonucleoproteins (→spliceosome, →spliceosomal RNP).
BERND-JOACHIM BENECKE

snurposome Term that was first used for granules containing *small nuclear ribonucleoproteins in germinal vesicles from frog and newt. Granules containing only U1 *small nuclear RNA (snRNA) were designated A snurposomes. B snurposomes contain all snRNAs involved in splicing except U7. A third class of snurposomes, called C snurposomes, was found, that contain snRNA U7. B and C snurposomes are often associated and have a spherical shape.

As later studies showed, B snurposomes contain the *RNA polymerase II *transcription and processing machinery, and the term *Pol II transcriptosome* was proposed for them. C snurposomes were shown to represent the main body of a cellular component called Cajal Body. Polymerase II components are supposed to be assembled in the Cajal Bodies and then sent to the B snurposomes. JÖRN WOLF

solid-phase synthesis of DNA and RNA →oligonucleotide synthesis

solid support →oligonucleotide synthesis

soluble RNA →transfer RNA

somatic gene transfer →gene transfer

somatic hypermutation →hypermutation

Southern blot describes the transfer of DNA onto a carrier (nitrocellulose, nylon membrane). The *blotting step is a part of the *Southern technique. IRENE DRUDE

Southern technique first described by Edwin Southern in 1975 and used for detection and quantification of DNA *fragments. DNA is first degraded into small fragments by *endonucleases and the fragments then are separated by size on a polyacrylamide or an agarose gel (→gel electrophoresis). The DNA is denatured directly on the gel by alkali, inducing the single-strand formation of DNA. After transfer of the DNA population from the gel onto a filter or a membrane by pressure, capillary power or an electric field, DNA is crosslinked onto the membrane by ultraviolet light or "baking" to fix deoxyoligonucleotides on the membrane. The membrane is incubated with a labeled DNA or RNA probe (→labeling of DNA and RNA) of known sequence, allowing a *hybridization of the probe to the *complementary sequence in the single-stranded DNA population. Bound fragments can be detected based on the labeling properties of the probe either radiographically when the probe is radioactive labeled, by fluorescence detection in the case of a *fluorescent dye incorporated into the probe or by chromogenic detection when a chromogenic dye is used (→blotting). IRENE DRUDE

South-western technique Technique for identifying and characterizing *DNA-binding proteins by their ability to bind to specific *DNA probes using a combination of *Southern and Western blot methods. Protein of the sample is separated by *electrophoresis on a sodium dodecylsulfate–polyacrylamide gel and blotted onto a membrane (typically nitrocellulose) for screening of DNA binding. After treatment of the membrane with a labeled DNA probe, *DNA-binding proteins can be detected based on the label properties of the *DNA probe (→blotting). IRENE DRUDE

spacer In general, a spacer molecule connects two different functional parts of a molecule. Usually, spacer molecules themselves are chemically neutral moieties, e.g. polyethylene glycol or alkyl chains. Spacer can be also an abbreviated term for *spacer DNA. Spacer DNA in eukaryotic and viral *genomes is usually made up of DNA regions separating two specific functional elements, e.g. protein target sites or functional *genes. They insure the correct and unbiased function of the two different DNA regions they separate. VALESKA DOMBOS

spacer DNA DNA segments between DNA that codes for functional RNA. In particular, the term is used for the long DNA sequences that separate *ribosomal *RNA (rRNA) genes in primary rRNA transcripts (→primary transcript), which appear in many copies in the *genome. Spacer DNA becomes transcribed together with the rRNA and is excised later during *rRNA processing. It can contain functional elements (→spacer, →spacer tRNA). JÖRN WOLF

spacer tRNA *transfer *RNA genes that are present in the spacer regions between *ribosomal RNA genes in *primary transcripts (→spacer DNA).

Spiegelmers Functional mirror-image RNA or DNA *oligonucleotides that are able to bind with high affinity and specificity to a given target molecule conceptually similar to antibodies that

aptamer · natural target

mirror-image · mirror-image
target   aptamer

mirror-image · aptamer
target

Spiegelmer · natural target

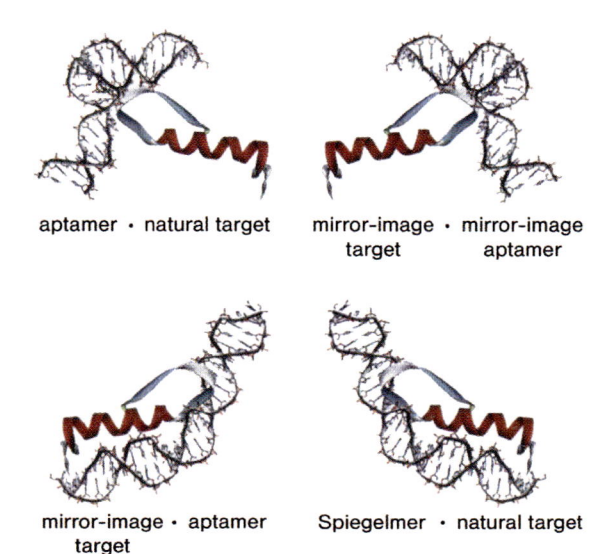

recognize antigens; Spiegelmers are also referred to as "mirror-image aptamers" (→aptamer). Spiegelmers can be identified through an adapted *SELEX process using chiral principles. (German: *Spiegel* = mirror.)

Due to the fact that Spiegelmers are totally composed of L-nucleotides (containing the non-natural sugars L-ribose for L-RNA and L-deoxyribose for L-DNA, respectively) and nature did not evolve appropriate enzymes to "handle" mirror-

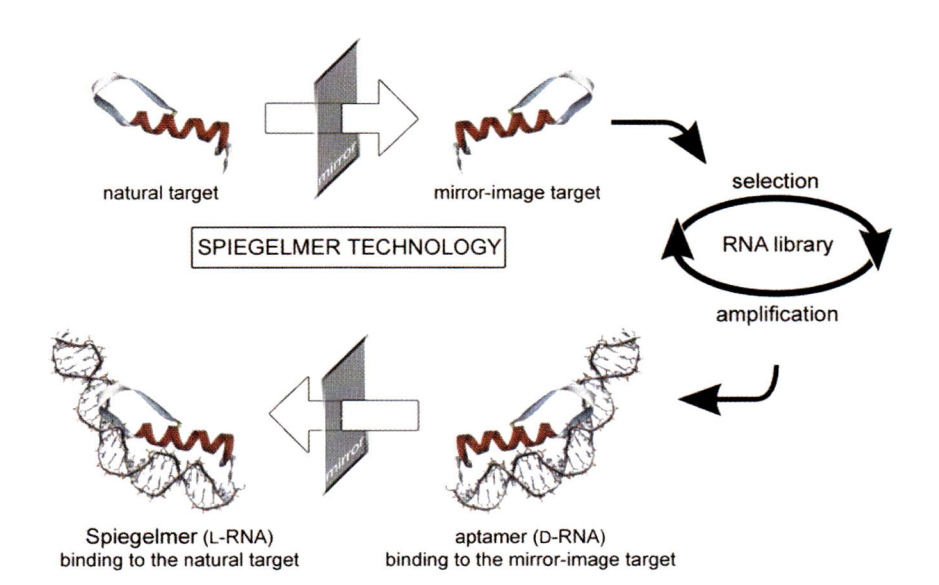

natural target → mirror-image target → selection

SPIEGELMER TECHNOLOGY

RNA library

amplification

Spiegelmer (L-RNA)
binding to the natural target

aptamer (D-RNA)
binding to the mirror-image target

image *nucleotides, Spiegelmers display an ultra-high biostability. However, the lack of Spiegelmer modifying enzymes prevents Spiegelmers from being introduced directly into the SELEX process. Active Spiegelmers can only be identified by introducing chiral inversion steps into the SELEX-process. Based on the fact that an aptamer binds its natural target, the mirror image of that aptamer will bind identically to the mirror-image of the natural target (mirror-image complex). This phenomenon is referred to as reciprocal chiral specificity. Therefore, if the process of aptamer selection is carried out against the mirror-image target, an aptamer against this unnatural mirror-image target will be obtained. The corresponding mirror-image nucleic acid (L-oligonucleotide) of the selected aptamer, the so-called "Spiegelmer", then binds to the natural target (i.e. the mirror-image of the mirror-image target) with similar binding characteristics as the aptamer itself.

Apart from their excellent biostability Spiegelmers seem to be non-immunogenic as well as non-toxic. For several different types of targets such as small molecules, peptides and proteins Spiegelmers have been identified. The efficacy of Spiegelmers in animal models also demonstrates their potency *in vivo*, underlining their potential as a new substance class for drug development. SVEN KLUSSMANN

spine of hydration Ordered hydrogen-bonded network of tightly arranged water molecules, partially substituted by Na$^+$ ions, in the *minor groove of *B-DNA.

spin label Small and stable organic molecules that carry an unpaired electron. These are covalently attached to the RNA to make the RNA accessible to *electron paramagnetic resonance spectroscopic methods or to quench those regions of a *nuclear magnetic resonance spectroscopy spectrum which are close to the spin label. Commonly used spin labels are nitroxides, because the unpaired electron is localized in the NO group and they are thermodynamically and kinetically stable. Therefore, they can be handled without problems in aqueous buffer solution and air, and can be stored in the frozen state for months without degradation. To be able to covalently attach nitroxides to RNA, they are derivatized with a *isocyanate, acetylene, iodoacetamide, acetylene, iodoacetamide or methanethiosulfonate group. Many of these nitroxides or their precursors are commercially available. Spin labeling is most powerful if it is done in a site-directed fashion, because then it is known where the spin label is attached. If the RNA is small enough to be synthesized on an automated RNA synthesizer it is possible to incorporate a labeled *phosphoramidite at the required position into the RNA (→oligonucleotide labeling, →oligonucleotide synthesis). In another approach *nucleotides with modifications that react selectively with a spin label are introduced into the RNA during the automated synthesis and than reacted with a specific label. Three examples are shown in the figure on next page.

Larger RNAs may be labeled by incorporating a thiobase enzymatically at the 3'-end and to subsequently let the thio group react with an iodoacetamide nitroxide. The 5'-end can be labeled after enzymatic elongation of the RNA with a nucleotide carrying a thiophosphate group which is than reacted with a iodoacetamide label. Labeling longer RNAs in the middle of the sequence is achieved by synthesizing a small RNA of the respective sequence, to label this and then to ligate this labeled RNA piece to the rest of the

Examples of different spin labels and the corresponding modifications on the RNA to conjugate them in a site-directed fashion: (a) thiouridine and a methane thiosulfonate nitroxide, (b) 2′-aminosugar and a isocyanate nitroxide and (c) 5-ioduridine and an acetylene nitroxide.

RNA sequence (→ligation). Generally, care should be taken that the label does not disturb the folding of the RNA (→RNA structures). In this respect it is very useful to be able to chose from a box of labels that bind at different sites of the RNA and where some are attached via a rigid linker to the RNA, which is good for PELDOR measurements (→electron paramagnetic resonance), whereas others are connected via

more flexible linkers, which might be better in the case of complicated *RNA folds.
OLAV SCHIEMANN

spliceosomal proteins → spliceosomal RNP, → spliceosome

spliceosomal RNA Group of *small nuclear RNAs (snRNA) involved in *pre-mRNA *splicing. This group consists of five *U-snRNA species (U1-, U2-, U4-, U5- and U6-snRNA). These U-snRNAs are of 106–187 nucleotides in length and share a common uridylic acid-rich Sm motif for binding of spliceosomal proteins. With the exception of U6-snRNA (synthesized by *RNA polymerase III), the other spliceosomal RNA molecules are synthesized by RNA polymerase II. Polymerase II transcribed U-snRNAs receive a special *cap structure, a *trimethylguanosine cap. In contrast, the 5'-end of RNA polymerase III U6-snRNA (like that of *7S K RNA) is modified by a *γ-monomethyl phosphate. Spliceosomal RNAs are fairly abundant and in mammals accumulate to about 10^6 copies per cell. In their functional state, spliceosomal RNAs are associated with a specific set of proteins, forming the *spliceosomal RNPs.
BERND-JOACHIM BENECKE

spliceosomal RNP *Small nuclear ribonucleoprotein (snRNP) particle composed of one *spliceosomal RNA and a set of 10 (U1-snRNP) to 19 (U2-snRNP) proteins. Those proteins range in size from 9 to more than 200 kDa. Basically, two groups of snRNP proteins can be distinguished. One group consists of seven Sm core proteins specifically recognizing the Sm motif of the spliceosomal RNA. The second group is formed by three to 12 additional proteins, specific for each snRNP particle. In their functional status, snRNPs may be monomeric (U1- and U2-snRNP) or may form hetero-oligomeric structures, such as the tri-*snurp of the U4/U6.U5-snRNPs.
BERND-JOACHIM BENECKE

spliceosome Abundant *ribonucleo protein complex that catalyzes the *splicing reaction of *pre-mRNA. The term spliceosome refers to the "major spliceosome" that removes U2-type *introns, as opposed to the *minor spliceosome removing U12-type introns (→atac splicing). The spliceosome is formed by the step-wise recruitment of *U-snRNPs to the *splice sites of pre-mRNA. In the first step, the U1-snRNP binds to the 5'-splice site of an intron via U1-snRNA/pre-mRNA base pairing. The resulting E-complex sometimes is designated as the "commitment complex" since from now on the pre-mRNA must go through the entire splicing reaction. In the second, this time ATP-dependent step, the U2-snRNP is bound to the *branch point region, via base pairing near the 3'-end of the intron. The resulting pre-spliceosome is called the A-complex and is characterized by spatial vicinity of the 5'- and 3'-splice sites, mediated by protein–protein interactions between the U1- and U2-snRNPs. Subsequently, the 25S tri-*snurp complex (consisting of the U4/U6.U5-snRNPs) enters the pre-spliceosome and builds the mature spliceosome. During spliceosome assembly, a variety of conformational changes takes place which require ATP consumption. Mostly, these conformational changes involve alternative systems of RNA base pairing (→edge-to-edge interaction of nucleobases) and, finally, result in C-complex formation which is the catalytically active form of the spliceosome (→ribozyme). Following *catalysis, the spliced *messenger RNA and the intron (in the form of a *lariat that is degraded subsequently) are released. The spliceosome disassembles with the free U-snRNPs being available for a new round of spliceosome assembly.

The two subsequent *transesterification reactions of the splicing reaction depend on an intricate catalytic RNA network. The 5′-splice site is defined by initial base pairing with U1-snRNA. Later, that interaction is replaced by base pairing with U5- and U6-snRNA. At the 3′-splice site the branch point A nucleotide is bulged out (and thereby activated) by base pairing of U2-snRNA to the adjacent *nucleotides of the pre-mRNA. In the tri-snurp complex, extensive base pairing exists between U4- and U6-snRNA. Prior to the first transesterification step, that interaction is broken and replaced by U6/U2-snRNA base pairing which constitutes an essential catalytic step. U5-snRNA has been shown to interact with pre-mRNA sequences both at the 5′- and 3′-splice sites. In summary, these RNA–RNA interactions prevent a premature release of both *exon ends and support properly directed introduction of the intermediates into the catalytic site of the spliceosome.

Aside from the integrated compounds of the snRNPs, another group of factors is required for spliceosome assembly. These 'non-snRNP spliceosomal factors' have been found by biochemical fractionation and complementation analyses during *in vitro* splicing reactions. As yet, however, these factors are much less well defined. BERND-JOACHIM BENECKE

splice site *Conserved sequence at the *intron–*exon boundary. The 5′-splice site (5′-end of the intron) is marked directly by U1-snRNA (→U-snRNA) via base pairing. The 3′-splice site (3′-end of the intron) is marked indirectly by binding of U2-snRNP to the *upstream *branch point (also via base pairing) and binding of the U2AF (U2 auxiliary factor) protein to the poly-*pyrimidine track between branch point and 3′-splice site. Some 3′-splice sites are "weak", meaning that U2-snRNP and U2AF binding is inefficient. In these cases additional proteins are required for efficient splicing. In particular, the *SR (serine/arginine-rich) proteins are involved. These SR proteins specifically recognize and bind to intronic or exonic splicing-enhancer sequences. BERND-JOACHIM BENECKE

splicing →RNA splicing

splicing mutation *Mutation within the sequence of *introns or *exons that prevents *splicing and thus may prevent protein biosynthesis. A splicing mutation can be (a) the mutation of a *splice site resulting in loss of function of that site, which may lead to a premature *stop codon, loss of an exon or inclusion of an intron; (b) the mutation of a splice site reducing specificity, which may result in variation in the splice sites, causing insertion or deletion of amino acids, or most likely a loss of the *reading frame; (c) the *transposition of a *splice site, resulting in inclusion or exclusion of more DNA than expected, which leads to longer or shorter hybrid exons. STÉPHANIE VAULÉON

splint ligation Enzymatic method for joining two nucleic acid *fragments. To achieve efficient ligation, the two fragments are hybridized with a short *complementary DNA *template (splint) that bridges the ligation site and thus pre-orients the two ends to be ligated (→DNA ligase, →ligase, →ligation). SABINE MÜLLER

split genes →mosaic genes

srp RNA →signal recognition particle, →7S L RNA

SR (serine/arginine-rich) proteins Regulatory proteins involved in *alternative splicing. SR proteins form a family of about 10 different proteins with a molecular mass between 20 and 70 kDa. The proteins contain two functional domains. The *ribonucleoprotein domain contains a RNA recognition motif (→RNA-binding proteins) and forms the RNA-*binding site. The RS domain is a cluster of arginine and serine residues and may be highly phosphorylated by specific *kinases. By binding to exonic and/or intronic *splicing enhancers (or silencers), SR proteins constitute an important tool for alternative splicing reactions. BERND-JOACHIM BENECKE

SSB →single-strand break

ssDNA short for single-stranded DNA

ssRNA short for single-stranded RNA

stacking →base stacking

start codon Starting point of protein *translation which defines the first amino acid to be incorporated into the growing polypeptide. Usually this is an AUG *codon, while CUG and UUG (also GUG and AUU in eukaryotes as well) operate at much lower frequency (in *Escherichia coli*: 90% AUG, 8% GUG, 2% UUG). All of them are decoded by the same *initiator tRNA (tRNA$_i$), a specialized tRNA^{Met} with a CAU *anticodon in prokaryotes becoming formylated after aminoacylation (→formylmethionyl-tRNA). Wobble base pairing (→wobble base pair) is crucial to do so and variation of the start codon is one strategy to regulate the level of protein biosynthesis (→translational control).

A start codon like AUG is necessary, but not sufficient, for *initiation of *translation – additional sequences *upstream of it are required to be present as well. Like in bacteria, the *purine-rich *Shine–Dalgarno sequence (e.g. AGGAGG) around 10 nucleotides upstream of the start codon allows for stabilizing RNA–RNA interactions with a *complementary sequence in the 30S *ribosomal subunit and subsequent ribosomal assembly. In eukaryotes the distance to the *messenger RNA 5'-cap (→cap structure) is one parameter for start codon identification as well as interaction with initiation *translation factors. This is different to the situation at a *stop codon and allows use of AUG as the only Met codon during *elongation as well (→ribosome). STEFAN VÖRTLER

stem–loop structure Structural motif that is formed when a single-stranded nucleic acid molecule loops back on itself to form a complementary double helix (stem) closed by a *loop. It is a very common secondary motif in RNA, mostly with the closing loop being a *tetraloop. It is also called a hairpin or *hairpin loop. SABINE MÜLLER

sticky ends →cohesive ends

stop codon Signal to stop the *translation of proteins. The universal *genetic code contains three *codons not assigned to a specific amino acid: the UAG (→amber), UGA (→opal) and UAA (→ochre) codons, identified by classical genetics after mutagenic treatment causing premature *termination. Although the nature of neighboring *nucleotides seems to affect the strength of the termination, a stop codon is as such sufficient to abort protein biosynthesis. Spontaneous *mutations therefore carry the danger of introducing a stop codon in the middle of a *coding sequence, prohibiting correct *gene expression (→nonsense mutation). Intriguing pathways insure rapid *messenger RNA (mRNA) degradation [termed nonsense-mediated decay (NMD)] parallel to the case of the degradation of incomplete in RNAs, which lack a stop codon at the 3'-end at the end of a mRNA. In the case of essential proteins cells have a rescue system based on *suppressor mutations like *suppressor transfer RNA (tRNA) to produce at least minimal amounts of protein and insure viability. They are derived from functional tRNA, but carry mutations in the *anticodon loop to allow *decoding of a stop codon. Their cellular concentration is too low to affect the correct termination at the end of an mRNA (particularly as several stop codons are present in the 3′-untranslated region (→ribosome). STEFAN VÖRTLER

strand polarity The two strands of DNA have antiparallel chemical polarity due to the opposite direction of their *sugar–phosphate backbones. The 5′→3′ direction has been assigned positive polarity, the 3′→5′ strand negative polarity. BETTINA APPEL

streptavidin Protein (53–60 kDa) composed of four subunits that binds *biotin with high affinity ($K_d \sim 10^{15} M^{-1}$). This strong interaction can be exploited in the lab, e.g. for *immobilization of nucleic acids carrying a biotin tag. Due to its lack of carbohydrate moieties and its isoelectric point being close to neutrality, streptavidin displays lower unspecific binding than, for example, avidin, which makes it especially suitable for laboratory use (→immobilization of nucleic acids). JÖRN WOLF

stripping Protocol that allows removal of a *nucleic acid probe or an antibody from a membrane used in a *Southern, *Northern or Western blot so as to re-use it with a new probe or antibody. GEMMA MARFANY, ROSER GONZÀLEZ-DUARTE

structural genes *Genes that encode any RNA or protein product other than a regulatory element.

structure mapping Approach to map the accessibility of *nucleotides of nucleic acids towards enzymes and chemicals. The method defines a set of single-stranded and double-stranded regions of nucleic acids. These data, combined with computer prediction and phylogenetic approaches, can derive *secondary structure models of RNA molecules (→chemical and enzymatic structure mapping of RNAs, →*in vivo* structure mapping of RNA). PASCALE ROMBY, PIERRE FECHTER

S-turn Common *RNA motif (also called Loop E motif) that is characterized by a number of *non-canonical *base pairs inducing a sharp bend in RNA helices. S-motifs have been also found to stabilize *internal *loops and *junction loops in *ribosomal RNAs. BETTINA APPEL, SABINE MÜLLER

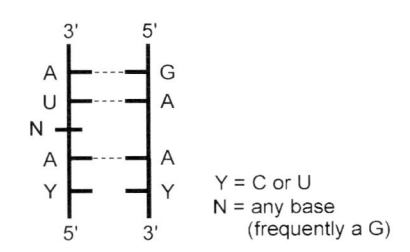

```
   3'        5'
A ┤-----├ G
U ┤     ├ A
N ┤     ├
A ┤-----├ A
Y ┤     ├ Y       Y = C or U
                  N = any base
   5'        3'   (frequently a G)
```

substitution mutation *Mutation caused by the replacement of one or more DNA (RNA) residues by an equal number of different residues (→deletion mutation, →insertion mutation). HANS-JOACHIM FRITZ

subunits of the ribosome →ribosome

succinimidyl ester Organic molecule containing a reactive N-hydroxysuccinimidyl ester that can be reacted with *oligonucleotides containing aliphatic amino groups. This reaction connects the organic

Reaction of succinimidyl ester with amine.

molecule to the oligonucleotide via an amide linkage. SNORRI TH. SIGURDSSON

sugar conformation In general, the conformation depends on the torsion angles for rotation around each bond of the sugar ring. *Ribose and *deoxyribose are five-membered rings and five torsion angles can be specified accordingly (→sugar pucker, →pseudorotation). KLAUS WEISZ

sugar edge →edge-to-edge interaction of nucleobases

sugar–phosphate backbone Backbone of nucleic acids consisting of alternating phosphate and pentose residues; the characteristic *bases may be regarded as side groups joined to the backbone at regular intervals. The *nucleotide units of both DNA and RNA are covalently linked through phosphate-group bridges. Specifically, the 5′-hydroxyl group of one nucleotide is joined to the 3′-hydroxyl group of the next nucleotide by a *phosphodiester linkage. The sugar–phosphate backbone of nucleic acids is hydrophilic. All the phosphodiester linkages in nucleic acid strands have the same orientation along the chain, giving each linear nucleic acid strand a specific polarity (→strand polarity), and distinct 5′- and 3′-ends. The conformational state of the sugar–phosphate backbone in DNA *secondary structure is characterized by the values of six torsion angles (→glycosidic torsion angle, →pseudorotation, →sugar conformation). NINA DOLINNAYA

sugar pucker As it is sterically and energetically unfavorable, a five-membered furanose ring will not be planar. It is rather puckered by pulling one or two atoms out of the plane to give envelope or twist forms. Sugar puckering modes can be characterized by the atoms displaced from the plane that is defined by three or four atoms. The two conformations with the 3′-carbon and the 2′-carbon out of the plane and on the same side as the base are called C3′-*endo* and C2′-*endo*, respectively. The sugar pucker in a five-membered ring can also be characterized by the pseudorotation phase angle P (→pseudorotation, →N-conformation, →S-conformation). KLAUS WEISZ

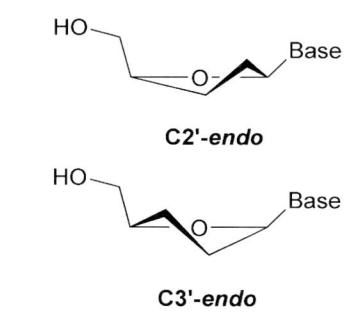

C2′-*endo*

C3′-*endo*

supercoil →supercoiled DNA

supercoiled DNA Compact state of closed circular *double-stranded DNA (dsDNA) (analogous to tertiary structure in proteins). Some natural DNAs exist as cyclized *macromolecules with no free 5′ and 3′-ends. Typically, these DNAs are present in *plasmids, bacteria and certain viral *genomic DNA. A phenomenon called supercoiling occurs with these molecules. Supercoiling is subject to some form of structural strain as a result of underwinding of the DNA in the closed circle. In other words, there are fewer *helical turns in closed-circular DNA than would be expected for the B-form structure (→B-DNA). The strain can be accommodated in two ways. (a) The two strands can simply separate over the distance corresponding to relaxed state (→relaxed DNA). (b) The DNA can form the supercoils. The resulting strained state of the DNA represents a form of stored energy. The underwound state can be maintained only if the DNA is a closed circle or if it is bound and stabilized by proteins such that the strands are not free to rotate about each other. If there is the break in one of the strands of *circular DNA, free rotation at that point will cause the underwound DNA to revert spontaneously to the relaxed state. The *linking number, Lk, of a DNA molecule specifies the number of *helical turns in a closed-circular DNA. Lk is a topological property because it does not vary when dsDNA is twisted or deformed in any way, as long as both DNA strands remain intact. Length-independent quantity, *superhelical density, σ, equal to $\Delta Lk/Lk_0$, where ΔLk is the change in linking number, and Lk_0 is the linking number in *relaxed DNA, measures the turns removed relative to those present in relaxed DNA. In cellular DNAs σ generally falls into the range of -0.05 to -0.07. The negative sign denotes that the change in linking number comes about as a result of underwinding the DNA. The supercoiling induced by underwinding is defined as negative supercoiling. Conversely, under some conditions DNA can be overwound and the resulting supercoiling is defined as positive. Two forms of a given circular DNA that differs only in a topological property such as Lk are referred to as topoisomers. Lk for the closed-circular DNA is always an integer. Lk can be broken down into two structural components called writhe (Wr) and twist (Tw). Wr may be thought of as a measure of the coiling of the helix axis and Tw as determining the local twisting or spatial relationship of neighboring *base pairs ($\rightarrow$twisting number, $\rightarrow$writhing number). When a change in *linking number occurs, some of the resulting strain is usually compensated by writhe (supercoiling) and some by changes in twist, giving rise to the equation: $Lk = Tw + Wr$. Twist and writhe are geometric rather than topological properties. They need not to be integers.

The degree of coiling is affected by various small molecules, particularly those that intercalate between the bases ($\rightarrow$intercalation). Underwinding DNA facilitates a number of structural changes in the molecule. Strand separation occurs more readily in underwound DNA. This is critical to the processes of *replication and *transcription, and represents a major reason why DNA is maintained in an underwound state. Underwinding a *DNA helix facilitates the formation of non-canonical conformations which decrease the structural strain: *Z-DNA, DNA *cruciform and *H-DNA. Supercoiling was shown to play a key role in the structural dynamics of nucleosomes. The enzymes that increase or decrease the extent of DNA underwinding are called *topoisomerases and the property of DNA they affect is Lk. These enzymes play an especially important role in processes such as replication and DNA packaging. There are two classes of topoisomerases. Type I topoisomerases act by transiently breaking one of the two DNA strands, rotating one of the ends about the unbroken strand and rejoining the broken ends; they change Lk in increments of 1. Type II topoisomerases break both DNA strands and change Lk in increments of 2. DNA topoisomerases are the enzymes responsible for maintaining the topological state of cellular DNA. These ubiquitous enzymes are involved in essential cellular processes, including transcription, *recombination, *chromosome condensation and segregation. NINA DOLINNAYA

super helix Superstructure adopted by DNA that allows for tight packing of DNA as a solenoid in the nucleus. Also, a superhelical *conformation can be adopted in order to react to superhelical tension caused by any protein, e.g. a *RNA polymerase, that moves along the strands ($\rightarrow$DNA superstructure, →super helix density). Jörn Wolf

super helix density (= specific linking difference) Quantity that defines to what extent a circular molecule is supercoiled (→supercoiled DNA). In the case of *circular DNA, it refers to the number of supercoils per 10 bp. Mathematically it can be described by the equation: $\sigma = (Lk - Lk_0)/Lk_0$, where σ is the super helix density, Lk_0 is the *linking number of the relaxed circular molecule and Lk is the linking number of the supercoiled molecule. For DNA, σ is usually negative (→DNA superstructure, →super helix density, →linking number). Jörn Wolf

suppression Occurrence of changes that eliminate the effects of a *mutation. Most mutants can undergo a second mutation that restores the original *phenotype. If the mutated *base pair is converted back to the original base pair, the process is called a reversion. If a mutation at a different site restores the original phenotype, it is called a suppression or second-site reversion. Suppression can occur within a *codon, intragenically or extragenically. An intragenic suppression may occur at a different codon, leading to an amino acid which restores the structure of the encoded protein impaired by the original mutation. Extragenic suppression is, for example, a mutation in a *transfer *RNA (tRNA) gene that enables the encoded tRNA to decode the *stop codon introduced by the original mutation. Beatrix Süß

suppressor gene *Gene that suppresses the phenotypic expression of another gene, particularly of a mutant gene.

suppressor mutation Classical genetic term defining a second *mutation that reverts the effect of the initial mutation. Spontaneous mutation of a sense to a *stop codon in an *open reading frame can be reverted by *suppressor tRNA, but as well by mutations in *ribosomal RNA, *termination factors or ribosomal proteins like S4, S5 and S12. Generally, in the latter cases, either hyperaccurate or error-prone ribosomes are generated. Stefan Vörtler

suppressor tRNA Specific *transfer RNA (tRNA) able to decode a *stop codon and to suppress *termination of protein biosynthesis (→translation) due to a *suppressor or second site mutation. It is a cellular rescue system counteracting initial *mutations of *sense codons to UAG, UGA or UAA stop codons in *open reading frames. Many suppressor tRNA are derived from *elongator tRNA by mutation of the *anticodons. Like su2, the most common suppressor in *Escherichia coli* derived from $tRNA_2^{Gln}$ by G36A mutation of the CUG anticodon now inserts Gln at UAG instead of the CAG codon. The level of suppressor tRNA is very low, just enough to support *translation of essential *genes, but not to interfere with appropriate termination at the end of translation, particularly as several stop codons are usually found at the 3'-untranslated region of mRNA. Increasing the suppressor tRNA level by overexpression or addition to *in vitro* translation systems is an important strategy in biotechnology to site-specifically introduce modified and unnatural amino acids. Stefan Vörtler

symmetric base pair A *base pair is called symmetric if a given base pairs with its most predictable pairing partner in a Watson–Crick-like fashion. Thus, *adenosine prefers to make *hydrogen bonds with *thymidine, *guanosine prefers to bind *cytidine (→Watson–Crick base pair). Matthäus Janczyk

syn conformation The *syn* conformation corresponds to a *glycosidic torsion angle κ in the range $0 \pm 90°$. The *syn* conformation can be favored by attaching a bulky group at the 8 position of *purines or the 6 position of *pyrimidines producing even more steric hindrance in the *anti* conformation. The *syn* conformations in *nucleotides are usually found with $\kappa = 45 \pm 45°$ ($\rightarrow$ glycosidic torsion angle). KLAUS WEISZ

synteny Conservation of the order of *genes between different species.

T

T →thymine

Tac promoter *In vitro* engineered *promoter, which consists of the –10 region of the *Lac promoter and the –35 element of the Trp promoter. In microbiology, the Tac promoter is often used to achieve a high expression yield of recombinant proteins in *Escherichia coli*. The presence of an additive *Lac operon region allows for the regulation of expression via isopropyl-β-D-thiogalactopyranoside.
IRENE DRUDE

TAFs Subunits of the eukaryotic *transcription factor TF$_{II}$D that assist the *TATA-binding protein in DNA binding and recruits other factors to the *transcription apparatus.
FRANZ NARBERHAUS

tandem repeat DNA sequences that are duplicated and joined head-to-tail along the DNA chain. The human *genome contains a high proportion of such sequences consisting of about 9–70 bp. These regions are called variable number of tandem repeats. The number of repeats varies between individuals from only a few copies to hundred of such copies. The *dinucleotide repeat such as the CA repeat is very common in the human genome and can be amplified by the *polymerase chain reaction and analyzed on polyacrylamide gels. In this way, individuals are distinguished by comparing the length of the amplified DNA *fragments. The CA repeat is also called *microsatellite DNA. TINA PERSSON

***Taq* DNA polymerase** *DNA-dependent DNA polymerase from the thermophile bacterium *Thermus aquaticus*. Due to its thermostability, *Taq* polymerase is often used for *polymerase chain reaction (PCR). The temperature optimum for the *Taq* polymerase is 72°C. *Taq* polymerase is able to amplify 1 *kb of DNA in 30 s. One of the drawbacks of *Taq* polymerase is the lack of *proofreading because of the absence of a 3'-*exonuclease activity. About 2×10^{-4} errors per *base are introduced into a DNA during a replication cycle. *Taq* polymerase adds an additional *adenine at the 3'-end, leading to polymerase chain reaction products with an adenine *overhang. As a result of this property, *Taq* amplified DNAs are very attractive for *cloning into a *vector with *thymidine overhang. IRENE DRUDE

TaqMan probes Dual-labeled hydrolysis probes (→nucleic acid probe) that utilize the 5'-*exonuclease activity of the *Taq* DNA polymerase for measuring the amount of target sequences in samples. TaqMan probes consist of 18–22 *nucleotides labeled with a fluorophore at the 5'-end and a quencher at the 3'-end (→oligonucleotide labeling). The TaqMan probe complementary to the target sequence is added to the *polymerase chain reaction (PCR) mixture. Whilst the probe is bound to the *template the fluorophore and quencher are in close proximity, and therefore there is limited fluorescence (the fluorescence is not completely quenched so there is background fluorescence). During polymerase extension the 5'-exonuclease

activity of the polymerase cleaves the probe thus releasing the fluorophore away from the vicinity of the quencher. As a result the fluorescence intensity of the reporter dye increases (→FRET). This process repeats in every cycle and does not interfere with the accumulation of PCR product.
DAVID LOAKES

T-arm *TΨC stem–loop of a *transfer RNA.

TAR RNA RNA structural motif that is essential for the *replication of HIV. It serves as a *recognition site for many *ligands, e.g. arginine, arginine amide, *amino glycosides, and other non-peptides as well as peptides. In HIV, the 59-nucleotide TAR RNA motif can be found at the 5′-end of all nascent *transcripts. It has two characteristic features. Its *bulge region interacts with the arginine and lysine-rich region of a protein called tat (trans activator protein). The other characteristic region, a *loop region, is involved in *transcription activation (→RRE RNA). JÖRN WOLF

TATA-binding protein (TBP) Subunit of the eukaryotic transcription factor $TF_{II}D$ that binds to the *TATA-box.

TATA-box Conserved AT-rich sequence about 25 bp *upstream of the *transcription start site of eukaryotic *genes transcribed by *RNA polymerase II.

tautomeric bases In the *nucleobases guanine, *uracil and thymine, the keto and enol form are in tautomeric equilibrium; in the nucleobases *adenine and *cytosine, the amino and imino form are in tautomeric equilibrium. Over 99.99% of the bases of *nucleosides exist in the keto or amino form, respectively. Tautomeric forms of the nucleobases may, if occurring during *replication, induce *mismatches and in the worst scenario lead to *mutations.
BETTINA APPEL

TBDMS method Allows for the chemical solid-phase synthesis of *oligoribonucleotides and is named after the nucleoside 2′-hydroxyl protection – the tert-butyldimethylsilyl (TBDMS) group. The method follows the *phosphoramidite method for the coupling of appropriately protected *nucleoside building blocks and was developed in the 1980s. The protection strategy employs two orthogonal protecting groups – an acid-labile (4,4′-dimethoxytrityl) group on the 5′-O-position and the fluorine ion-labile silyl ether on the 2′-O-position. The 3′-O-position is derivatized with a 2-cyanoethoxy-(N,N-diisopropylamino)phosphinyl residue as commonly used in automated DNA solid-phase synthesis (→oligonucleotide synthesis). The originally used *nucleobase protecting groups were N^6-benzoyl for *adenosine, N^4-benzoyl or acetyl for *cytidine, and N^2-isobutyryl for *guanosine. Advanced amino protecting groups such as N-phenoxyacetyl (pac) or N-tert-butylphenoxyacetyl groups (tac) enable very mild *deprotection of the RNA. The use of potent activators, such as 5-(benzylthio)-1H-tetrazole (BTT) for the actual coupling step increases coupling efficiencies and rates. These improvements are responsible for holding the TBDMS method its own and for still having the widest application beside the recently developed *TOM method and *ACE method. The structures of TBDMS nucleoside phosphoramidites are depicted above.

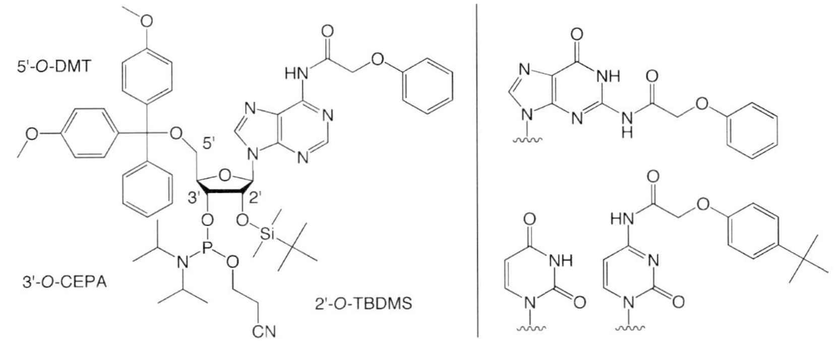

Oligonucleotide assembly proceeds with coupling yields higher than 98.5% and with coupling rates slightly slower than the other methods mentioned above. After strand assembly, the detachment of the sequence from the solid support and the removal of the acyl *nucleobase and *phosphodiester *protecting groups are carried out under basic nucleophilic conditions, preferably with a 1:1 mixture of concentrated aqueous ammonia and 8 M ethanolic methylamine. Importantly, when using methylamine, cytidine must not be protected with N^4-benzoyl, otherwise substitution at the carbon in position 4 is observed. Removal of the remaining 2'-O-TBDMS protecting groups is preferably achieved by using triethylamine tris(hydrofluoride) in N-methyl pyrrolidone. RONALD MICURA

T-DNA (transferred DNA) Part of the *Ti plasmid in the virulent Agrobacterium tumefaciens. Upon infection of a plant cell this DNA part is transferred to and stably integrated into the *genome of the host causing permanent expression and proliferation into tumors. SABINE MÜLLER

TDP Abbreviation for thymidine diphosphate (→nucleoside phosphates).

tecto nucleic acids Assembly of RNA molecules held together by loop–loop interactions. Tecto-RNA molecules are defined as artificial RNA molecules able to *self-assembly to form nanoscale RNA objects (→DNA nanotechnology, →DNA nanoarchitectures). BETTINA APPEL

tecto-RNA →tecto nucleic acids

tel-DNA →telomeres

telomerase Large *ribonucleoprotein (RNP) complex that universally adds specific DNA sequence repeats at the ends of *chromosomes. It is a *RNA-dependent DNA polymerase, an enzyme containing a RNA molecule as *template for the elongation of the G-rich strand. In principle, telomerase activity universally complements the *DNA polymerase and plays an important role to replicate the chromosomal ends ensuring full genetic integrity in actively dividing cells. The enzyme is essentially made up of two components. One is a functional RNA component that serves as a template for DNA synthesis and the other is the catalytic component with *reverse transcriptase activity.

During normal growth, telomerase activity is strictly regulated. There are reports indicating that telomerase activity is not regulated during the cell cycle, but the activity is downregulated during the end of cell cycle and differentiation. Deregulation of telomerase expression or activity is often associated with cancer – telomerase activity has been observed in more than 90% of cancerous cells. In contrast, it has been reported to be absent in normal human somatic cells. Telomerase activity can be seen in a few cell types like germ cells or lymphocytes.

The deregulated activity of telomerase would lead to progressive loss of the telomeric cap with each cycle of cell division resulting in critically shortened *telomeres. The mechanisms behind telomerase activity regulation are not yet fully understood. RAJESH SINGH

telomeres DNA sequences bound by telomere-associated proteins (tel-DNA). They define the ends of *chromosomes. These caps function to protect chromosomes from degradation. Telomeres consist of several *tandem repeats of the GT-rich hexanucleotide sequence $(TTAGGG)_n$.

The length of telomeric DNA segments varies with chromosomes, species, type and age of a cell. It has been reported that in humans, telomeres end in a 3'-single-stranded *overhang of 75–300 *nucleotides. Telomeres are often associated with cancer and aging, since they protect chromosomes from *recombination and ensure the complete *replication of chromosomes during the cell division cycle.

There are reports mentioning that with each cell division cycle, telomeres are shortened by 50–200 bp. When telomeres become critically short, cells enter a phase of growth arrest often called cellular senescence (→telomerase). RAJESH SINGH

temperature-induced riboswitching
→RNA thermometer

template Molecule that acts as a mold for the synthesis of another molecule, thereby determining the structure of the synthesized molecule, e.g. a RNA serves as template strand for a DNA strand during *reverse transcription. VALESKA DOMBOS

template-independent DNA polymerase
Enzyme that catalyzes the *polymerization of *deoxynucleotides without using a *template. The template-independent DNA polymerase *terminal deoxynucleotidyl transferase catalyzes the repetitive addition of deoxynucleotides to the 3'-hydroxyl terminus of deoxyoligonucleotides, *single-stranded or *double-stranded DNA molecules. VALESKA DOMBOS

template-independent RNA polymerase
Enzyme that catalyzes in a *template-independent way the *polymerization of *nucleotides, like the poly(A) RNA polymerase which catalyzes the addition of *adenine monomers to the 3'-end of a *messenger RNA (→polyadenylation, →poly(A) tail). VALESKA DOMBOS

terminal deoxynucleotidyl transferase (TdT) *Template-independent enzyme which catalyzes addition of *nucleotides to the 3'-ends of DNA. This specialized *DNA polymerase needs cobalt as cofactor (→template-independent DNA polymerase). VALESKA DOMBOS

terminal labeling →end labeling

terminal loop →hairpin loop

terminal transferase →terminal deoxynucleotidyl transferase

termination One of the three steps of *translation in which a *stop codon leads to peptide/protein release and ribosomal recycling. Termination refers also to the process of *transcription. Transcription stops if a terminator sequence is formed in a protein-dependent or -independent manner. STEFAN VÖRTLER, SABINE MÜLLER

termination codon → stop codon

termination factor → translation factor

ternary complex Tripartite molecular ensemble, e.g. a *ribozyme–*ligand–substrate or ribozyme–substrate1–substrate2 complex.

tertiary structure Three-dimensional arrangement of secondary structure elements like helices, loops, *junctions, or *bulges found in *biomolecules.
SLAWOMIR GWIAZDA

tether → linker

tetracycline riboswitch *In vitro* selected, tetracycline-binding *aptamer that can be used for conditional *gene expression when inserted in untranslated regions of eukaryotic *messenger RNAs (mRNA). It consists of two main stems forming the scaffold. They are separated by a *bulge and a *loop which form the tetracycline-*binding pocket. Both regions cooperate and tetracycline binding reinforces an intramolecular connection within the *aptamer structure, thereby inhibiting binding of the *small ribosomal subunit when placed directly behind the *trimethylguanosine cap, interfering with formation of the 80S *ribosome for *start codon proximal insertion or inhibiting mRNA *splicing when inserted near the 5'-*splice site of an *intron. *In vitro*, one molecule of tetracycline is bound by one molecule of pre-structured apo-RNA with an exceptional low dissociation constant of 770 pM. BEATRIX SÜß

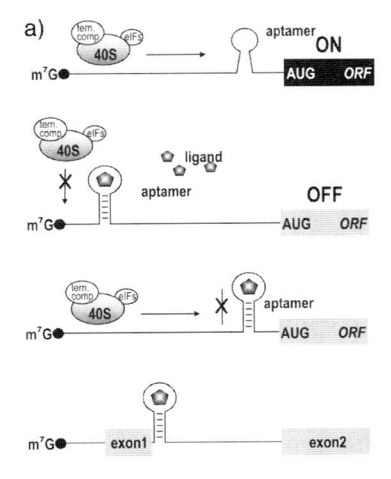

Model of tetracycline riboswitch-mediated control of gene expression. (a) Addition of the ligand tetracycline facilitates the formation of a tetracycline–aptamer complex which interferes with the binding of the 43S pre-initiation complex, the scanning or pre-mRNA splicing.

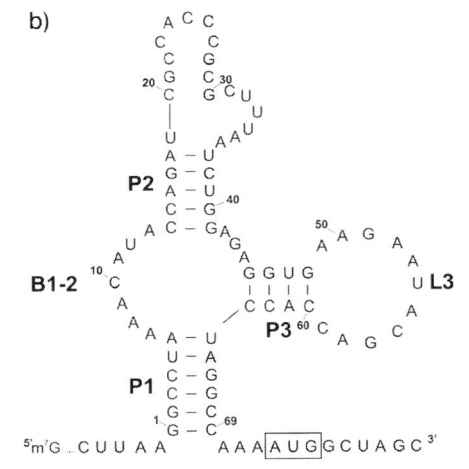

(b) Predicted secondary structure of the tetracycline riboswitch in the context of a mRNA (the start codon is boxed). Important elements are indicated as stem = pedestal (P), bulge (B) and loop (L). Positions determined as important for tetracycline binding are mainly located in B1–2 and L3.

Tetrahymena ribozyme *Ribozyme named after the ciliated protozoan *Tetrahymena thermophila* from which it was originally isolated in the 1980s by Thomas R. Cech. It was found in a *group I intron of a *ribosomal RNA *transcript and was the first ribozyme described. It consists of approximately 400 nucleotides and catalyzes its own excision through two *transesterification reactions. In the first splicing step, the 5'-*splice site *downstream of a GU *wobble base pair in helix P1 [formed by base pairing between the 5'-*exon and the *internal guide sequence (IGS) of the *intron] is cleaved by an exogenous guanosine lofted within the intron. After cleavage, helix P1 is displaced to allow formation of the P10 helix between the IGS and 4 nucleotides downstream of the 3'-splice site. In the second splicing step, the 3'-hydroxyl of the terminal *uridine of the 5'-exon is ligated to the phosphate upstream of the 3'-exon. A truncated form of this ribozyme without 5'-exon can *trans*-splice a sequence attached to the ribozyme 3'-end to a target mimicking the 5'-exon (→*trans*-splicing). This was used for developing RNA-based *RNA repair techniques. STÉPHANIE VAULÉON

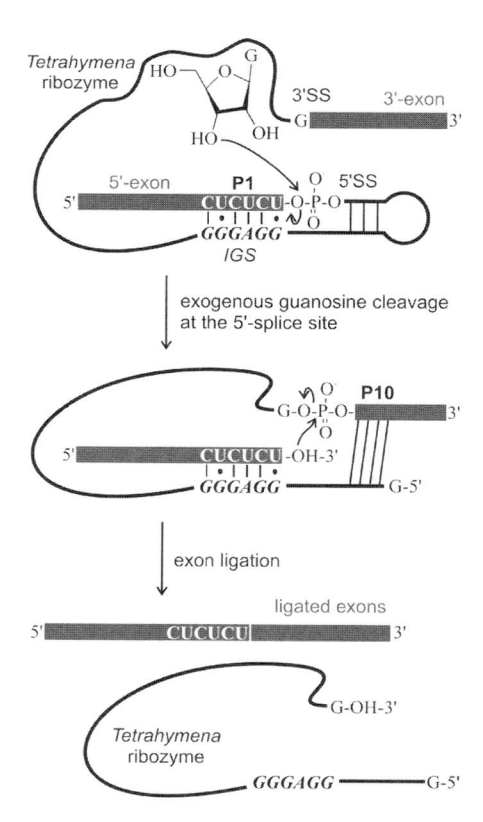

tetraloop Composed of four unpaired nucleotides and the most abundant of all RNA *loops. According to current estimates, more than 50% of all loops are tetraloops. Tetraloops are found to act as nucleation sites for *RNA folding, and are also important in RNA–RNA and RNA–protein recognition. Tetraloops are classified according to their nucleotide sequence. Three main classes of tetraloops are known; they have the sequences GNRA, UNCG and CUUG, and their thermodynamic parameters are given below.

The cUUCGg tetraloop is a representative of the YNMG motif. UNCG tetraloops are involved in long-range tertiary RNA–RNA interactions via the second unpaired uridine, which is able to bind an unpaired *adenosine nucleotide remote in sequence.

In the light of the three-dimensional structure the remarkable stability can be explained as follows. Nucleobases uracil 1 and guanine 4 form a *mismatch *base pair because guanine 4 adopts the *syn* conformation around the *glycosidic bond. This base pair is stabilized by *hydrogen bonds between the imino and amino group of guanine and the oxygen atom O2 of the uracil base [imino group: G4(N1)···H···(O2)U1, amino group and G4(N2)···H···(O2)U1]. Furthermore, the pairing is stabilized by an unusual hydrogen bond between the

Thermodynamic parameters for RNA tetraloops with different loop sequences

Sequence		T_m (°C)	ΔH (kcal mol^{-1})	ΔG (kcal/mol^{-1})
YNMG	UUCG	76.2	−55.9	−6.3
	UACG	73.8	−53.6	−5.7
Non-YNMG	GCUU	70.9	−45	−4.4
	UUUG	70.3	−44	−4.2
	UUUU	69.9	−44.3	−4.2

[1] In each case the remaining nucleotides are: 5'-GGACLLLLGUCC-3' (L = loop nucleotides listed in the table). Values are determined in 1 M sodium chloride, 0.01 M sodium phosphate, 0.1 mM EDTA at pH 7.

hydroxyl moiety of the *ribose ring of uracil 1 and the base oxygen O6 of guanine 4. Although the pair is classified as a mismatch, the *Watson–Crick edge of guanine 4 is completely hydrogen bonded to its counterpart uracil 1.

The base of the second nucleotide uracil 2 is flipped out and points towards the

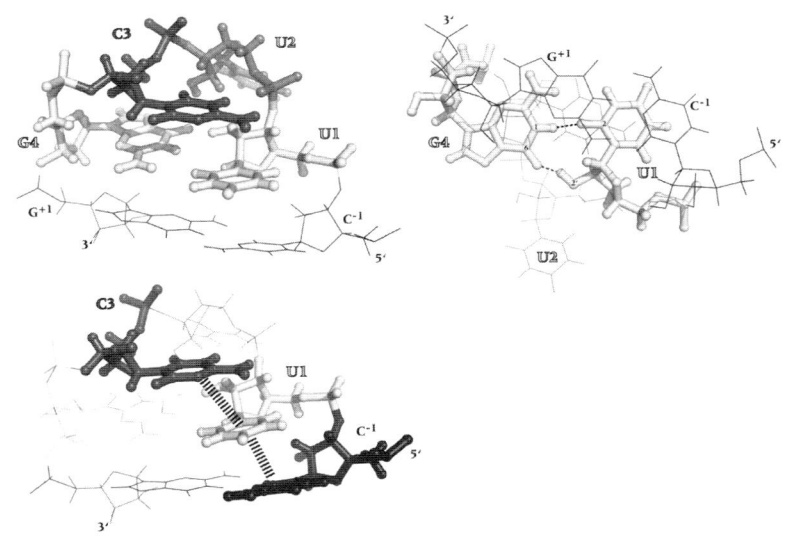

Three-dimensional structure of a cUUCGg tetraloop as revealed by X-ray crystallography (PDB code: 1F7Y): upper left: arrangement of the loop nucleotides, where the nucleotides U1 and G4 form a mismatch GU base pair (light grey), the bulged-out nucleotide U2 pointing into the solvent is colored in dark grey, the nucleotide C3 positioned above the mismatch base pair is shown in black grey, the closing base pair named C^{-1} and G^{+1} are indicated as black lines; upper right: bottom view of the cUUCGg tetraloop highlighting the mismatch between G4 and U1 that is stabilized by hydrogen bonds between G4(N1)···H···(O2)U1 and C3(O2')···H···(O2)C3, furthermore the bulged-out position of U2 becomes obvious; lower left: the favorable stacking interaction between the nucleotides C3 (dark grey), U1 (light grey) and C^{-1} is highlighted as indicated by the connecting lines.

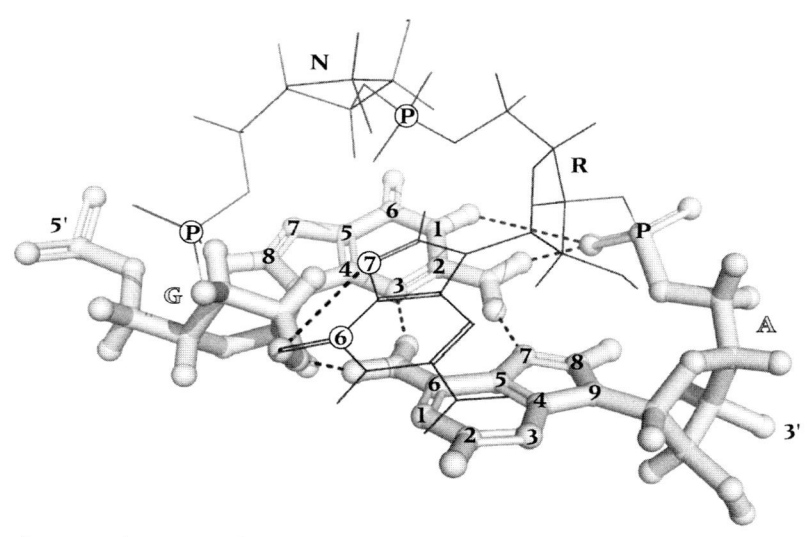

Three-dimensional structure of a GNRA tetraloop as revealed by NMR spectroscopy (PDB-Id.: 1ZIF, 1ZIG, 1ZIH): the hydrogen bonds that are formed by the nucleotides G1, R3 (in the depicted case G3) and A4 are indicated as dashed black lines; the nucleotides forming the base pair are shown in stick-ball representation, whereas the nucleotides N/R are shown as black lines; numbers are indicating atom nomenclature; P indicate phosphorous atoms in the backbone, the base of nucleotide N2 is omitted for clarity.

solvent. The *ribose moiety of this nucleotide and of the following nucleotide (cytosine 3) adopts the C2′-endo conformation and thereby facilitates the 180° turn of the *nucleic acid backbone.

Between the nucleobases of cytosine 3 and uracil 1, there is an additional *base stacking interaction that might be extended in the case of the cUUCGg tetraloops by additional stacking to the cytosine closing base. Furthermore, a side-on hydrogen bond between the amino group and one of the phosphate oxygens [C3(N4)···H···(^{proR}O)C3] and one between the hydroxyl group and the base [C3(O2′)···H···(O2)C3] stabilize the structural arrangement.

The stability of the other stable family of tetraloops described as containing the *consensus sequence GNRA of nucleotides can also be explained by the three-dimensional conformation.

In GNRA tetraloops, a network of heterogeneous hydrogen bonds is found as the central structural feature of these *secondary structure elements. An extensive stacking of nucleobases within the loop furthermore complements these intramolecular interactions between several hydrogen bond donor and acceptor sides. As revealed by high-resolution *nuclear magnetic resonance (NMR) spectroscopy structures of three different GNRA tetraloops, the major change in direction of the molecule is made between the first and the second nucleotide, and therefore results in an asymmetric loop structure where the first nucleotide, i.e. the nucleotide G, stacks on the 5′-side of the stem and nucleotides 2–4, i.e. N, R and A, stack onto the 3′-side of the helical stem. All of the nucleotides are found in the *anti conformation and so a sheared anti–anti GA base pair (→sheared base pair) between the first and the last loop nucleotide is facilitated. In these *non-canonical base pairs, the *minor groove side of nucleotide G1 is paired via hydrogen bonds to the *major groove side of nucleotide A4. Interestingly, this base pair continues the stacking of the helix because A4 is moved towards the minor groove, shortening the phosphate-to-phosphate distance by 6.2 Å compared to a canonical Watson–Crick GC base pair.

A network of up to seven hydrogen bonds stabilizing the loop surrounds the described central GA pair. It should be mentioned that the loop hydrogen-bonding network is dynamic, it differs between individual *conformations of the structure ensemble. The variation may reflect a conformational fluctuation that is best described by a network of heterogeneous hydrogen bonds. This stacking of nucleobases plays a stabilizing role and is reflected not only in the extension of the helix by the GA pair, but also by the fact that with decreasing ability to perform stacking, the order of the tetraloop is decreased.
BORIS FÜRLING, HARALD SCHWALBE

tetraloop receptor →GAAA tetraloop receptor

tetraplex →G-quadruplex

TΨC arm →TΨC stem–loop

TΨC stem–loop One branch in the *cloverleaf structure of a *transfer RNA containing the conserved *nucleotides *ribothymidine (T), *pseudouridine (Ψ) and *cytidine (C) at positions 54, 55 and 56. Conserved *base pairs between the TΨC and the D or *dihydrouridine stem–loop nucleotides G19 and G20 fold the cloverleaf into its characteristic L-shaped three-dimensional structure. STEFAN VÖRTLER

TFO →triplex-forming oligonucleotides

TGS →transcriptional gene silencing

thermal denaturation profile →melting curve

thermal melting Method for determining the stability of *oligonucleotide duplexes and higher-order structures. Duplexes (and other higher-order structures) are held together by hydrogen bonding interactions (→hydrogen bond) between *complementary *base pairs. Whilst most base pairs conform to the normal Watson–Crick hydrogen-bonding mode (→Watson–Crick base pair), there are a number of other hydrogen-bonding modes which include *Hoogsteen, *reversed Watson–Crick and *wobble base pairs. Other factors affecting duplex stability include *stacking interactions between adjacent nucleobases, hydrophobic effects (→hydrophobic interaction) and salt concentration. As a general rule, GC base pairs are stronger than AT(U) base pairs, thus a GC-rich duplex will melt at a higher temperature than an AT duplex. Thermal melting experiments are carried out by slowly raising the temperature of the duplex (or other structure) and observing either the ultraviolet *absorbance of the duplex (→absorption spectra of nucleobases, →hyperchromic shift) or the optical rotation by *circular dichroism (CD) spectroscopy. As a duplex is heated so the duplex will start to come apart to give two *random coils, and as a consequence both the absorbance intensity will increase (as measured by ultraviolet) and the optical rotation will decrease (as measured by CD). The *melting temperature (T_m) is defined as the point at which 50% of the DNA or RNA is duplex and 50% is random coil. There are a number of methods for calculating T_m available in the literature, but it is important to remember that T_ms are dependent on salt concentration.
DAVID LOAKES

thiamine pyrophosphate (TPP) riboswitch One of the most abundant classes of *riboswitches. It regulates *genes involved in thiamine (vitamin B$_1$) biosynthesis, phosphorylation and transport. The *aptamer *consensus motif has been found in all three domains of life. It is most widespread in bacteria where it controls *gene expression on either transcriptional or translational level. Examples of functional TPP aptamers in fungi and in plants seem to suggest regulatory roles in *splicing or *messenger RNA stability.

The TPP aptamer is composed of two extended *stem–loop domains flanking the central part formed by the P1 stem and L2–4 loop. The P3 stem is highly variable in length, comprising more than 100 *nucleotides in some eukaryotic sequences. Biochemical and crystal structure data show that most conserved nucleotides reside in regions that selectively recognize different parts of TPP: loop L3–2 binds to the *pyrimidine moiety, while residues in and around stem P4 make contacts to the *pyrophosphate group. No specific interactions are found for the thiazole group, which is the key functional part in the role of TPP as protein cofactor in carbonyl transfer reactions. Consistent with this finding, the TPP aptamer binds with high affinity to the thiamine antagonist pyrithiamine (PT), which contains a *pyridine instead of the thiazole ring. It is furthermore remarkable that the aptamer, although made of RNA with its polyanionic character, not only binds to a *ligand that carries a negatively charged pyrophosphate group, but efficiently discriminates against metabolic precursors thiamine and its *monophosphate by more than 1000−fold.
RÜDIGER WELZ

thiophosphate →phosphorothioate oligonucleotides

4-thiouridine (S4U) Analog of *uridine containing a sulfur in 4 position Some bacterial *transfer RNA contain 4-thiouridine. 4-Thiouridine can be used in photochemical *crosslinking experiments or as site for oligonucleotide functionalization (→oligonucleotide labeling).
BETTINA APPEL

TPP riboswitch aptamer consensus sequence

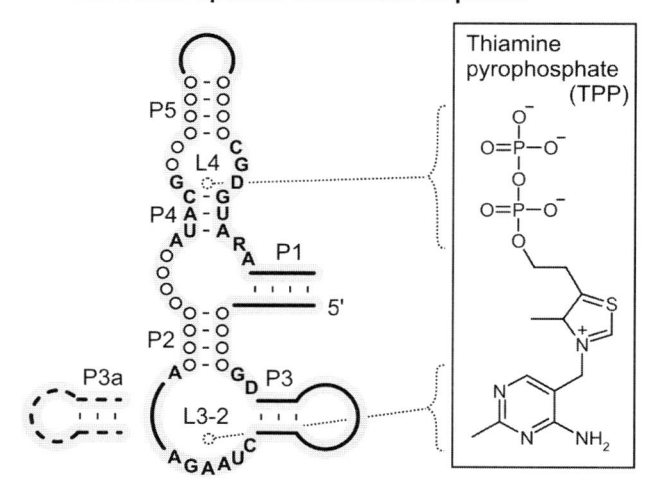

4-thiouridine

three-stem junction →three-way junction

three-way junction Structural element of DNA or RNA. It describes a point in the *secondary structure of a single molecule or a multi-molecular arrangement where the structure branches into three double-helical stems. SLAWOMIR GWIAZDA

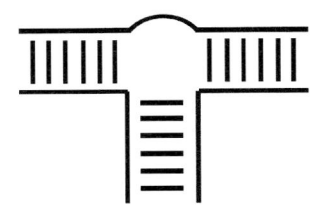

three-way-junction-based sensors *Three-way junctions are characterized by hydrophobic pockets defined by unstacked *base pairs on the ends of individual double helical stems. These pockets are similar to other hydrophobic receptors, e.g. cyclodextrans. Derivatization of three-way junctions with *fluorescein yields sensors that can report the binding of hydrophobic molecules, such as steroids, through a change in fluorescence. MILAN STOJANOVIC, DARKO STEFANOVIC

threose nucleic acid (TNA) Analog of RNA based on α-L-threofuranosyl units joined by 3′,2′-*phosphodiester linkages. TNA can be assembled by natural enzymes. BETTINA APPEL

thymidine Natural building block of DNA (→nucleosides).

TNA

thymidine dimer DNA is significantly damaged by exposure to ultraviolet irradiation (280–320 nm). One of the major effects of this type of irradiation is the formation of cyclobutane pyrimidine dimers (CPD). Among these thymidine–thymidine dimers are formed to a much larger extent than cytidine–cytidine dimers. The basis for the dimer formation is a photochemically allowed $[2\pi-2\pi]$ cycloaddition of the C5–C6 double bonds as part of two adjacent *pyrimidines. The irradiation of two thymidines which are located next to each other in single-stranded oligonucleotides yields thymidine dimers in two different isomers (*cis-syn* and *trans-syn*). Due to the steric situation, in *duplex DNA only the *cis-syn* configured thymidine dimer is formed, which represents the major *DNA lesion upon exposure of DNA to ultraviolet light. DNA *photolyase catalyzes the repair of the *cis-syn* configured cyclobutane thymidine dimers by utilizing the energy of visible light to break the cyclobutane ring. The thymidine dimers represent stable species inside the *DNA double helix, and are thought to initiate skin cancer and/or induce cell death through blocking of *replication and *translation pathways. Even though they are formed to a much lesser extent, cytidine dimers are considered even more highly mutagenic since they cause a variety of mutagenic reactions like deamination. The latter conversion leads ultimately to a C →T *mutation (→ DNA repair). HANS-ACHIM WAGENKNECHT

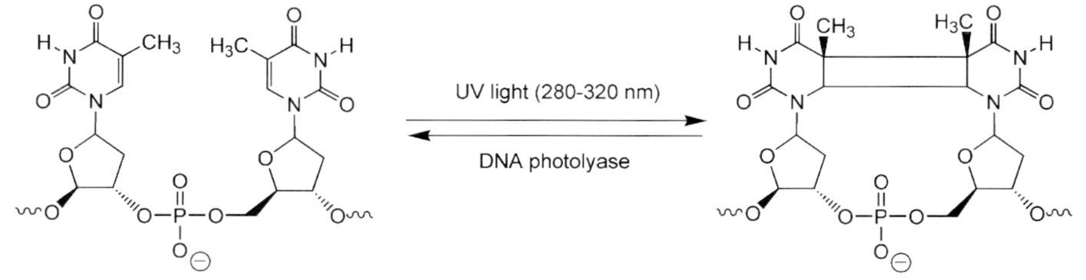

thymidine diphosphate (TDP) →nucleoside phosphates

thymidine monophosphate (TMP) →nucleoside phosphates

thymidine phosphates →nucleoside phosphates

thymidine triphosphate (TTP) →nucleoside phosphates

thymine →nucleobase

thymine arabinonucleoside →arabinonucleosides

thymine dimer →thymidine dimer

thymine ribonucleoside →nucleosides

thymine xylonucleoside →xylonucleosides

Ti plasmid Stable *episome in *Agrobacterium* that carries the *genes in the T region responsible for tumor induction (crown gall disease) in infected plants.

T-loop →TΨC stem–loop of a *transfer RNA

T_m →thermal melting

TMP Abbreviation for thymidine monophosphate (→nucleoside phosphates).

TNA →threose nucleic acid

TOM method Allows for the chemical solid-phase synthesis of oligoribonucleotides and named after the nucleoside 2′-hydroxyl protection – the [(triisopropylsilyl)oxy]methyl (TOM) group. First introduced in 1998, the method follows *phosphoramidite method for the coupling of appropriately protected nucleoside building blocks and was developed under the aspect that a sterically less-hindered 2′-O-protection should result in high efficiency of RNA synthesis. The protection strategy employs two orthogonal *protecting groups – an acid-labile (4,4′-dimethoxytrityl) group on the 5′-O-position and a fluorine ion-labile silyl ether on the 2′-O-position. The 3′-O-position is derivatized with a 2-cyanoethoxy-(N,N-diisopropylamino)phosphinyl residue as commonly used in standard DNA and in the TBDMS RNA synthesis method (→oligonucleotide synthesis, →TBDMS method). The chemical structures of TOM nucleoside phosphoramidites are depicted below.

Oligonucleotide assembly proceeds with coupling yields higher than 99% and with coupling rates as fast as one nucleotide per 2 minutes. After the assembly, the detachment of the sequence from the solid support and the removal of the acetyl *nucleobase and *phosphodiester *protecting groups are carried out under basic nucleophilic conditions with MeNH$_2$/water/ethanol, followed by removal of the remaining

2′-O-TOM protecting groups with tetra-n-butylammonium fluoride (TBAF). The deprotection step using TBAF can be carried out in tetrahydrofuran containing up to 20% water. This opens up an increased variety of modifications that can be incorporated into RNA and that are otherwise labile in fluorine ion solutions, e.g. O^6-trichloroethyl guanosine-modified RNA. Complete removal of the TBDMS group, in contrast, occurs only in the absence of water and requires drying of the TBAF solutions.

Using commercial synthesizers, the TOM method enables routine preparation of oligoribonucleotides up to 80 bases in length. The TOM method provides crude oligoribonucleotides of high sequence integrity, excellent purity and biological activity. The TOM method produces milligram quantities of more than 90% pure RNA and has been successfully commercialized (Xeragon Inc.). A further strength of the method is that it can be easily combined with the existing large pool of nucleoside labeling and marker building blocks developed for the DNA synthesis and the *TBDMS RNA synthesis methods. RONALD MICURA

topoisomerase →DNA topoisomerase

topology →DNA topology, →RNA topology

TPP riboswitch →thiamine pyrophosphate riboswitch

T7 promoter Promoter sequence 5′-TAATACGACTCACTATA-3′ that regulates the *expression of *downstream *genes in phage T7. This sequence is recognized by the T7 phage, coding *T7 RNA polymerase, which in turn catalyzes the synthesis of T7 phage RNA. IRENE DRUDE

***tra* genes** *Genes that carry the information for the conjugation transfer of *Ti plasmid between *Agrobacterium* cells.

trailer In the context of an *messenger RNA, the 3′-untranslated region (UTR), as opposed to "leader" (→leader sequence), the *5′-untranslated region. The more generic untranslated region terms are now more widely used. GEMMA MARFANY, ROSER GONZÀLEZ-DUARTE

***trans*-activator** *Gene (usually encoding for a *transcription factor) that regulates the *expression of other genes. Transcription factors regulate the binding of *RNA polymerase and the initiation of transcription (→transcription initiation). MAURO SANTOS

trans-active As opposed to *cis-active, proteins (or genetic elements located at a certain distance) that stimulate (enhancer) or suppress the *transcription of eukaryotic genes. In the context of *RNA catalysis, *trans*-active refers to an intermolecular reaction. MAURO SANTOS, SABINE MÜLLER

transcript RNA molecule synthesized by *RNA polymerase in the process of *transcription.

transcriptase *DNA-dependent, *RNA polymerase, an enzyme responsible for RNA synthesis according to the information encoded in DNA *genes. Typical bacteria, such as *Escherichia coli*, possess one type of RNA polymerase, responsible for the synthesis of all RNA molecules in the cell. Eukaryotic cells have three nuclear RNA polymerase species, numbered I, II and III. RNA polymerase I transcribe genes coding for the large *ribosomal RNA precursor. RNA polymerase II is responsible for *messenger RNA synthesis and for the synthesis of several types of small *non-coding RNA. RNA polymerase III synthesizes various small RNA, such as 5S *ribosomal RNA, *transfer RNA, U6 *small nuclear RNA and several other. Transcriptase is distinct from *reverse transcriptase or *RNA-dependent DNA polymerase. PETR V. SERGIEV

transcription Process of DNA-dependent RNA *polymerization from *nucleoside 5′-triphosphates. It is carried out by an enzyme *RNA polymerase (→transcriptase). The process of transcription can be subdivided into three steps: *initiation, *elongation and *termination of transcription. During initiation RNA polymerase recognizes the *promoter, a specific part of DNA, necessary and sufficient for RNA polymerase binding and transcription start. Initiation of transcription requires *transcription factors. For the typical bacteria *Escherichia coli*, a specific RNA polymerase subunit, the *sigma factor is required for promoter recognition and initiation of transcription. While σ^{70} is responsible for the recognition of the majority of *E. coli* promoters, other sigma factors are required for initiation of transcription from the minor classes of promoters, active at various stress conditions. In addition to a sigma factor, a number of DNA-binding *transcription factors are necessary for activation or inhibition of the *transcription initiation process. For eukaryotic cells, instead of the single sigma factor, a number of so-called basal transcription factors are necessary for transcription start. For the RNA polymerase I these are SL1 and UBF factors. For the RNA polymerase II, TFIIA, TFIIB, TFIID, TFIIF, TFIIE and TFIIH factors are required. In addition, a large number of specific transcription factors are necessary for transcription of each particular *gene. RNA polymerase III needs the TFIIIB factor for all types of promoters, TFIIIA and TFIIIC for transcription of the 5S *ribosomal *RNA genes, and TFIIIC for transcription of the *transfer RNA genes and Oct-1, STAF and SNAPc factors for various small RNA synthesis.

During promoter recognition RNA polymerase forms a closed complex, where DNA is still double stranded. Following that, DNA is locally melted to form an open complex. After the synthesis of an around 9-nucleotide long RNA, the RNA polymerase enters the elongation step and from this time initiation factors are no longer necessary. During elongation, the RNA polymerase catalyzes the nucleophilic attack of the 3′-hydroxyl of the growing

RNA chain onto the α-phosphate atom of the incoming *nucleoside-5'-triphosphate, complementary to the base of the DNA *template. RNA polymerase interacts with various elongation factors that are necessary for the regulation of elongation. Particularly, in bacteria, elongation of transcription is a subject of regulation by the mechanisms of *antitermination and *attenuation (→transcriptional attenuation).

Transcription stops at specific sequences, the terminators. *E. coli* terminators could be divided into Rho dependent, relying on the RNA *helicase Rho factor, and Rho independent, which are composed of an RNA hairpin followed by several uridine residues. PETR V. SERGIEV

transcriptional attenuation Mechanism of *attenuation, involving a highly structured RNA leader sequence, e.g. a signature sequence named the T-box, frequently found for *genes encoding *aminoacyl-tRNA synthetases and amino acid biosynthesis enzymes in Gram-positive bacteria. In the presence of limiting amounts of the appropriate amino acid, *transcription *antitermination is mediated by uncharged *transfer RNA (tRNA), which acts as a positive regulator. Uncharged tRNA directly interacts with the leader *messenger RNA at a specifier (the *codon for the appropriate amino acid) and at the T-box sequence to stabilize the antiterminator stem and promote transcription antitermination. The T-box transcriptional attenuation mechanism is involved in the regulation of many amino acid biosynthetic *operons and tRNA synthetase genes in Gram-positive bacteria. BEATRIX SÜß

transcriptional gene silencing →gene silencing, →RNA interference, →RITS

transcription bubble →DNA bubble

transcription control One of the most important mechanisms of the regulation of *gene expression. Since *transcription is the very first step of gene expression, the decision to transcribe or not to transcribe a particular *gene is the most effective tool of regulation. Transcriptional control, especially in eukaryotes, is combinatorial. It is a result of the combination of activity of the large number of *transcription factors, often regulated by a set of different intra- and extracellular signals. Transcriptional control also includes epigenetics, or cell memory, encoded by DNA modification (→modified DNA) and *histone modifications. Such modifications influence the condensation of *chromatin, and thus DNA accessibility to various transcription factors and *RNA polymerase. Usually, transcriptional activation is synergistic, i.e. several regulatory mechanisms act together to activate transcription significantly better than the sum of individual activation mechanisms. PETR V. SERGIEV

transcription factors *DNA-binding proteins affecting the *transcription initiation. In eukaryotes, transcription factors are divided into two groups: basal transcription factors, necessary for the transcription of all *genes, and specific transcription factors, which are required for transcription of a subset of genes, usually involved in the same or interrelated processes. Basal transcription factors for the *RNA polymerase I are SL1 and UBF. For the RNA polymerase II there are TFIIA, TFIIB, TFIID, TFIIF, TFIIE and TFIIH basal transcription factors. RNA polymerase III needs TFIIIB for all types of *promoters. Specific transcription factors could either activate (activators) or inhibit (repressors) *transcription. Proteins that affect transcription, but cannot bind DNA and rely on other *DNA-binding proteins to be directed to a particular gene, are called coactivators or *corepressors. DNA-binding transcription factors often belong to common families, sharing similar folds. Among the typical DNA-binding folds of transcription factors one can list *helix–turn–helix, *helix–loop–helix, leucine zipper and zinc finger (→DNA-binding proteins). PETR V. SERGIEV

transcription initiation Initiation of *transcription starts from the recognition of the *promoter, a specific part of DNA that is necessary and sufficient for *RNA polymerase binding and transcription start. Binding of the RNA polymerase to a promoter usually is facilitated by *transcription factors. Recognition of the promoter results in a closed complex formation. In this complex, DNA is still double stranded. Local *melting of DNA results in the formation of open complex. After DNA melting, the RNA polymerase starts RNA synthesis, which is initiated with a single *nucleoside-5′-triphosphate *in vivo* and can be initiated with *nucleoside-5′-monophosphate or even *dinucleotide *in vitro*. This initiation mechanism is sharply distinct from those of *DNA polymerase, which can only use RNA or DNA *primer, but not the nucleoside-5′-triphosphate to initiate *polymerization. Before the RNA chain reaches a certain length, usually 9 nucleotides, RNA synthesis in unstable and the synthesized RNA can be lost. After polymerization of 9 nucleotides, the RNA polymerase loses contact with the promoter and transcription initiation factors, and enters *elongation. PETR V. SERGIEV

transcription regulation Regulation of *gene expression is achieved on the level of *transcription, *RNA processing, RNA transport, *translation, protein maturation, post-translational modification and protein or *messenger *RNA degradation. Most efficient is regulation at the stage of transcription, since it assumes practically no waste of materials. Regulation of transcription usually is achieved at the stage of initiation (→transcription control). More rarely, transcription is regulated at the stage of *elongation. In prokaryotes, regulation at the stage of elongation is achieved by either *antitermination or *attenuation. PETR V. SERGIEV

transcription systems Model systems for *in vitro* transcription. The most popular *in vitro* transcription system utilizes bacteriophage *T7 RNA polymerase. Less common is use of phage T3 and SP6 RNA polymerases for *in vitro* transcription. Transcription systems based on eukaryotic RNA polymerases are more complex, but also available. If the goal is protein synthesis, *in vitro* transcription is often coupled with *in vitro* translation. Such systems are called coupled transcription–translation system. PETR V. SERGIEV

transcription termination →transcription

transduction Transfer of *genetic information. Depending on the specific context, this term refers to DNA transfer between two cells by means of bacteriophage or viral infection, or it refers to the signal transduction cascade from a stimulus to the final protein effectors within the cells. GEMMA MARFANY, ROSER GONZÀLEZ-DUARTE

transesterification In nucleic acids, conversion of one *phosphodiester into another by cleavage of a phosphoester bond and subsequent formation of a new one. SABINE MÜLLER

transfection Introduction of exogenous DNA into eukaryotic cells, most commonly through *plasmids.

transfer RNA (tRNA) Ubiquitous RNA that transfers an amino acid to the nascent polypeptide chain during *translation. tRNAs consist of 70–85 *nucleotides including a number of modified bases (→rare bases). Each of the 20 proteinogenic amino acids has at least one specific tRNA per cell. Since organelle and type specificity contribute to the multiplicity of tRNAs, the true number of tRNA species per cell is estimated to be about 50–70. A short nomenclature has been introduced in order to differentiate between specific tRNAs. tRNA$^{Ala}_{yeast}$, for example, is the alanine specific tRNA from yeast. All tRNAs adopt the characteristic *cloverleaf structure. It is composed of four intramolecular helices separated by three *loops. The *acceptor stem carries the 3'-terminal *overhang CCA (→CCA-tail), where the amino acid is attached via an ester bond to one of the adenosine hydroxyl groups. This 3'-terminal A is always 21 nucleotides apart from the first nucleoside of the *TΨC loop. The *anticodon loop contains the *codon *recognition site that is specific for each tRNA and which is responsible for recognition of the corresponding codon in the *messenger RNA (mRNA). The third loop is the *DHU loop, which together with the two other loops is involved in protein, mRNA and *ribosome recognition processes and interaction (→major tRNA, →minor tRNA, →ribosome, →wobble hypothesis, →aminoacyl-tRNA, →elongator tRNA, →translation). SABINE MÜLLER

transformation Process of the genotypic modification of an acceptor cell by introduction of external DNA. Several bacteria have a natural *competence for the uptake of external DNA. Others, like *Escherichia coli*, should be treated by different methods in order to obtain cells that are able to accept DNA. In eukaryotic cells, transformation is the inheritable conversion of a normal cell to a cancer-like state after treatment with viruses or *carcinogens. ANNEGRET WILDE

transgene *Gene that has been transferred from one organism to a different one.

transition Event that results in the exchange of a *purine base with a different purine base or a *pyrimidine base with a different pyrimidine base in a nucleic acid strand. Transitions may cause *gene mutations (→transversion, →mutation, →point mutation). SABINE MÜLLER

transition state Particular configuration of a molecular assembly along the reaction path. It is defined as the highest energy state of the reaction.

transition state analog Molecule that mimics the *transition state of a reaction.

transition state binding Biochemical *catalysts such as proteins and *ribozymes function by binding more tightly to the *transition state than to the substrate or product by using multiple binding interactions. The free binding energy of the ribozyme–substrate transition state complex partly compensates the required activation energy of the reaction, thus stabilizing the transition state complex and lowering the activation barrier. DENISE STROHBACH

translation Process of protein biosynthesis occurring in the *ribosome (→translational attenuation, →translation control, →translation system, →translation factor).

translational attenuation Mechanism of *attenuation, involving a highly structured RNA leader sequence, found for *genes encoding amino acid biosynthesis enzymes (Trp operon). In the presence of limiting amounts of the appropriate amino

acid, synthesis of a leader peptide leads to *ribosome stalling in the leader region waiting for the tryptophanyl *transfer RNAs. This leads to the formation of an *antiterminator hairpin and allows *RNA polymerase to proceed with *transcription.
BEATRIX SÜß

translation control One mechanism to regulate *gene expression and to adapt the cellular protein content (the proteome) to different environmental conditions. Primarily, *gene regulation occurs at the transcriptional level through control of *messenger RNA (mRNA) synthesis during the interplay between regulatory DNA regions (→promoters) and *transcription factors (→transcription control, →transcription regulation). Translational control concerns all regulatory events leading to protein biosynthesis, while post-translational control comprises all forms of protein modification and processing. In contrast to *transcription, both have the advantage of very short response times and allow the stability, location and activity of *gene products to be governed directly.

In prokaryotes all control processes are intertwined as they occur parallel in the same compartment on the same mRNA. This is exemplified by *attenuation, a mechanism, for example, regulating the biosynthetic tryptophan pathway: translation of a *polycistronic RNA from the Trp *operon coding for three proteins is discontinued due to amino acid starvation and lack of Trp-tRNA^{Trp}, causing *ribosomes to stall on the mRNA. This blocks *secondary structure formation with a sequence further *downstream, which would otherwise form a transcriptional termination structure (→transcription) and block gene expression. In this way the *RNA polymerase will continue to transcribe the downstream gene leading to protein expression, increase of Trp biosynthesis, decrease of ribosomal stalling and mRNA *secondary structure formation of the leader with the downstream sequence shutting the pathway off. In eukaryotes compartmentalization requires organized transport of mRNA and its protein products which offer alternative points of control.

There are a number of known major effectors of translational control. (a) The time a synthesized mRNA becomes available for *translation as in several instances it is kept dormant for later developmental steps, a situation termed masked mRNA. (b) The way translation proceeds with respect to *internal ribosomal entry sites shortcutting conventional eukaryotic *initiation by using an internal *start codon, −1/+1 *frame shifts and hops of up to 50 *nucleotide along the mRNA discontinuing or alternating the *reading frame, *recoding to introduce selenocysteine and pyrrolysine at *stop codons or translational enhancer sequences which aid initiation in prokaryotes. (c) How fast and therefore with how many protein copies per *template the gene is expressed. Often this is related to the *codon usage. One example is the use of different *start codons all of which are decoded by the *initiator tRNA, but with different efficiency. (d) How long translation proceeds is a matter of mRNA stability. It can be affected by RNA secondary structure but as well by proteins binding to the 3′-untranslated region (UTR) and blocking access to *nucleases. On example of the latter is a regulatory element for iron homeostasis involving translation of transferritin, a membrane iron receptor important for iron uptake. At low iron levels the iron regulatory proteins IRP1 and IRP2 adopt a conformation which is able to bind to mRNA hairpins called iron responsive elements (IRE). Five of which are located at the 3′-UTR blocking nuclease degradation. However, at high iron levels IRP1 acts as functional aconitase enzyme employed in the tricarboxylic acid cycle and is unable to bind the transferritin mRNA, while IRP2 is prone to increased protease degradation – both effects lead in turn to m*RNA degradation, decreased transferritin level and reduced iron uptake. (e) The location where translation occurs. Targeted transport and immobilization of the mRNA gives rise to strictly localized translation and in turn formation of a protein gradient which is crucial for a broad set of developmental processes like pattern formation, definition of body axes by Vg1 or VegT genes in oozytes, initiation of head and thorax formation by *Drosophila* Bicoid *bcd* and Nanos *nos* genes, segregation between germ and somatic cells by *Drosophila* Oskar *osk* gene, determination of mating types between mother and daughter cells by the *Saccharomyces cerevisiae* ASH1 gene as well as synaptic plasticity associated with formation of neuronal connections important to learning. (f) Post-translational control mechanisms including covalent addition or removal of low-molecular-weight compounds like formylmethionine, carbohydrates, and acetyl, biotinyl and phosphoryl groups. Additional mechanisms are the formation of cysteine bridges, specific cleavages to remove leader peptides, internal sequences (inteins) or the processing of precursors like in the case of insulin or the protease trypsin. STEFAN VÖRTLER

translation factor Collective term for all factors that assist ribosomal protein synthesis (→translation). The peptidyl transferase reaction is based on RNA and proceeds even on isolated 50S *ribosomal subunits – albeit a three-orders of magnitude reduced rate compared to the 20–50 peptide bond formations per second observed *in vivo* under optimal growth conditions. Translation factors ensure this high rate by supporting specificity and processivity while reducing errors due to *proofreading. All of them are proteins, which is of interest for the evolution of the translational system and its interplay with the RNA-made *ribosome. As the main reaction partner during translation is also *transfer RNA (tRNA), many translation factors interact with RNA or RNA-*binding sites, and were proposed to mimic RNA in structure and/or function. This view falls short to describe the complex molecular interplays, but was an important intellectual step away from a simple rigid *polymerization platform as ribosomes and translation was considered before the 1980s.

Translation factors are classified according to the stage in the translational cycle in which they participate. Functional homologs exist in all branches of the tree of life underlining the conservation of the translation mechanisms. A small "e" represents eukaryotic and a "a–e" archeal factors of the eukaryotic type. However, for historic reasons the nomenclature is confusing. Similar numbers do not represent functional or structural homologs and eukaryotic factors outnumber prokaryotic factors 10-fold. This large number is a consequence of (a) a more demanding and complex initiation, (b) parallel pathways for regulation and control at different stages of translation with similar factors being present in duplicate or even more and (c) formation of multi-factor complexes, which might stay assembled and act as supramolecular units reducing again the number of active players.

Initiation factors (IF/eIF/a-eIF) help to identify translational start sites, recruit ribosomal subunits and match the initial tRNA (→initiator tRNA, →formylmethionyl-tRNA) with the correct *start codon. This is important as it determines the *reading frame translated. Prokaryotes use three monomeric initiation factors, whereas in the eukaryotic systems at least 12 factors are made up from 23 different polypeptides. IF1/eIF1A binds to the small subunit *A-site, possibly blocking an early entry of tRNA. IF2/eIF5B is a GTPase facilitating initiator tRNA binding to the *P-site as well as subunit joining. IF3 (with no direct homolog, but likely eIF1 fulfilling the role in eukaryotes) assists IF2 to find the correct start site, but seems also to keep ribosomes dissociated to avoid premature assembly and aid ribosomal recycling after termination. With this protein, translational steps are clearly interconnected. The additional eukaryotic factors are involved in alternative pathways to prohibit premature subunit association or assist re-dissociation of the subunits, and particularly identify and load a *messenger RNA (mRNA) to the assembling ribosome. Prokaryotes do not have the latter problem as translation and *transcription are coupled processes, and as soon as a *Shine–Dalgarno sequence appears, ribosomes will assemble on it. Eukaryotic mRNA is an isolated, processed, likely structured and protein-covered molecule. This requires, for example, 7-methylguanosine and poly(A) identification [→cap structure, →poly(A) tail] as proof of complete processing or RNA unwinding and removal of m*RNA-binding proteins to prepare for small subunit loading.

Elongation factors (EF/eEF/a-eEF) support the processive polymerization once the ribosomal P-site is loaded with an initiator tRNA. There is higher similarity among the factors involved compared to the other steps, which is taken as sign of the universal mechanism of the peptidyl transferase reaction. Newly incoming *elongator tRNA, which are all tRNA not involved in initiation or suppression, bind as s ternary complex with EF-1A·GTP to the A-site. The bacterial factor was historically named EF-Tu, "T" for transfer and "u" for unstable, to differentiate it from "s" for stable, as it was first isolated in complex with Ts which split during purification. The protein is among the most ubiquitous in a translationally active cell (around 100 μM; 5–10% of cellular protein) with a nanomolar binding affinity for *aminoacyl-tRNA protecting it from deacylation. Correct *codon–*anticodon interaction triggers GTP hydrolysis, dissociation of EF-1A and leads into peptidyl transfer to the aminoacyl-tRNA. To continue polymerization, A-site-bound *peptidyl-tRNA has to be translocated to the P-site, P-site tRNA into the E-site and the ribosome has to move one *triplet further on the mRNA – all of which is catalyzed under GTP expense by EFG (eEF2, a-eEF2).

Termination factors insure the controlled ending of protein synthesis at the appropriate *stop codons, release of the nascent polypeptide and start of ribosomal recycling, as a result they are also called release factors. In prokaryotes, RF1 and RF2 decode UAG/UAA and UGA/UAA, respectively, while just one eRF1 in eukaryotes or a-eRF1 in Archea recognizes all three *stop codons. These are called class I termination factors. All of them require a stop codon presenting A-site, induce at the *peptidyl transferase center the hydrolysis of P-site bound peptidyl-tRNA like catalyzed by the ribosome itself and remain bound to the A-site until release under GTP expense by a class II release factor (RF3, eRF3 or aRF3).

Finally, recycling of the GDP-bound factors to the GTP-forms is required. Separate guanine exchange factors (GEF) perform this task and allow for regulation of the G-proteins. GEF can be proteins like EF-Ts for EF-1A, multi-factor complexes like eEF1B formed by eEF1α and eEF1β (the latter formerly known as eEF1γ) for eEF1A or even the *ribosome as suggested for RF3/eRF3. STEFAN VÖRTLER

translation systems *In vitro* reconstitution of the protein biosynthesis in the test tube was a goal long aimed for, starting in the late 1940s when radioactive amino acids became available for incorporation and tracer studies. In comparison to direct injection into animals or addition to the growth medium, *in vitro* systems offer control over the conditions and the specific activity of isotopes. Moreover, they add the possibility to use modified or altered components like *translation factors, mutated *ribosomes or unnatural amino acids charged to *transfer RNA. With this one can study translational mechanisms or, maybe even more importantly, obtain difficult to isolate proteins, screen in parallel large numbers of different proteins or produce toxic and artificial polypeptides containing novel amino acids. This avoids bacterial fermentation for classical *overexpression and subsequent laborious purification since translation systems contain just a subset of the more than 3000 proteins of live bacteria. Affinity purification using tags on the desired protein makes the isolation a matter of hours rather than days.

One approach uses semipurified preparations of ribosomes and a mix of enzymes and translation factors, while a different strategy relies on reconstitution from isolated components. First translation systems were purified from tissues and cells with high protein synthesis rates like liver, reticulocytes, germinating plant seeds and bacteria in their logarithmic growth phase. Up until the 1990s yields were rather moderate, but the application of continuous *dialysis supported translation for days rather than a few hours and allowed yields in the milligram range in just 1 mL of lysate. During dialysis energy-providing compounds as well as amino acids are replenished while potential inhibiting reaction products are removed. Careful adjustment of the reaction parameters like the nature of salts, buffer components and their respective concentrations increased yields even for non-dialysis or batch systems as well. It seems that not a singular effect stabilizes *in vitro* translation systems, but the sum of many small improvements makes the difference. Several systems are commercially available, even in a transcription–translation form where just a *polymerase chain reaction product or a *vector with the cloned *gene of interest needs to be added to induce *messenger RNA *transcription by endogenous *RNA polymerases leading subsequently to protein synthesis (→*in vitro* translation). STEFAN VÖRTLER

translation termination One of the three steps of *translation in which a *stop codon leads to peptide release and ribosomal recycling.

translocation Stepwise movement of the *ribosome along a *messenger RNA by simultaneously transferring the *peptidyl-tRNA from the *A-site to the *P-site of the ribosome. Intrachromosomal or interchromosomal shifting of *chromosome segments is also referred to as translocation. Sometimes these chromosomal shifts lead to diseases, e.g. chronic myelogenous leukemia. VALESKA DOMBOS

transposable element Discrete DNA sequence, present in the prokaryotic and eukaryotic *genome, able to change its location and insert itself in a variety of non-homologous genomic targets. *Transposons fall in two classes: those that mobilize via DNA and the others that propagate themselves through a RNA intermediate, which is reverse transcribed (→reverse transcription) and then inserted into a new genomic location. MAURO SANTOS

transposition Copying a stretch of DNA from one locus to another with or without loss of the transposed sequence at the original location. Transposition typically requires pre-formed DNA elements of specific modular organization such as IS (insertion sequence) elements or *transposons. HANS-JOACHIM FRITZ

transposon →transposable element.

***trans*-splicing** Rare intermolecular *splicing reaction that ligates *exons from different *primary transcripts. *Trans*-splicing has been observed in lower eukaryotes such as fungi, trypanosomes and nematodes. A well-studied example features three different actin *messenger RNAs (mRNAs) found in the nematode *Caenorhabditis elegans*. At their 5'-end, these actin mRNAs contain a common 22-nucleotide *leader sequence that is not encoded by the actin *gene. Rather that leader originates from another 100-nucleotide RNA molecule transcribed from a different gene. BERND-JOACHIM BENECKE

transversion Event that results in the exchange of a *pyrimidine base with a *purine base or *vice versa* in a nucleic acid strand. Transversions may cause *gene mutations (→transition, →mutation, →point mutation). SABINE MÜLLER

tree alignment →RNA alignment and structure comparison

tree editing →RNA alignment and structure comparison

trimethylguanosine cap *Cap structure found in various *U-snRNA and *small nucleolar RNA species transcribed by RNA polymerase II. Chemically this cap forms a $m_3^{2,2,7}$GpppN nucleotide sequence element at the 5'-end of those small RNAs. BERND-JOACHIM BENECKE

trinucleotide repeat *Nucleotide motif, consisting of 3 nucleotides, which is consecutively repeated in a *genome. Too many repeats can lead to trinucleotide repeat disorders (→genetic diseases), e.g. Huntington's disease. VALESKA DOMBOS

triple helix Three-stranded nucleic acid structure that is formed by binding a third strand in the *major groove of a *double helix through the formation of specific *hydrogen bonds with the Watson–Crick bases of the duplex (→Watson–Crick base pair, →Hoogsteen base pair). Triple helices can be classified according to the composition and orientation of the third strand. In the pyrimidine (parallel) motif, a homopyrimidine third strand binds parallel to the purine strand of a homopurine–homopyrimidine duplex domain to form T·AT and C$^+$·GC base triads through Hoogsteen hydrogen bonds. In the purine (antiparallel) motif, a purine-rich third strand binds antiparallel to the Watson–Crick purine strand of the duplex forming G·GC, A·AT or T·AT base triplets via reverse Hoogsteen hydrogen bonds (→parallel orientation, →antiparallel orientation, →triplex-forming oligonucleotides). KLAUS WEISZ

triplet Three-nucleotide stretch in *messenger RNA forming one *codon of the *genetic code.

triplet code →genetic code

triplex-forming oligonucleotides (TFOs) While *antisense oligonucleotides act on the post-transcriptional level and hybridize to a specific *messenger RNA, TFOs modulate *gene expression at an early

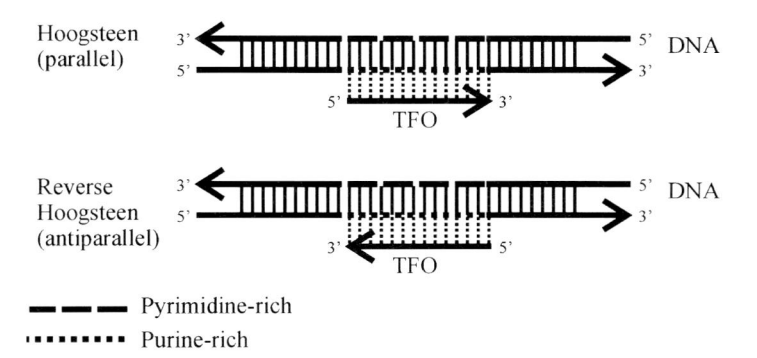

```
Hoogsteen      3' ←||||||||||||[|||||||||||||||]|||||||||||||→ 5' DNA
(parallel)     5' ────────────────────────────────────────── 3'
                        5' ····················→ 3'
                                   TFO

Reverse        3' ←||||||||||||[|||||||||||||||]|||||||||||||→ 5' DNA
Hoogsteen      5' ────────────────────────────────────────── 3'
(antiparallel)          3' ←···················  5'
                                   TFO

━━━  Pyrimidine-rich
·······  Purine-rich
```

stage by binding to *double-stranded DNA. This approach has therefore been named antigene strategy in analogy to the term *antisense strategy for the former method. TFOs specifically recognize oligopyrimidine–oligopurine sequences and bind to the *major groove of the *purine-containing strand of the *DNA double helix. TFOs form *hydrogen bonds in the unusual Hoogsteen geometry (→ Hoogsteen base pair), either in parallel or in reverse (antiparallel) orientation.

TFOs can be employed to interfere with various biological functions of DNA in a sequence-specific manner. Inhibition of *gene expression by preventing *initiation or *elongation of *transcription is the most frequently reported application. Interestingly, TFOs can also enhance transcription, e.g. by competition with a *repressor factor. In addition, they have been used to induce a permanent modification of the DNA at the triplex site. For this purpose, reactive molecules, like photo-inducible *crosslinker or DNA-cleaving agents, were conjugated to the oligonucleotide and delivered to the intended site of action by the TFO. Further fields of application include the induction of site-directed genomic *mutations in somatic cells as well as the stimulation of *homologous recombination.

As can be seen from this listing, TFOs can be employed for a broad range of purposes, and they have already been proven to be suitable for cell culture and *in vivo* experiments. However, improvement of their efficiency is still one of the key challenges for successful applications. To this end, various *modified nucleotides with enhanced binding affinity have been introduced into the oligonucleotides. JENS KURRECK

tRNA → transfer RNA

tRNA gene *Gene that is transcribed into *transfer RNA

tRNA nucleotidyl transferase Enzyme that adds the *CCA-tail to the 3'-end of *transfer RNAs generating the functional 3'-*acceptor arm (→ RNA processing).

T7 RNA polymerase *DNA-dependent RNA polymerase from the bacteriophage T7. The 99-kDa protein catalyzes the *polymerization of RNA in 5' → 3' direction in a magnesium-dependent manner. T7 polymerase is extremely *promoter specific (→ T7 promoter) and starts *transcription directly after the promoter. Apart from the non-transcribed promoter sequence, it requires an adjacent sequence motive which starts with a 5'-cytidine (mostly 5'-CCCTCT-3'). This motive is always transcribed together with the *downstream *gene information, wherefore all T7 polymerase transcribed RNAs contain a guanine (mostly 5'-GGGAGA-3' hexanucleotide motif) at their 5'-terminus. T7 polymerase is also able to introduce ^{32}P-labeled *nucleotides or 5'-modified nucleotides into a RNA, which makes it an attractive tool for *radioactive labeling or labeling of RNA with a *fluorescent dye. IRENE DRUDE

tRNA processing → RNA processing

tRNA splicing Removal of *introns from *transfer RNA (tRNA) precursor molecules. About 10% of the *transfer *RNA genes in yeast contain an intron invariantly located 3' to the *anticodon. The removal of these 14- to 46-nucleotide long *intervening sequences follows a different mechanism than that of "normal" *RNA splicing. Here, the tRNA intron is removed in a two-step reaction: cleavage and *ligation are two separated steps. Base pairing between the intron and the anticodon sequence is essential for the reaction. BERND-JOACHIM BENECKE

TTP Abbreviation for thymidine triphosphate (→nucleoside phosphates).

tumor gene →oncogene

tumor suppressor genes Endogenous *genes of eukaryotic cells that play a surveillance role on the integrity of the genetic material (DNA) or the general cellular state and which can trigger cell apoptosis if the cell appears to be excessively damaged or not responding to regulatory control signals. MAURO SANTOS

turnover number Number of substrate molecules which are turned over to product by one enzyme molecule per unit of time – under conditions of maximum activity of the enzyme. The turnover number corresponds to the k_{cat} value in the *Michaelis–Menten equation (→Michaelis–Menten model). According to that equation at saturating substrate concentrations, i.e. far above the K_M value, the maximum rate is reached which equals $v_{max} = k_{cat} \cdot c_E$. Obviously, for calculation of the turnover number (k_{cat}) the enzyme concentration must be known (→ K_M determination). CHRISTIAN HERRMANN

twin ribozyme RNA molecule that is composed of two *ribozyme units. For example, mobile *group I introns (→mobile genetic elements) containing two ribozyme domains and a homing endonuclease (termed twin ribozyme intron organization) can integrate by *reverse splicing into the *small subunit *ribosomal RNA of bacteria and yeast. A different type of twin ribozymes has been developed by *RNA engineering. These twin ribozymes are derived from the *hairpin ribozyme and

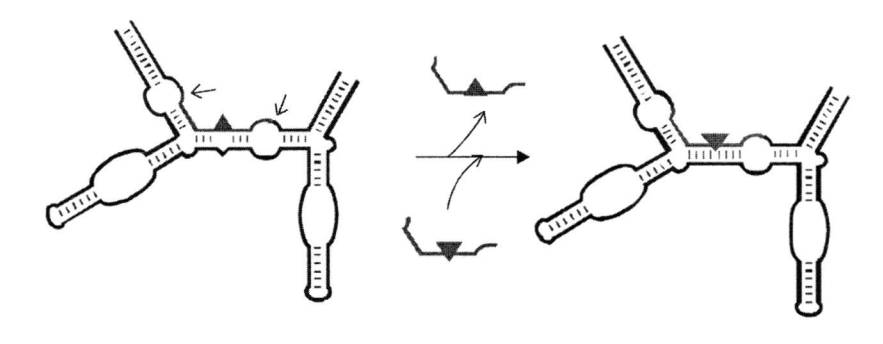

catalyze in a strictly controlled fashion two chain cleavage events and two *ligations. In this manner, they mediate the site-directed and patchwise exchange of RNA fragments, and thus are potential tools for *RNA repair strategies in *gene therapy.
SABINE MÜLLER, STÉPHANIE VAULÉON

twist →helical twist

twisting number (Tw) In circular DNA, the ratio between the number of *base pairs and the number of base pairs per *helical turn (→supercoiled DNA, →DNA topology). VALESKA DOMBOS

U

U →uracil

UDP Abbreviation for uridine diphosphate (→nucleoside phosphates).

ultraviolet crosslinking Direct covalent binding of non-modified nucleic acids to protein, support or other polymers upon ultraviolet irradiation. As example, short wave ultraviolet light causes the nitrogenous bases in RNA, mostly *uracil, to become highly reactive and to form covalent linkages to amine groups on the surface of the membrane. Care must be taken not to under or overexpose the RNA to ultraviolet light — both of which will decrease *hybridization signals. Usually a 1-minute exposure with 254 nm light or 3 minutes with 302 nm light is sufficient. Ultraviolet light is a zero length *crosslinker and is believed to produce less perturbation of the complex than chemical crosslinkers. Ultraviolet crosslinking may take place in the case of photo-affinity labeling of macromolecules. For example, arylazide derivatives have become especially popular among photo-activatable nucleic acid analogs due to their high reactivity. *Nucleosides containing arylazido groups can be introduced into nucleic acid by both enzymatic and chemical methods (→oligonucleotide synthesis). The main disadvantages of arylazide include the lack of selectivity in the binding to proteins and large size of the arylazido group, which can lead to disturbances in the local structure of a nucleic acid–protein complex and to *crosslinking with remote protein fragments. Nevertheless, this type of modified nucleic acid is widely used in the study of structures of protein–nucleic acid complexes. Among other commonly used photo-activatable nucleic acid derivatives, mention can be made of 5-halogenopyrimidine and 8-halogenopurine nucleosides. Photolysis of these photo-activatable nucleic acids produces stable radicals which react with electron-donor amino acid residues of proteins, particularly with aromatic and sulfur-containing side-groups. In the case of bromo derivatives, the optimum wavelength of the ultraviolet light necessary for the abstraction of the halogen atom is 308 nm, while that for iodo derivatives is 325 nm. The increase in the wavelength reduces photo-induced damage of proteins and nucleic acid, but decreases the yields of the photo-addition products. By virtue of their small size (the van der Waals radii of iodine and bromine atoms correspond approximately to that of the methyl group), halogen atoms do not induce any significant changes in the structure of nucleic acid and do not prevent protein – nucleic acid recognition. This makes halogen derivatives of nucleic acid highly attractive tools for probing protein–nucleic acid contacts in the *binding sites of *biopolymers (→crosslinking of DNA, →ultraviolet laser crosslinking).

TATIANA S. ORETSKAYA, VALERIY G. METELEV

ultraviolet laser crosslinking Exhibits several advantages over techniques using low-intensity light sources. These advantages are determined by the high amount of photons delivered by the laser in a single nanosecond or picosecond pulse.

The laser-induced generation of radical cations exhibits a high quantum yield, which leads to a higher efficiency of *crosslinking (exceeding by close to two orders of magnitude that obtained with conventional ultraviolet light sources). This makes possible the formation of *crosslinks that cannot be induced by conventional ultraviolet lamps. The crosslinking reaction itself is completed in much less than 1 μs, which avoids the possibility of unwanted crosslinking of ultraviolet-damaged molecules and permits the trapping of rapid dynamic changes in *protein–DNA interactions. Ultraviolet laser crosslinking has been used *in vitro* for studying different aspects of protein–DNA interactions including measurements of binding constants, the determination of protein–DNA contact points and the size of the protein–nucleic acid complexation site (→crosslinking of DNA, →ultraviolet crosslinking).
TATIANA S. ORETSKAYA, VALERIY G. METELEV

ultraviolet spectroscopy Technique that uses light in the ultraviolet region of the electromagnetic radiation to promote electronic transitions in molecules. The ultraviolet spectrophotometer measures the intensity of light passed through a sample (I) and compares it to its initial intensity (I_0). Based on the strong *absorbance of nucleobases at 260 nm (→absorption spectra of nucleobases) and a linear change in absorbance with concentration, measuring the absorbance at 260 nm by ultraviolet spectroscopy is the most commonly used technique for determining nucleic acid concentrations (→Beer–Lambert law). In contrast, temperature-dependent ultraviolet experiments on nucleic acids exploit the *hyperchromic effect, i.e. an increase in extinction coefficient upon unstacking of nucleic acid bases (→base stacking). Thus, structural transitions that change the fraction of stacked bases also change the extinction coefficient. Extinction coefficients depend on both the magnitude of the intrinsic transition moment of each base and the relative orientations of adjacent bases.
KLAUS WEISZ

UMP Abbreviation for uridine monophosphate (→nucleoside phosphates).

uncharged tRNA *transfer RNA carrying no amino acid.

unidentified reading frame (URF) →open reading frame

universal bases *Base analogs capable of forming *base pairs with each of the natural DNA/RNA bases without discriminating between them. The easiest way to determine whether *nucleobase analogs can act as a universal base is by measuring the thermal *melting temperature (T_m) (→thermal melting) of *oligonucleotide duplexes with the analog opposed to each of the natural bases (A, C, G and T/U). If an analog behaves as a universal base then each of the four duplexes would exhibit almost identical T_ms. The earliest and most widely used universal base described is *hypoxanthine, which has proven very useful in *polymerase-based applications, but in practice is a poor example because duplexes containing it opposite natural bases show a wide T_m discrimination, whilst polymerases preferentially recognize it as a *guanine residue.

There are now many nucleobase analogs described which act as a universal base, and each possesses a series of common features, i.e. they are usually hydrophobic, non-polar, aromatic and non-hydrogen bonding. As they usually possess no hydrogen bonding functionality they stabilize base pairs with natural nucleobases in oligonucleotide duplexes by enhanced *stacking interactions. As universal bases stabilize principally by stacking interactions, as a general rule, the larger the surface area, the more stable the duplex containing it. An additional feature of such aromatic nucleobase analogs is that they preferentially form more stable self-pairs within a duplex. A universal base self-pair is probably more stable than a base pair between a universal base and a natural nucleobase because natural nucleobases need to be desolvated (loss of water molecules) in order to base pair with the hydrophobic universal base. To date there are no reported examples of universal bases that are capable of forming hydrogen-bonding base pairs to each of the natural bases.

Hypoxanthine   3-Nitropyrrole   5-Nitroindole

Common universal base analogs

As each of the described universal base analogs are hydrophobic, aromatic nucleobases lacking hydrogen-bonding capability they are poorly tolerated in enzymatic reactions. For example, in polymerase-based reactions they cause stalling of the growing chain. They are therefore most effective in applications which rely only on the *hybridization properties of the oligonucleotides containing them. They can be used in *primers for polymerase reactions (→polymerase chain reaction, →sequencing) to aid stabilization provided the universal base is not close to the 3'-end of the primer. For example, universal bases can be used to stabilize short primers for thermal cycling reactions or they can be complementary to positions in the *template whose precise sequence is unknown. They can also be used in probes (→nucleic acid probe) in in situ hybridization applications and in *single-nucleotide polymorphism detection. The most widely used universal base analogs are those derived from 3-nitropyrrole and 5-nitroindole. These analogs have proven useful in a number of applications as described above. The larger 5-nitroindole derivative is more stabilizing in a duplex than the smaller 3-nitropyrrole, although it still causes destabilization compared to native DNA. DAVID LOAKES

universal primer A *primer is a short *oligonucleotide which is complementary to a DNA *template and which is to be used for DNA *amplification by, for example, the *polymerase chain reaction. Primers can be either specific to a particular DNA sequence or they can be universal. A specific primer (or pair of primers) is used to amplify specific DNA template sequences. A universal primer is complementary to *nucleotide sequences which are very common in a particular set of DNA molecules, and thus they are able to bind to a wide variety of DNA templates. Thus, if there are a number of different nucleic acid sequences each complementary to the universal primer, then each will be amplified during the polymerase step.
DAVID LOAKES

5'-untranslated region (5'-UTR) Region of *messenger RNA containing the *cap structure usually a 7-methylguanosine (→trimethylguanosine cap), the *start codon for the protein synthesis and the *binding site for the *ribosome. In prokaryotic organisms the binding site for the ribosome is called the *Shine–Dalgarno sequence. Recently, genetic control elements such as *riboswitches have been localized in the non-translated region of the 5′-UTR in prokaryotes. TINA PERSSON

up mutation *Mutation causing the strengthening of a *phenotype; often a mutation in a regulatory DNA sequence such as a *promotor (→down mutation, →null mutation). HANS-JOACHIM FRITZ

upstream To the 5′-side of a particular position in a directional nucleic acid strand.

uracil →nucleobase

uracil arabinonucleoside →arabinonucleosides

uracil DNA glycosylase (UDG) Enzyme involved in *DNA repair. Any *uracil residue in DNA resulting from either misincorporation or deamination of *cytosine is removed, leaving behind an *abasic site to be further processed by *DNA repair enzymes. SABINE MÜLLER

uracil ribonucleoside →nucleosides

uracil xylonucleoside →xylonucleosides

URF Abbreviation for unidentified reading frame (→open reading frame).

uridine Natural building block of RNA (→nucleosides).

uridine diphosphate (UDP) →nucleoside phosphates

uridine monophosphate (UMP) →nucleoside phosphates

uridine phosphates →nucleoside phosphates

uridine triphosphate (UTP) →nucleoside phosphates

U-snRNA Uridylic acid-rich *small nuclear RNA (snRNA; 60–215 nucleotides). When initially identified and characterized, those snRNA molecules revealed a much higher portion of *U residues than other stable RNA molecules known at that time. Today, U-snRNA mostly refers to two groups: one is the subgroup among snRNA which is involved *pre-mRNA *splicing, i.e. the *spliceosomal RNAs. The second group consists of the large number of *small nucleolar RNAs which are involved in the maturation of *ribosomal RNA (→RNA processing). In addition to these two groups, another representative of U-snRNA is the U7 RNA involved in 3′-end formation of non-polyadenylated *histone mRNAs. BERND-JOACHIM BENECKE

U-snRNP *Ribonucleoprotein complex, associated from *U-snRNA and spliceosomal proteins (→spliceosomal RNP).

UTP Abbreviation for uridine triphosphate (→nucleoside phosphates).

5′-UTR →5′-untranslated region

3′-UTR short for 3′-untranslated region

U-turn A well-characterized and common *RNA motif involved in *tertiary interactions. It was first observed in the structure of *transfer RNA^{Phe}, and is also prominent in the crystal structures of the *hammerhead ribozyme *active site.

This stable structure element facilitates a sharp change in the strand direction, and can close a RNA hairpin. Due to the fact that the motif exposes the *nucleotides located at the 3′-side of the turn towards the solvent, it creates an anchor for long-range tertiary interactions. Typically, the sequential signature of a U-turn motif is 5′-UNR-3′ and it is closed by YY, YA or GA pairs. As revealed by comparative sequence analysis, the U-turn is a very abundant motif; it is identified 33 times in a large set of *ribosomal RNAs (library of more than 7000 of different species).

BORIS FÜRTIG, HARALD SCHWALBE

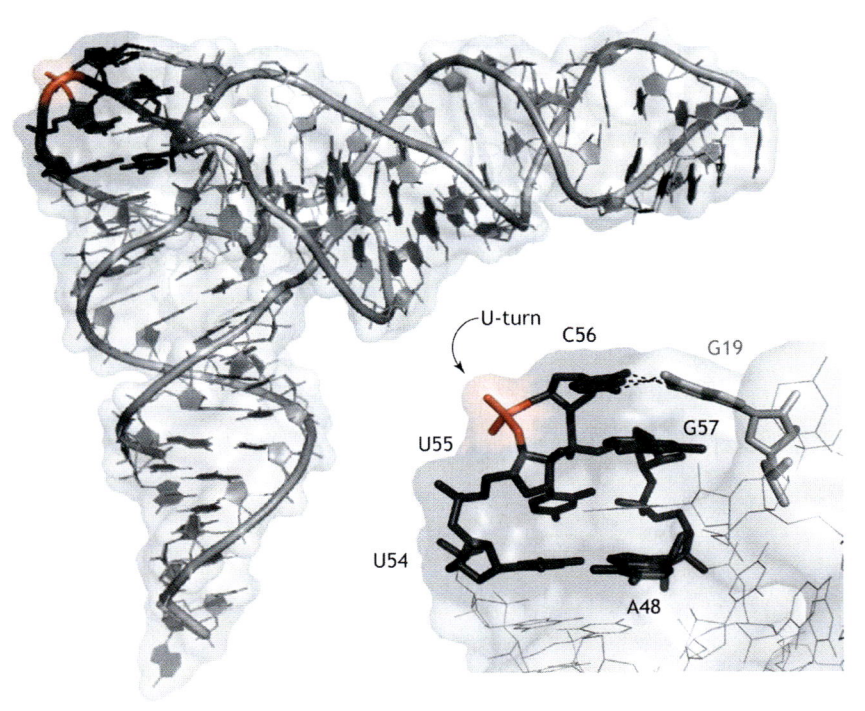

U-turn motif as seen in the tRNA^{Phe} (4TRA); The nucleotides forming the U-turn motif are colored in black, in a stick model, the turn in strand direction that is made around a backbone phosphate group is indicated by red color. Left panel displays the whole molecule, the anticodon-loop pointing right, the CCA-tail down; Right panel is a close-up of the U-turn motif, in the context of the tRNAPhe molecule it facilitates a long-range interaction between the nucleotide C56 directly 3′ of the turn and the nucleotide G19 (also shown in stick representations, grey).

V

van der Waals interaction very weak intermolecular force which arises from the polarization of molecules into dipoles. A transient dipole of one atom induces a dipole at another atom which results in an electrostatic attraction of the two atoms. Polarization increases with rising size of an atom. The interaction is very weak and has only a very low range. MATTHÄUS JANCZYK

variable loop Structure element of a *transfer RNA located between the *anticodon and *TΨC stem–loops in the *cloverleaf structure. As implied by the name, it contains a variable number of extra residues (designated by an "e") after *nucleotide 45 of a canonical 76 nucleotide tRNA. The variable loop can harbor 4–24 nucleotides, with the vast majority having 4–5 nucleotides (called class I tRNA) in contrast to those with 10–24 nucleotides (class II tRNA). The latter comprise usually tRNA^{Leu}, tRNA^{Ser} and tRNA^{Tyr} of eubacteria and some organelles. STEFAN VÖRTLER

Varkud satellite (VS) ribozyme Small ribozyme component within transcripts of mitochondrial satellite plasmids. It was originally identified by R. A. Collins in the Varkud isolate of the fungus *Neurospora*. Its function is autocatalytic processing of multimeric precursor transcripts (resulting from a *rolling-circle amplification of the VS plasmid) in order to release monomeric forms of this RNA. With a length of about 150 nucleotides, it is the largest ribozyme with a cleavage mode leading to 2′,3′-*cyclic phosphate and 5′-hydroxyl ends.

V-arm → variable loop

Since no crystal structure is available yet, current models rely on structure probing and mutagenesis experiments. The global architecture of the VS ribozyme is organized by two three-way helical junctions (II–III–VI and III–IV–V, → three-way junction), with helices III, IV and VI assumed to form a *coaxial stack (a and b). The VS ribozyme is unique in that it recognizes its substrate hairpin domain I mainly by *tertiary interactions, including a "kissing complex" (→ kissing loops) which consists of 3bp between the *loops of helices Ib and V. Formation of this loop–loop interaction induces a major rearrangement of the base-pairing pattern in domain I, creating a single nucleotide *bulge on the 3′-side of helix Ib and a metal-binding *secondary structure motif near the cleavage site. The current *tertiary structure model (b) is consistent with the finding that ribozyme activity is sensitive to *mutations in loop V and changes in the length of helices III and V, but rather insensitive to truncations of helices IV and VI.

Mutation analyses indicate that the bases in the central bulge of helix VI, especially A756, are of particular importance for *catalysis (the base numbering of the VS ribozyme starts with U617 and ends with U783, due to its origin in the 881-nucleotide VS transcript). The *active site is thought to be formed by juxtaposing this bulge and the cleavage site (b). The *catalytic core structure also accepts RNA substrates added *in trans* (→ *trans*-active): an independent *transcript carrying the VS helix Ib and the 5′-part of helix Ia hybridizes to the corresponding part of helix Ia of a 5′-truncated

Nucleic Acids from A to Z: A Concise Encyclopedia. Edited by Sabine Müller
Copyright © 2008 WILEY-VCH Verlag GmbH & Co. KGaA, Weinheim
ISBN: 978-3-527-31211-5

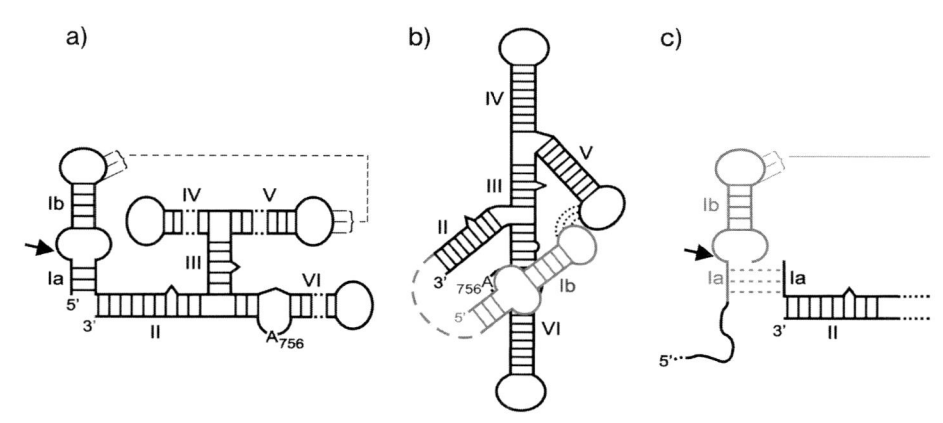

The VS ribozyme. (a) Basic secondary structure; the tertiary base pairing between helix Ib and helix V is indicated by the dashed line. The catalytically important residue A756 is indicated. (b) Secondary structure presentation based on the current tertiary structure model of the VS ribozyme, with the stem–loop I in grey and positioned in front of helix VI. (c) Set-up for *trans*-cleavage; the stem–loop Ib module can be fused to the 3'-end of a RNA of interest (black line) for tailored 3'-end cleavage of transcripts by the *trans*-acting VS ribozyme moiety. The cleavage position is indicated by an arrow in (a) and (c).

VS ribozyme version. This substrate is efficiently cleaved at the canonical site between helices Ia and Ib (c). In the *cis* format (→ *cis*-active), the cleavage position may be preceded by a single 5'-nucleotide only, which deletes helix Ia. Apparently, in the natural context in which the substrate module is covalently linked to helix II, the tertiary contacts of *stem–loop Ib to loop V and helix VI are sufficient for efficient and precise cleavage. ROLAND K. HARTMANN, MARIO MÖRL, DAGMAR K. WILLKOMM

vector Self-replicating DNA molecule (e.g. *plasmid, virus) that carries cloned DNA fragments (→cloning vector, →plasmid).

vector DNA →vector

very short patch repair Special case of *short patch repair. Historically, the name is derived from experiments of genetic *recombination and refers to the repair of T/G *mismatches in *Escherichia coli* as they result from the hydrolytic deamination of 5-me-C residues. HANS-JOACHIM FRITZ

VS ribozyme →Varkud satellite ribozyme

W

Watson–Crick base pairing →Watson–Crick base pairs

Watson–Crick base pairs *Purine– *pyrimidine complementary pairs AT, TA, GC and CG, linked by *hydrogen bonds: two bonds between *adenine and *thymine, but three between *guanine and *cytosine. Watson–Crick base pairs are able to vicariously replace one another in *double-stranded DNA because they are isomorphous – the same size, a similar shape; the *glycosidic bonds (the links between sugar and base) occur in identical orientation with respect to the helix axis. This feature allows all four bases to occur on both chains and so any sequence of bases can fit into the *double helix. Other pairing of bases tend (to varying degrees) to destabilize the double-helical structure.
NINA DOLINNAYA, BETTINA APPEL

Watson–Crick edge →edge-to-edge interaction of nucleobases

Watson–Crick helix Right-handed DNA double helix consisting of two intertwined *sugar–phosphate backbones, with the heterocyclic DNA bases projecting inwards from each of the two strands. The bases are arranged in *purine–*pyrimidine pairs, *adenine with *thymine, *guanine with *cytosine, linked by *hydrogen bonds, and these *base pairs are stacked on top of each other along the helix axis at a distance 0.34 nm apart (→base stacking). The secondary repeat distance of about 3.4 nm is accounted for by the presence of 10 (now 10.5) nucleotide pairs in each complete turn of the double helix. The two antiparallel polynucleotide chains (→antiparallel orientation) are not identical in either base sequence or composition. Instead, they are complementary to each other (→complementary strand).
NINA DOLINNAYA

Watson–Crick model Three-dimensional model of DNA molecule postulated by James Watson and Francis Crick in 1953. It consists of two helical DNA chains coiled around the same axis to form a right-handed double helix (→right-handed DNA). The hydrophilic *backbone of alternating *deoxyribose and negatively charged phosphate groups are on the outside of the double helix, facing the surrounding water. The *purine and *pyrimidine bases of both strands are stacked inside the double helix, with their hydrophobic and newly planar ring structures very close together and perpendicular to the axis of the helix. Each base of one strand is paired in the same plane with a base of the other strand. *Watson–Crick base pairs (AT, TA, GC and CG) fit best within the structure, providing a rationale for the specific A=T and G=C base equivalence discovered by Chargaff. The Watson–Crick model accounted not only for *Chargaff's rule, but also for the X-ray diffraction data, and for the other chemical and physical properties of DNA. Moreover, the model immediately suggested a mechanism for the transmission of *genetic information. NINA DOLINNAYA

wobble base Base at the third position of degenerated *codons (→wobble base pair, →wobble hypothesis, →genetic code).

wobble base pair Unconventional non-*Watson–Crick base pair required to form between the third position of a *messenger RNA *codon and the first nucleotide of a *transfer RNA *anticodon (position 34) to enable proper *decoding of the *genetic code as stated by the *wobble hypothesis. One example is the controlled *frame shift during *translation of bacterial RF2 genes. STEFAN VÖRTLER

wobble hypothesis Hypothesis put forward in 1965 by Francis Crick, trying to explain why a specific *transfer RNA can recognize more then one *codon of the *messenger RNA. *Decoding of the *genetic information occurs by codon–*anticodon interaction, with the 5′-base of the codon pairing with the 3′-base of the anticodon. The 3′-base of the anticodon and the following central base form standard *Watson–Crick base pairs with the *complementary bases of the codon. The third base (5′-base) of the anticodon however, has more play in choosing its pairing partner – it can "wobble", meaning that it unconventionally pairs with a variety of bases. For example, the alanine-specific transfer RNA (tRNA^{Ala}) has the anticodon 3′-CGI-5′ and can bind three different codons (GCU, GCC and GCA), forming one CG and one GC Watson–Crick base pair, and a third *wobble base pair between *inosine and either *uracil, *cytosine or *adenine. In the same manner, uracil as the 5′-base of the anticodon can pair with either adenine or guanine, and guanine as the 5′-anticodon base can pair with uracil or cytosine in the codon. SABINE MÜLLER

writhing number (Wr) The number of times a DNA helix crosses itself during supercoiling (→supercoil, →supercoiled DNA).

X

X-chromosome One of the two *chromosomes that determines the sex in many animal species, the other is the *Y-chromosome. This form of sex determination is usually labeled as XX/XY. Thus, in mammals for instance, females have two X-chromosomes and are called the homogametic sex, while males have and a X- and a Y-chromosome and are the heterogametic sex. In certain insects and nematodes the females are XX, but the males have just a singe X and no other sex chromosome. This system is labeled XX/X0. In other species like birds and butterflies the females are the heterogametic sex and the males the homogametic sex. This system of sex determination is labeled ZW/ZZ. The hallmark of both XX/XY and ZW/ZZ systems is that there is a 1:1 ratio of females and males.

Genetic disorders that are due to *mutations in *genes located on the X-chromosome are called X-linked. A famous example is the transmission of hemophilia in the pedigree of Queen Victoria and some of her descendants: daughters Princess Alice (wife of Grand Duke Ludwig IV of Hessen-Darmstadt) and Princess Beatrice (wife of Prince Henry Maurice of Battenberg), and son Prince Leopold that died at age 31 of hemorrhage after a fall; granddaughters Princess Alix (later Queen Alexandra, wife of Tsar Nicholas II of Russia), Princess Alice (wife of Alexander, Prince of Teck) and Princess Victoria Eugenié of Battenberg (wife of King Alfonso XIII of Spain), and grandsons Prince Leopold of Battenberg (died at age 33, presumably of hemorrhage, after surgery) and Prince Maurice of Battenberg (died at age 23, in Battle of Ypres). The present English and Spanish royal families are free of hemophilia. Since the mutations causing hemophilia are recessive and their frequency is low, XX women are normally protected because they usually carry a normal *allele in one of the X-chromosomes. On the other hand, the Y-chromosome has no gene to mask the mutation that causes the disease and XY men that inherited the X-chromosome with the mutation from their mother are hemophilic. MAURO SANTOS

xDNA (expanded DNA) Designed, synthetic genetic encoding system analogous to DNA, but in which all *base pairs are replaced by new pairs having size-expanded geometry. The enlarged xDNA bases are related to the natural *bases, but have their *Watson–Crick edges shifted outward by 2.4 Å (the width of a benzene ring, see figure). When the xDNA bases are attached to *deoxyribose, the novel expanded *nucleosides are abbreviated dxA and dxG (*benzo*purines), and dxC and dxT (*benzo*pyrimidines).

As in natural DNA sequences (where each position is occupied by an A, T, C or G), xDNA relies on Watson–Crick hydrogen bonding (→Watson–Crick base pairs) and shape complementarity to govern stable and selective duplex assembly – thereby adhering to the *purine–*pyrimidine rule. However, in xDNA, *benzo*pyrimidines are paired with purines and *benzo*purines are paired with pyrimidines. This gives rise to eight possible information-encoding base pairs – more storage potential than nature's four-base system. A theoretical xDNA nucleosides with size-expanded bases. An xDNA base pair is shown on the right, below a natural base pair.

functioning genetic system based on xDNA would be orthogonal to the natural genetic system.

Sequences of size-expanded base pairs can assemble into fully expanded, right-handed, antiparallel duplexes, structurally resembling natural B-form DNA (→B-DNA). However, these xDNA duplexes have increased diameter (13 versus 10Å), widened *major and *minor grooves, and a smaller helical twist than B-DNA (involving about 12 bp per turn). In addition, xDNA helices are more thermally stable than a natural helix of analogous sequence. This is due to enhanced *stacking ability of the xDNA bases (→base stacking), which have increased aromatic surface area. The increase in size also renders the xDNA bases fluorescent, a property virtually absent in natural DNA bases.

The xDNA base analogs are able to selectively recognize their natural Watson–Crick partners in assembly (i.e. xA prefers to be paired with T, over C, G and A), thereby retaining the ability to selectively encode *genetic information. The ability of xDNA to store information as well as resemble natural DNA structurally illustrates the importance of consistent base pair geometry and uniform helical structure in the evolution of nature's genetic system. In addition, the observed base pair *mismatch discrimination and unique fluorescent properties of xDNAs makes them promising tools in nucleic acid detection.
ANDREW T. KRUEGER, ERIC T. KOOL

X-ray analysis of nucleic acids
→crystallization of nucleic acids

xylonucleosides Structural analog of *ribonucleosides in which ribose is replaced with *xylose.

xylose Pentose monosaccharide ($C_5H_{10}O_5$; 150.13 g mol^{-1}) that exists in two enantiomeric conformations, D(−)-xylose and L(+)-xylose.

Y

YAC library Library of human *genomic DNA (or, exceptionally, murine genomic DNA) cloned into *yeast artificial chromosomes.

Yarus inhibitor *Transition-state analog which was used in crystallization studies of the *ribosome. It binds to the catalytic center of the ribosome and is a strong inhibitor of the peptidyl transferase reaction in the ribosome. In selection studies, the Yarus inhibitor was used to screen RNAs that strongly bind to it in order to find RNAs that could catalyze peptide bond formation (→ artificial ribozymes). It consists of CCdApPuromycin. JÖRN WOLF

Y-chromosome One of the two *chromosomes that determines the sex in many animal species (→ X-chromosome). In humans, the Y-chromosome contains only 78 *genes and most of its DNA is heterochromatic. This is a typical situation in most species, i.e. the Y-chromosomes contain the fewest genes of any other of the chromosomes. SRY (sex-determining region) lies on the *Y-chromosome in humans and other primates, and encodes the testis-determining factor. MAURO SANTOS

yeast artificial chromosome *Artificial chromosome containing human *genomic DNA and which is meant to be episomically maintained in yeast. The average insert stretches from 100 kb to 1 Mb (→ YAC library). GEMMA MARFANY, ROSER GONZÀLEZ-DUARTE

Z

Z-DNA Left-handed form of the *double helix. Z-DNA can be formed by certain sequences containing alternating *purine and *pyrimidine bases: d(GC)$_n$ > d(GT/AC)$_n$ >> d(AT)$_n$. When deoxycytidines in d(GC)$_n$ sequence are 5-methylated, Z-DNA can form under physiological salt conditions. The existence of Z-DNA was first suggested by optical studies demonstrating that a *polymer of alternating dG and dC residues produced a nearly inverted circular dichroism spectrum in a high salt solution (→circular dichroism spectroscopy). An atomic-resolution crystallographic study of d(GC)$_3$ revealed a left-handed double helix, which maintained *Watson–Crick base pairs. The Z-DNA helix is built from a dinucleotide repeat with the dCs in the *anti* conformation while dGs are in the unusual *syn* form (→*syn* conformation). In contrast to *B-DNA, where the repeating unit of the helix is a single *base pair, in Z-DNA the sugar residues of dCs and dGs are in different puckering modes. In Z-DNA, there is a single narrow groove that corresponds to the *minor groove of B-DNA. There is no *major groove. The *backbone follows a zigzag path, giving rise to the name Z-DNA. The transition from B- to Z-conformation involves "flipping" the base pairs upside down. Z-DNA is less stable than B-DNA at physiological salt concentration. In addition, the formation of B–Z junctions, each of which has a free energy ΔG near +4 kcal mol^{-1}, is a significant energetic barrier to left-handed helix formation. *In vivo* Z-DNA is stabilized by the *negative supercoiling generated by DNA *transcription. Negative supercoiling arises behind a moving *RNA polymerase as it ploughs through the *DNA double helix. Unlike B-DNA, Z-DNA is highly immunogenic, and polyclonal and monoclonal antibodies were prepared in order to recognize this conformation in eukaryotic systems. Individual *genes were assayed by *crosslinking the antibody to DNA using a 10-ns exposure of a laser at 266 nm (→ultraviolet laser crosslinking). These experiments made it possible to determine which regions of a *gene form Z-DNA. The amount of the Z-form was found to depend on DNA negative torsional strain. It increased dramatically as transcription increased, but was largely unaffected by *DNA replication. Formation of Z-DNA could facilitate *recombination of homologous chromosomal domains by relieving topological strain that arises when intact duplexes are intertwined. Z-DNA formation could effect the placement of nucleosomes as well as the organization of chromosomal domains. Discovery of proteins that bind to Z-DNA with high affinity and specificity helps establish a biological role for this shape. Z-DNA-binding domains are found in the *RNA editing enzyme, *double-stranded RNA adenosine deaminase and tumor-associated protein DLM-1. Recognition of Z-DNA by RNA editing protein may block the gene from further transcription until editing of the RNA is complete. Z-DNA binding is a common feature of E3L gene products in DNA-containing poxviruses which replicate in the cytoplasm of infected cells. It was shown that binding to the Z-conformation is necessary for E3L biological activity (a protein that does not bind Z-DNA is not pathogenic, but a mutation that creates Z-DNA binding makes a lethal virus). Z-form regions associated with transcriptional activity might exist for only a very short period in a living cell.

NINA DOLINNAYA